Transmission Techniques for Emergent Multicast and Broadcast Systems

Transmission Techniques for Emergent Multicast and Broadcast Systems

Mário Marques da Silva
Américo M.C. Correia
Rui Dinis
Nuno Souto
João Carlos Silva

CRC Press
Taylor & Francis Group
Boca Raton London New York

CRC Press is an imprint of the
Taylor & Francis Group, an **informa** business
AN AUERBACH BOOK

Auerbach Publications
Taylor & Francis Group
6000 Broken Sound Parkway NW, Suite 300
Boca Raton, FL 33487-2742

Printed in the United States of America on acid-free paper
10 9 8 7 6 5 4 3 2 1

International Standard Book Number: 978-1-4398-1593-9 (Hardback)

Visit the Taylor & Francis Web site at
http://www.taylorandfrancis.com

and the Auerbach Web site at
http://www.auerbach-publications.com

Contents

Preface

This book presents a detailed and comprehensive description of the most important enhancements from a requirements perspective, transmission techniques, and receiver designs using multiresolution schemes for modern broadband wireless systems. It covers several hot topics that are key in the development of emergent services. In order to support the requirements of new services, this book addresses the means that allow increasing the bit rate per user and, as a result, increasing the capacity of digital cellular radio networks due to improved spectral efficiency.

The purpose of this book is to concentrate in a single place several important research and development (R&D) activities currently taking place in the field of wireless communications for multimedia broadcast and multicast service (MBMS). These aspects are normally split into different references, and thus the proposed book reduces the time and cost required to learn and improve skills and knowledge in the field. Moreover, this book presents a compilation of the latest developments in the area that is the outcome of several years of research and participation in many international activities and projects. The focus is on the key requirements of emergent services, with a special emphasis on the MBMS. The purpose is to cover several subjects that allow reaching such key requirements, providing the corresponding description of fundamentals and theory.

The transmission, detection techniques, and schemes presented in this book are relevant to many digital communication systems (wireless, cellular, satellite, etc.). Nevertheless, a special focus is placed on the multimedia services using multicast and broadcast techniques, which is the main subject of this book.

With such an approach, this book covers a wide range of potential readers: it can be used either by an engineer with a BSc degree to learn more about the latest R&D wireless activities for the purpose of an MSc or PhD program, or for business activities; this book can also be used by academic, institutional, or industrial researchers in order to support the study, planning, design and development of prototypes and systems.

Although the subjects associated with MBMS covered in this book are wide and generic, applicable to unicast, multicast, and broadcast, the final and concluding chapter focuses on a system-level evaluation of MBMS using different transmission techniques. Owing to the demanding Quality-of-Service requirements (bandwidth, bit error rate, capacity, latency, jitter packet loss, etc.), in order to allow the deployment of the emergent services, such as video broadcast, HDTV over the Internet, video-on-demand, mobile television, and so on, several enhancements to MBMS were introduced. The combination of the enhancements is accomplished by adaptive transmission techniques. The concluding chapter of this book presents the combination of these techniques and enhancements for the purpose of implementing the evolved MBMS (E-MBMS).

Authors

Mário Marques da Silva [marques.silva@ieee.org] is a researcher at the Portuguese Instituto de Telecomunicações (Institute for Telecomunications) and the Centro de Estudos em Sistemas de Informação e Tecnologias Informáticas (CESITI). He is a professor at the Universidade the Autónoma de Lisboa and leads an Engineering Branch at Information Technologies and Telecommunications Directorate, Portuguese Navy. He received his BSc in electrical engineering in 1992, and MSc and PhD degrees in electrical engineering and mobile communications, respectively, in 1999 and 2005, both from the Instituto Superior Técnico, Technical University of Lisbon, Portugal.

Between 2005 and 2008, he was with the NATO Air Command Control Management Agency (NACMA) in Brussels, Belgium, where he managed the deployable communications of the new Air Command and Control System Program. He has been involved in several telecommunications projects, either as a researcher or project leader, including research, architecture, development, analysis, simulation and testing of networking, HF, V/UHF, satellite and cellular/UMTS communications systems. He is the author of more than 30 papers and a reviewer of several international publications including journals and conferences. He is a member of the Institute of Electrical and Electronics Engineers (IEEE) and the Armed Forces Communications and Electronics Association (AFCEA). His research interests include networking (e.g., mobile ad hoc networking, TCP/IP) and mobile communications, including block transmission techniques (OFDM, SC-FDE), WCDMA, multiuser detection, interference cancelation, space–time coding, MIMO systems, smart and adaptive antennas, channel estimation, and software-defined radio.

Américo M.C. Correia received his BSc degree in electrical engineering from the University of Angola in 1983, and his MSc and PhD degrees from the Instituto Superior Técnico (IST-UTL), Lisbon, Portugal, in 1990 and 1994, respectively. From 1991 to 1999 he was with IST as an assistant professor. He is currently a full professor with the Instituto Superior de Ciências do Trabalho e da Empresa (ISCTE-IUL), Lisbon, Portugal. He visited the Nokia Research Center from September to December 1998 as a visiting scientist. From September 2000 to August 2001 he worked for Ericsson EurolabNetherlands. His current research topics include wideband CDMA, OFDMA, MIMO, radio resource management, and multimedia broadcast/multicast services.

Rui Dinis received his PhD degree from the Instituto Superior Técnico (IST), Technical University of Lisbon, Portugal, in 2001. From 2001 to 2008 he was a professor at IST. Since 2008 he has been teaching at FCT-UNL (Faculdade de Ciências e Tecnologia da Universidade Nova de Lisboa). He was a researcher at the Centro de Análise e Processamento de Sinais (CAPS)/IST from 1992 to 2005; from 2005 to 2008 he was researcher at ISR/IST (Instituto de Sistemas e Robótica); and in 2009 he joined the research IT center (Instituto de Telecomunicações).

Nuno Souto graduated in aerospace engineering–avionics branch in 2000 from the Instituto Superior Técnico, Lisbon, Portugal. From November 2000 to January 2002 he worked as a researcher in the field of automatic speech recognition for the Instituto de Engenharia e Sistemas de Computadores, Lisbon, Portugal. He received his PhD in electrical engineering from the Instituto Superior Técnico in 2006. He currently works as an assistant professor at the Instituto Superior de Ciências do Trabalho e da Empresa (ISCTE), Lisbon, Portugal. He has participated in several European research projects and has published several articles in international journals. His research interests include wideband CDMA systems, OFDM, channel coding, modulation, channel estimation, and MIMO systems.

João Carlos Silva received his BS degree in aerospace engineering from the Instituto Superior Técnico (IST), Lisbon Technical University, in 2000. From 2000 to 2002 he worked as a business consultant for McKinsey & Company. He also completed his PhD thesis at IST in 2006, focusing on spread spectrum techniques, multiuser detection schemes, and MIMO systems. From 2006 to 2007 he was a professor at the Escola Superior Tecnologia e Gestão (ESTG) in Leiria (teaching physics and electronics), and has been a professor since 2007 at the Instituto Superior de Ciencias do Trabalho e da Empresa (ISCTE), in Lisbon (teaching computer networks). He is the author of more than 40 papers and a reviewer of several international publications. His research interests include computer networks and mobile communications, such as block transmission techniques (OFDM, SC-FDE), WCDMA, multiuser detection, interference cancelation, space–time coding, MIMO systems, smart and adaptive antennas, channel estimation, and software-defined radio.

Chapter 1

Introduction

Future wireless systems will be (IP) internet protocol-based and will provide acceptable quality of service (QoS) in terms of speed, bit error rate (BER), end-to-end packet loss, jitter, and delays for different types of traffic. Many technological achievements were made in the last few years in the area of communications and others are planned for the future to allow new, emergent, and future services. Whereas in the past, new technologies pushed new services, today reality can be the opposite: end users want services to be employed on a day-to-day basis, whatever the technology supporting it. Users want to browse the Internet, have e-mail access chat, and establish a video conference session, regardless of the technology used (e.g., WiMAX, Wi-Fi, 3G, HSPA, etc.). It is most important to have the required/desired service available, and with quality. Thus, services must be delivered following the concept of "anywhere" and "anytime." Figure 1.1 presents the bandwidth (BW) requirements for each different service.

Multimedia broadcast and multicast service (MBMS) is intended to use spectrum-efficient multimedia services, such as video broadcast or mobile television (by transmitting data over a common radio channel), both in the core network and, above all, in the air interface, where the bottleneck in terms of spectrum efficiency is placed to a large group of users. MBMS allows downlink data to be transmitted from a single source to multiple recipients in broadcast or multicast modes and can be offered via existing global system for mobile communication (GSM) and universal mobile telecommunications system (UMTS) cellular networks. In addition, MBMS includes point-to-point (PtP) and point-to-multipoint (PMP) modes. Whereas the former allows individual retransmissions, the latter does not.

1.1 Requirements for MBMSs

There is still a lot of ongoing investigation on ways of improving the delivery of multimedia information. The multimedia paradigm has put pressure on resources optimization, and sharing channels is one of the most important aspects in network optimization. Efficient network resource usage should be the leverage for upcoming multimedia applications. Besides this, in order to guarantee scalability, multiresolution schemes have to be considered in current UMTS and emergent cellular standards.

One of the most important properties of MBMS is resource sharing among many user equipments (UEs). This means that many users should be able to listen to the same MBMS channel, at the same time. Therefore, power should be allocated to these MBMS channels for arbitrary UEs in the cell to receive this service. This approach goes against the traditional power control concept used in unicast, where the power should be controlled to provide the required QoS [typically a minimum signal to noise ratio (SNR)] to a user located at the edge of the cell, while avoiding high interference levels to users in the same cell or in adjacent cells.

Services may be classified by the type of cast, namely unicast, multicast, or broadcast. In unicast, a single source transmits the data to a single destination. Multicast service consists of a node that sends the same content to multiple (specific) destinations. In wireless medium, broadcast is the basic mode, as the data are sent to all nodes. MBMS is a system that enables mobile networks to efficiently deliver data from a single content

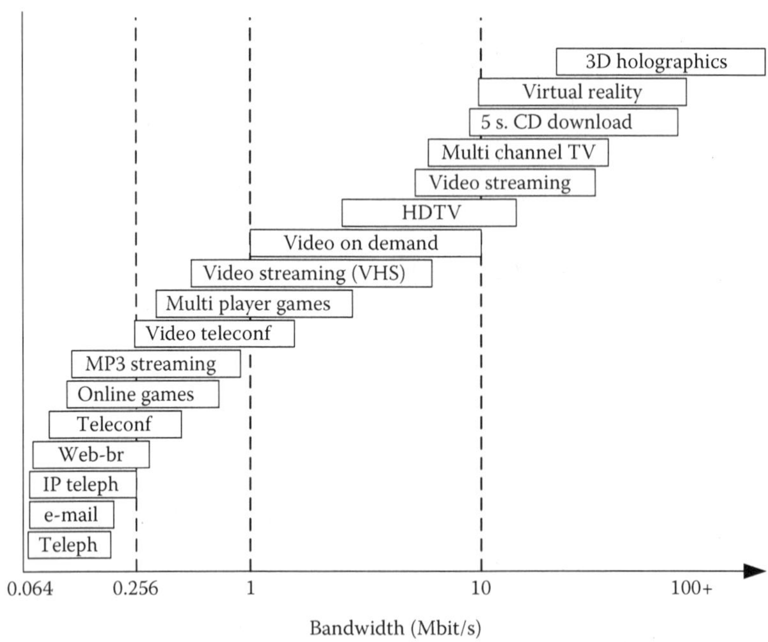

Figure 1.1 BW requirements of the different services.

provider source to multiple users by sharing radio and transport network resources. Multimedia services can be delivered in either multicast or broadcast mode. In multicast mode multimedia data can be transmitted to a specific group of users (MBMS user groups) in a specific area (MBMS service area). In broadcast mode data are transmitted in a specific area (MBMS service area), but all users in the specific MBMS service area will be able to receive the transmitted MBMS data.

For broadcast and multicast transmissions in a mobile cellular network, depending on the communication link conditions some receivers will have better SNR than others, and thus the capacity of the communication link for these users is higher. In broadcast transmissions, it is possible to exchange some of the capacity of the good communication links to the poor ones and the trade-off can be worthwhile. Figure 1.2 shows typical cellular topology, where a Node B (i.e., a BS) is the central node of each cell.

The specific feature for multiresolution consists of Fast Link Adaptation: instead of compensating the variations of downlink radio conditions by means of power control, the transmitted power is kept constant and the modulation and coding of the transport block are chosen at every transmission time interval (TTI). This is called adaptive modulation

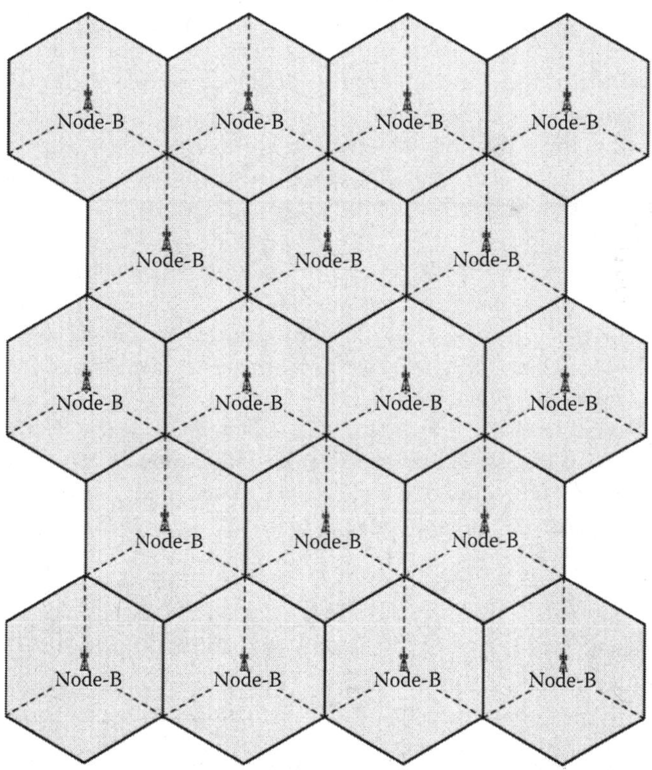

Figure 1.2 Typical cellular topology.

and coding (AMC). To users in good conditions, 16-QAM (quadrature amplitude modulation) or even 64-QAM can be allocated to maximize throughput, whereas users in bad conditions are penalized on throughput, reaching a point to which the service can be denied. In wideband code division multiple access (WCDMA), up to 30 different channel quality indicators (CQIs) can be employed associated with corresponding different bit rates. M-QAM hierarchical constellations and multiple input–multiple output (MIMO) along with the orthogonal frequency division multiple access (OFDMA) system can also be seen as multiresolution schemes for the enhanced MBMS network that are not yet standardized [Bingham 1990; Falconer et al. 2002]. This combination of techniques is specified for use by long-term evolution (LTE).

The comparison between the provision of MBMS based on WCDMA as adopted for UMTS and OFDMA LTE is not straightforward because of the difference of BW and intersite distances (ISDs). This comparison is performed in the concluding chapter of this book.

1.2 Transmission Techniques to Support Wireless Advances for MBMS Requirements

The challenge facing the mobile telecommunications industry today is how to continually improve the end-user experience, to offer appealing services through a delivery mechanism that offers improved speed, service attractiveness, and service interaction. Furthermore, to deliver the required services to the users with the minimum cost, the technology should allow better performances, higher throughputs, improved capacities, and higher spectral efficiencies. The following sections describe several measures to reach such desiderates.

The BW requirements for MBMS present a considerable challenge since multipath propagation leads to severe time-dispersion effects. In this case, conventional time-domain equalization schemes are not practical. Block transmission techniques, with appropriate cyclic prefixes (CPs) and employing frequency-domain equalization (FDE) techniques, have been shown to be suitable for high data rate transmission over severely time-dispersive channels [Falconer et al. 2002], and therefore present advantages for use with emergent MBMS. The OFDM technique is the most popular modulation based on this technique and has been selected for LTE, as opposed to WCDMA, which is the air interface technique that has been selected by European Telecommunications Standard Institute (ETSI) for UMTS. Single-carrier (SC) modulation using FDE is an alternative approach based on this principle [Sari et al. 1994; Falconer et al. 2002]. Because of the lower envelope fluctuations of the transmitted signals [and implicitly a lower peak-to-mean envelope power ratio (PMEPR)], SC FDE equalization (SC-FDE) schemes [also designated as SC-frequency division multiple access (SC-FDMA)] are especially interesting for the uplink transmission [Falconer et al. 2002]. Furthermore, MIMO schemes enhanced with state-of-the-art receivers are also normally associated with MBMS in order to improve the overall system performance in terms of capacity, spectral efficiency, and coverage.

For MBMS services, it makes sense to have two or more classes of bits with different error protection, to which different streams of information can be mapped. Depending on the propagation conditions, a given user can attempt to demodulate only the more protected bits or also the other bits that carry the additional information. By using nonuniformly spaced signal points in hierarchical modulations, it is possible to modify the different error protection levels [Vitthaladevuni and Alouini 2001; Vitthaladevuni and Alouini 2003]. These techniques are interesting for applications where the data being transmitted are scalable, that is, they can be split into classes of different importance. For example, in the case of video transmission data from the video source encoders may not be equally important. The same happens in the transmission of coded voice. The nonuniform QAM constellation concept has already been incorporated in the Digital Video Broadcast—Terrestrial (DVB-T) standard.

Other aspects, such as synchronization, alongside advanced equalizers and turbo-decoders also come into play, in order to obtain a correct estimate of the medium, and sort all interferences between the different links.

Since typical FDE receivers are coherent, we need accurate channel estimates at the receiver. These estimates can be obtained with the help of pilots multiplexed with data, in either time or frequency [Hoher et al. 1997]. As an alternative, we can employ implicit pilots (also known as pilot embedding or superimposed pilots) [Orozco-Lugo et al. 2004], where pilots or training signals are added to data so as to save BW. The major problem associated with superimposed pilots is the interference between data and training signals: on the one hand, channel estimates are corrupted by the data signal; on the other hand, detection performance will be degraded due to interference from the training block. With advanced receivers it is possible to jointly perform detection and channel estimation, leading to a good trade-off with improved performance and improved spectral efficiency, while keeping complexity at a low level [Marques da Silva et al. 2009b]. This is especially advantageous when combined with multiantenna systems [Marques da Silva et al. 2009c] and/or when hierarchical constellations are employed [Souto et al. 2007b], situations where channel estimation can present a significant challenge.

1.2.1 Wideband Code Division Multiple Access

First-generation (1G) cellular networks (1980–1992) were analogues (NMT, AMPS, TACS, etc.). These systems were of low reliability, low capacity, low performance, and without the ability to perform roaming between different networks and countries. The multiple access technique adopted was FDMA, where different signals were transmitted in different (orthogonal) frequency bands.

The second generation (2G), such as GSM—(1992–2002/3), introduced digital technology in the cellular environment, with much better performance, better reliability, higher capacity, and with the ability to perform roaming among operators, due to its normalization and technology. The multiple access technique used by GSM was time division multiple access (TDMA), where different signals originating from different users were transmitted in different (orthogonal) TSs.

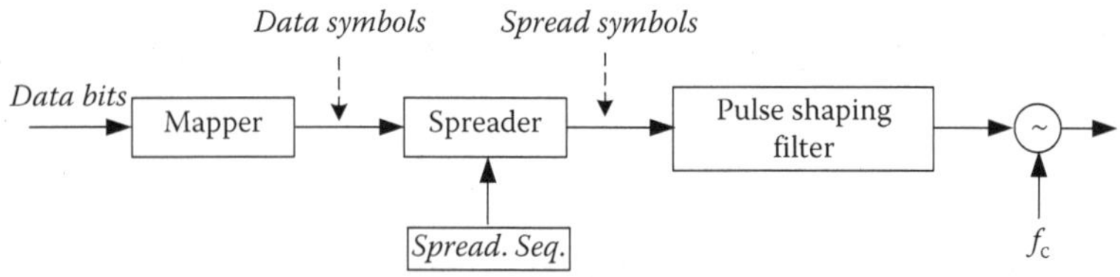

Figure 1.3 Generic block diagram of a spread spectrum transmitter.

The narrowband code division multiple access (CDMA) system was adopted in the 1990s by IS-95 standard, in the United States. Afterward, UMTS proceeds with the utilization of CDMA, in this particular case WCDMA.

The CDMA system consists of different spread spectrum transmissions [Ojanperä and Prasad 1998; Holma and Toskala 2000], each one associated with a different user's transmission using a different (ideally orthogonal) spreading sequence.

Figure 1.3 depicts a simplified block diagram of a spread spectrum transmitter, where the first block consists of a symbol modulator (mapper), responsible for the conversion of source bits into symbols. The resulting signal is then sampled for spreading purposes. The spreader consists of a block that performs multiplication of sampled versions of the symbols with samples of the spreading sequence. Finally, the signal is passed through a band-limited pulse shaping filter and carrier modulator. At the receiver, the signal is convoluted again with exactly the same spreading sequence.

The result of the operations performed by a CDMA transmitter can be seen from Figure 1.4 applied to a binary phase shift keying (BPSK) modulated signal, where T_S and T_C stand for the symbol and chip period, respectively and S_F stands for the spreading factor.

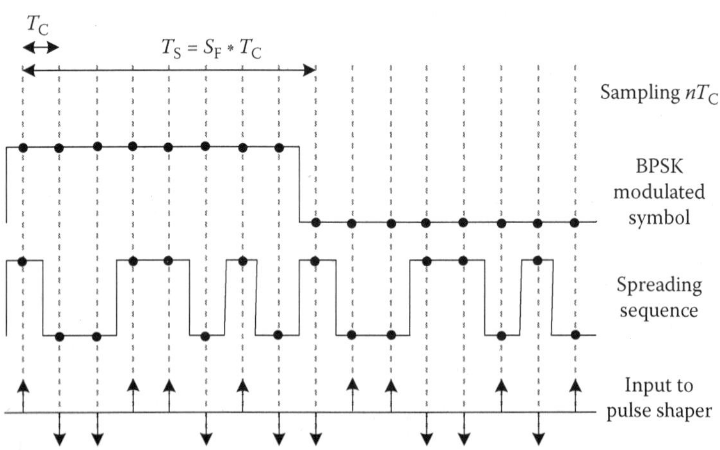

Figure 1.4 Generation of a BPSK spread spectrum signal (spreading factor = 8).

The main characteristics of spread spectrum systems are the additional resistance to interference and the possibility of taking advantage of the multipath channel (MC) in order to exploit multipath diversity. This leads to improved performance and spectral efficiency, as compared to narrowband signals. The relationship between the power spectral densities of a spread spectrum signal and that of the original signal corresponds to the spreading factor. Figure 1.5 depicts the exposed concept.

Assuming that different signals use orthogonal spreading sequences, the corresponding signals are also orthogonal and the same spectrum can be shared among different signals. Moreover, WCDMA consists of a CDMA system whose spread BW is typically higher than the coherence BW of the channel. This allows a better exploitation of multipath diversity, but requires higher spectrum availability.

The typical WCDMA receiver is a RAKE receiver [Glisic and Vucetic 1997], which has several fingers to detect different multipaths of the channel. In a multipath environment, since each finger of the RAKE receiver discriminates a different multipath, the combination of signals from different fingers with a maximal-ratio combining (MRC) tends to achieve a performance improvement compared to a single decorrelator. For this reason, it is normally stated that a WCDMA system jointly with a RAKE receiver is able to exploit multipath diversity. As the spreading factor increases, the resolution of the RAKE receiver also increases, allowing better discrimination of the several propagation paths, increasing the diversity order and, potentially, improving the performance.

Assuming correct synchronization, the received signal at the RAKE receiver output is the original signal plus higher frequency components, which are not part of the original signal, and are to be filtered. If there is any undesired interfering signal at the receiver, the spread signal will affect it just as it did to the original signal at the transmitter, spreading

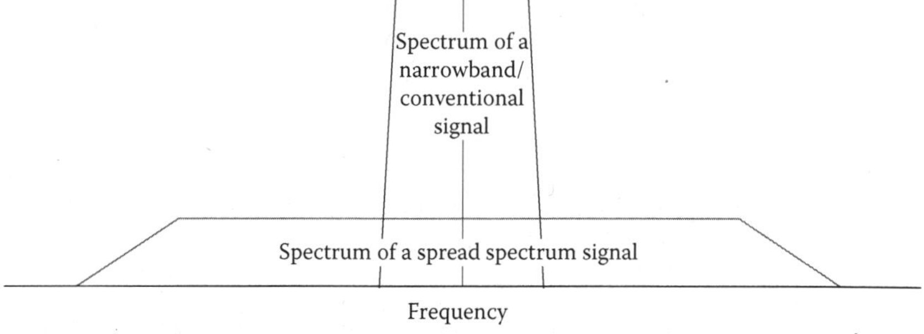

Figure 1.5 Spectrum of spread spectrum signal versus narrowband signal.

it to the BW of the spread signal. Thus, the neglecting effect of the interfering signal is less powerful than in conventional narrowband signals. Furthermore, if the spreading sequences of the different signals that share the same spectrum are not perfectly orthogonal at the receiver side (due to lack of orthogonality itself, due to lack of synchronization, due to the multipath effect, etc.), then the resulting signal is composed of the desired signal plus the noise and a component called multiple access interference (MAI), which can be mitigated by use of a multiuser detector (MUD) for WCDMA signals [Marques da Silva and Correia 2003a; Marques da Silva 2005; Marques da Silva et al. 2005a].

1.2.2 Orthogonal Frequency Division Multiplexing

OFDM is a transmission technique adopted by many high data rate communication systems as the standard IEEE 802.11n (IEEE stands for Institute of Electrical and Electronics Engineers), IEEE 802.16e (2005), and so on, being well suited for frequency-selective fading channels [Cimini 1985; Bingham 1990; Liu and Li 2005]. In opposition to conventional SC transmission techniques where information symbols are transmitted in a single stream, the OFDM technique splits the symbols into several lower rate streams, which are then transmitted in parallel subcarriers. As a consequence, the symbol period is increased, making the signal less sensitive to intersymbol interference (ISI). To avoid interference between subcarriers, the several streams should be transmitted in orthogonal subcarriers.

It is known that sinusoids with frequencies spaced by $1/T$ form an orthogonal basis set in a T-duration interval, and a periodic signal with period T can be represented as a linear combination of the orthogonal sinusoids. Therefore, the orthogonality between subcarriers is assured by using discrete Fourier transform (DFT) and inverse DFT (IDFT). In practice, OFDM is normally implemented through an efficient technique called fast Fourier transform (FFT) and inverse FFT (IFFT). Therefore, OFDM signals are commonly generated by computing the N-point IFFT, where the input of IFFT is

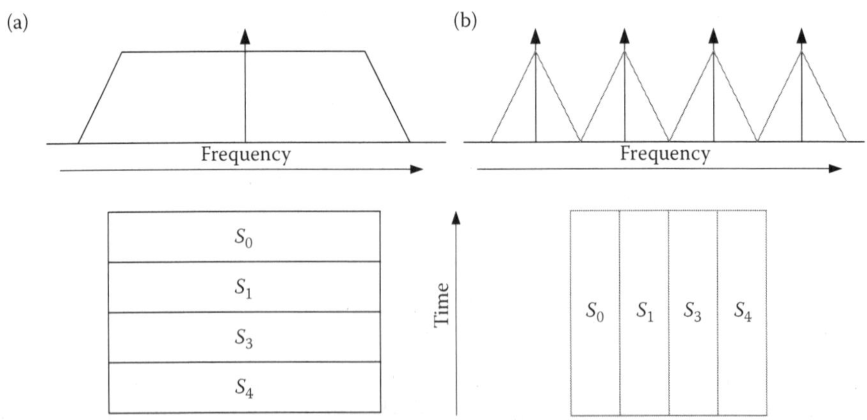

Figure 1.6 Spectrum of (a) SC signal versus (b) OFDM signal.

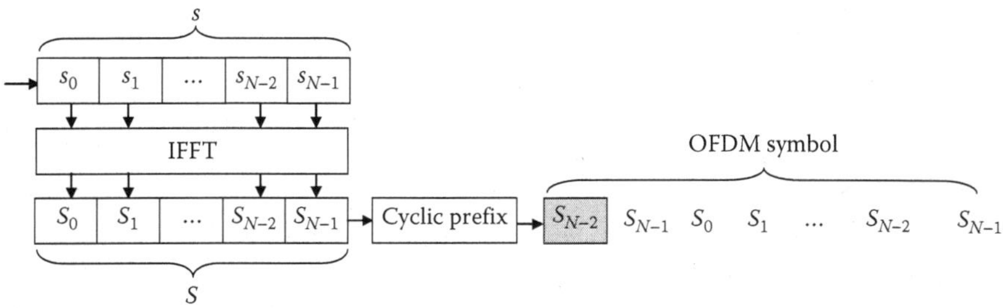

Figure 1.7 CP insertion.

the frequency-domain representation of the OFDM signal. The output of IFFT is the time-domain representation of the OFDM signal; the N-point IFFT output is defined as a useful OFDM symbol. This "time-domain" OFDM signal is composed of N subcarriers, as depicted in Figure 1.6b, as opposed to an SC signal also depicted in the same figure.

Although with OFDM signals the symbol stream is split into several parallel and lower rate substreams (each one associated with a different subcarrier), ISI can still occur within each substream. To mitigate the effect of ISI caused by channel delay spread, each block of N IDFT coefficients is typically preceded by a CP or a guard interval consisting of N_g samples, such that the length of the CP is, at least, equal to the time span of the channel (channel length). The CP is simply a repetition of the last N_g time-domain symbols. The prefix insertion operation is illustrated in Figure 1.7.

Subcarrier spacing is determined by the IDFT size N and input sampling rate of the IDFT. A subset of active subcarriers is mapped with the data modulation symbols and pilot symbols. The remaining subcarriers are left inactive prior to the IDFT; the inactive subcarriers are determined based on the spectral constraints, such as the BW allocation and the spectral mask.

If the CP length is greater than the length of the channel, the linear convolution of the transmitted sequence of IDFT coefficients with the discrete-time channel is converted into a circular convolution. As a result, the effects of ISI and intercarrier interference (ICI) are completely and easily removed. After removal of the guard interval, each block of N received samples is converted back to the frequency domain using a DFT. Each of the N frequency-domain samples are processed with a simple one-tap frequency-domain equalizer and applied to a decision device to a metric computer.

In conventional time-domain signals, the equalization process consists of a series of convolution operations, whose length is proportional to the time span of the channel. This is very complex from the hardware requirements or time consuming, and its effectiveness is always a function of the number of symbols considered. In OFDM, equalization is performed as a simple multiplication of the OFDM signal spectrum with the frequency response of the channel. This represents a great advantage in terms of processing requirements and effectiveness, as compared to the equalization process normally employed in time-domain signals.

Multiple access is achieved in OFDMA (multiuser scenario) by assigning subsets of OFDM subcarriers to individual users, allowing simultaneous low data rate transmission from several users.

1.2.3 Multiple Input–Multiple Output

The use of multiple antennas at both the transmitter and the receiver aims to improve the performance or symbol rate of systems, but it usually requires higher implementation complexity. Moreover, the antenna spacing must be larger than the coherence distance to ensure independent fading across different antennas [Foschini 1996; Foschini and Gans 1998; Rooyen et al. 2000].

The various configurations, shown in Figure 1.8, are referred to as multiple input–single output (MISO), single input–multiple output (SIMO), or MIMO. On the one hand, the SIMO and MISO architectures are a form of receive and transmit diversity schemes, respectively. On the other hand, MIMO architectures can be used for combined transmit and receive diversity as well as for the parallel transmission of data or SM. When used for SM, MIMO technology promises high bit rates in a narrow BW; therefore, it is of great significance to spectrum users. MIMO systems transmit different signals from each transmit element so that the receiving antenna array receives a superposition of all the transmitted signals. Receive and transmit antennas must be sufficiently separated in space and/or polarization to create independent propagation paths.

MIMO schemes are used in order to push the performance or capacity/throughput limits as high as possible without an increase in spectrum BW, although there is an obvious increase in complexity [Rooyen et al. 2000; Hottinen et al. 2003; Marques da Silva and Correia 2001; Marques da Silva and Correia 2002a; Marques da Silva and Correia 2002b; Marques da Silva and Correia 2003b]. Figure 1.9 presents a generic diagram of a MIMO scheme.

For M transmit and N receive antennas, we have the capacity equation [Telatar 1995; Foschini and Gans 1998]

$$C_{EP} = \log_2 \det\left(I_N + \frac{\beta}{M}\mathbf{HH}' \right) \text{ b/s/Hz}, \tag{1.1}$$

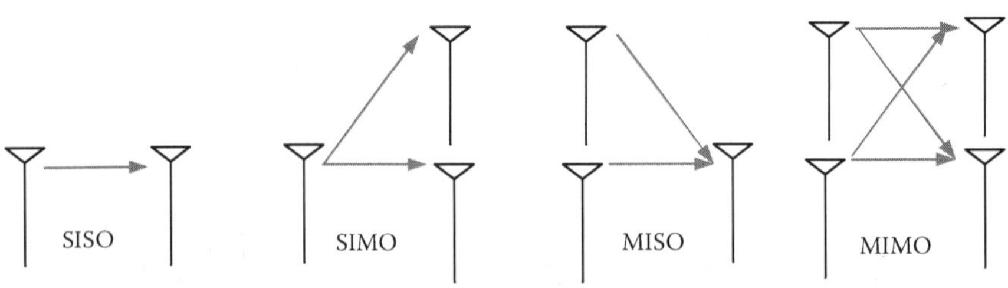

Figure 1.8 Multiple antennas configurations.

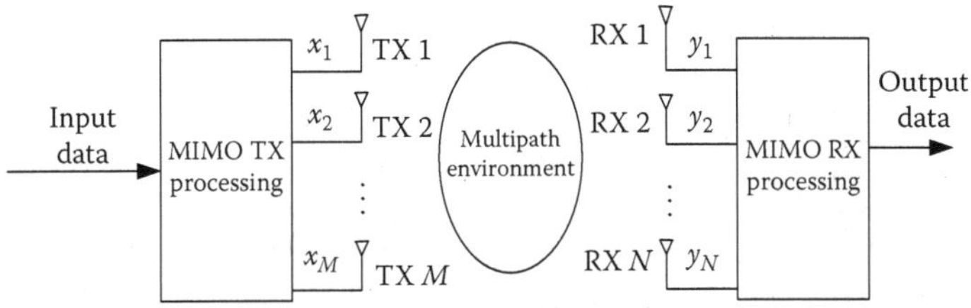

Figure 1.9 Generic diagram of a MIMO scheme.

where I_N is the identity matrix of dimension $N \times N$, **H** is the channel matrix, $\mathbf{H}'$ is the transpose-conjugate of **H**, and β is the SNR at any receive antenna. Foschini [Foschini 1996] and Telatar [Telatar 1999] both demonstrated that capacity grows linearly with $m = \min(M, N)$ for uncorrelated channels. Therefore, it is possible to employ MIMO as a multiresolution distribution system where concurrent data streams are transmitted by M and received by $N(M \leq N)$ different antennas. The downside to this is receiver complexity, sensitivity to interference, and correlation between antennas, which is more significant as the antennas are closer. For a UMTS system, it is inadequate to consider more than two or four antennas at the UE/mobile receiver.

MIMO schemes are implemented based on multiple-antenna techniques. These multiple-antenna techniques are

- Space–time block coding (STBC)
- Multilayer transmission
- Space division multiple access (SDMA)
- Beamforming

Although STBC is essentially a MISO system, the use of receiver diversity makes it a MIMO, which corresponds to the most common configuration for this type of diversity. STBC-based schemes focus on achieving performance improvement through the exploitation of additional diversity, while keeping the symbol rate unchanged [Alamouti 1998; Tarokh et al. 1999]. Symbols are transmitted using an orthogonal block structure, which enables simple decoding at the receiver [Alamouti 1998; Silva et al. 2004a; Marques da Silva et al. 2009a].

Multilayer transmission and SDMA belong to the same group, entitled spatial multiplexing (SM), whose principles are similar but whose purposes are quite different.

The goal of the MIMO based on the multilayer transmission scheme is to achieve higher data rates in a given BW, whose increase rate corresponds to the number of transmit antennas [Foschini 1996; Foschini and Gans 1998; Nam and Lee 2002]. An example of the multilayer transmission scheme is the Vertical—Bell Laboratories Layered Space–Time (V-BLAST). In this case, the number of receive antennas must be equal to or higher than the number of transmit antennas. The increase in symbol rate is achieved by "steering" the receive antennas to each one (separately) of the transmit antennas, in order to receive the corresponding data stream. This is achieved through the use of the nulling algorithm. With a sufficient number of receive antennas it is possible to resolve all data streams, as long as the antennas are sufficiently spaced so as to minimize the correlation [Marques da Silva et al. 2005b].

The goal of the SDMA scheme is to improve the capacity (more users per cell), while keeping the spectrum allocation unchanged. It is usually considered in the uplink, where the transmitter (UE) has a single antenna while the receiver (BS) has several antennas. Figure 1.10 depicts an SDMA configuration applied to the uplink. SDMA assumes that the number of antennas at the receiver is higher than the number of users that share the same spectrum. With such an approach, the receiver can decode the signals from each transmitter, while avoiding the signals from the other transmitters. Similarly to the decoding performed in multilayer transmission, this can be achieved through the use of the nulling algorithm.

In SM, the symbol with the highest SNR is first detected using a linear nulling algorithm such as zero forcing (ZF) or minimum mean square error (MMSE) [Foschini 1996]. The detected symbol is regenerated, and the corresponding signal portion is subtracted from the received signal vector using typically a successive interference cancelation (SIC). This cancelation process results in a modified received signal vector with fewer interfering signal components left. This process is repeated until all symbols are detected. According to the detection-ordering scheme in Ref. [Foschini 1996], the detection process is organized so that the symbol with the highest SNR is detected at each detection stage.

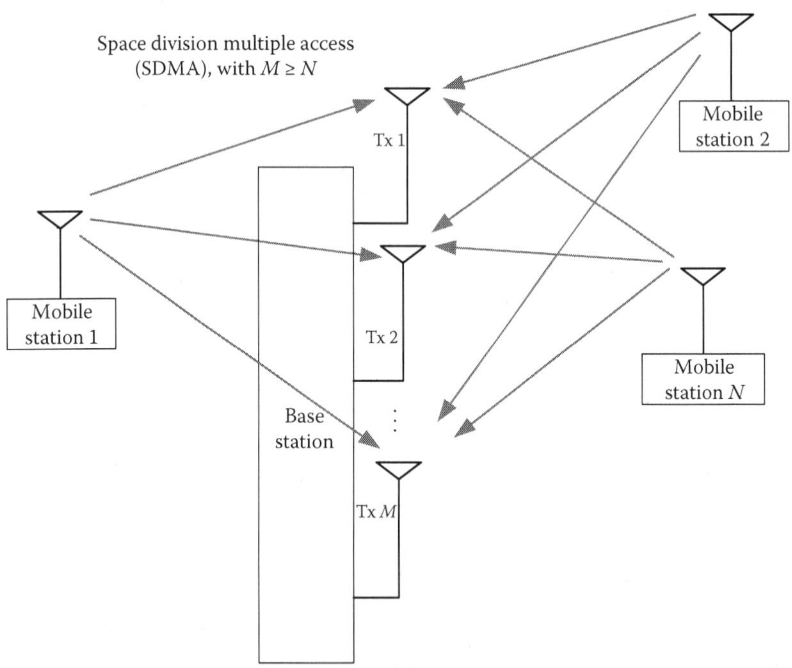

Figure 1.10 Example of the SDMA scheme applied to the uplink.

In opposition to the STBC and SM MIMO schemes where those antenna elements that form an array are usually widely separated in order to form a transmit diversity array with low correlation among them, beamforming is implemented by an antenna array with certain array elements at the transmitter or receiver being closely located to form a beam array with antenna elements spacing typically half wavelength. This scheme is an effective solution to maximize the SNR, as it steers the transmit (or receive) beam toward the receiver (or transmitter) [Marques da Silva et al. 2009a]. As a result, improved performance or coverage is achieved with beamforming, that is, the number of required cells to cover a given area is reduced.

MIMO schemes normally require some level of additional processing and complexity from the receiver. Different receivers were analyzed in Refs. [Silva et al. 2005a; Silva et al. 2005b], in order to establish the trade-off between performance and complexity for such systems. An alternative approach that minimizes the level of computation from the receiver, transferring it to the transmitter, while maintaining the ability to exploit diversity, has been proposed in Refs. [Marques da Silva 2005; Marques da Silva et al. 2008b; Marques da Silva et al. 2009a], which consists of a preprocessing scheme for MIMO systems.

1.2.4 Hierarchical QAM Constellations

The limited spectrum resources available are one of the major restrictions for achieving high bit rate transmissions in wireless communication networks. M-QAM is considered an attractive technique to achieve this objective due to its high spectral efficiency and has been studied and proposed for wireless systems by several authors [Webb and Hanzo 1994; Webb and Steele 1995; Goldsmith and Chua 1997]. In fact 16-QAM modulation has already been standardized for the HSDPA mode of the UMTS by the Third Generation Partnership Project (3GPP) [3GPP04E].

M-QAM constellations can be constructed in a hierarchical way so as to provide multiresolution and improve the efficiency of the network in broadcast/multicast transmissions, as discussed in Ref. [Cover 1972]. In this case, the constellations can be referred to as hierarchical, embedded, or multiresolution M-QAM [we will denote them as multiresolution hierarchical QAM (M-HQAM)].

To present an example of that, 16-QAM hierarchical constellations can be constructed using a main QPSK constellation where each symbol is, in fact, another QPSK constellation, as shown in Figure 1.11. This construction procedure results in two classes of bits with different error protection that can be modified by using nonuniformly spaced signal points.

The bits used for selecting the symbols inside the small inner constellations are called weak bits and the bits corresponding to the selection of the large outer QPSK constellation are called stronger bits. The idea is that the constellation can be seen as a 16-QAM constellation if the channel conditions are good or as a QPSK constellation otherwise. In the latter situation, the received bit rate is reduced to half. Some alterations to the physical layer of the UMTS system to incorporate these modulations have already been proposed in Refs. [Souto et al. 2005a; Souto et al. 2005b; Souto et al. 2007b].

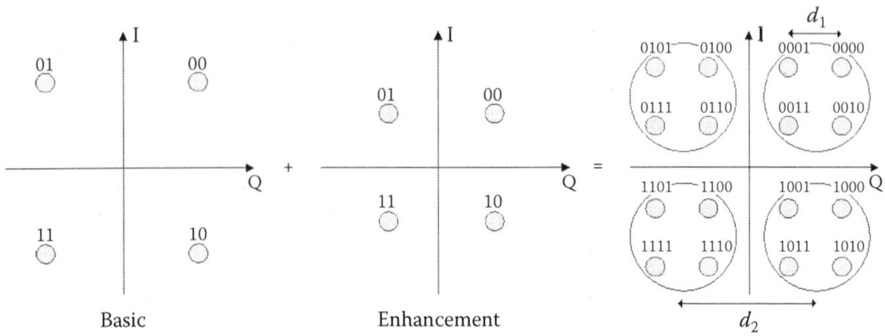

Figure 1.11 Signal constellation for 16-QAM nonuniform modulation.

Figure 1.12 shows an example of the usage of a 64-HQAM constellation in a cellular system. Depending on the position in the cell, users will demodulate the received signal as 64-QAM, 16-QAM, or QPSK.

These techniques are interesting for applications where the data being transmitted are scalable, that is, they can be split into classes of different importance. For example, in the case of video transmission, the data from the video source encoders may not be equally important. The same happens in the transmission of coded voice. Several authors have studied the use of hierarchical constellations for this purpose. In Refs. [Ramchandran et al. 1993] and [Wei 1993], hierarchical QAM constellations were employed for the transmission of digital high-definition television (HDTV) signals. Furthermore, [Engels and Rohling 1998] compares the performance of 64-QAM and 64-DAPSK hierarchical

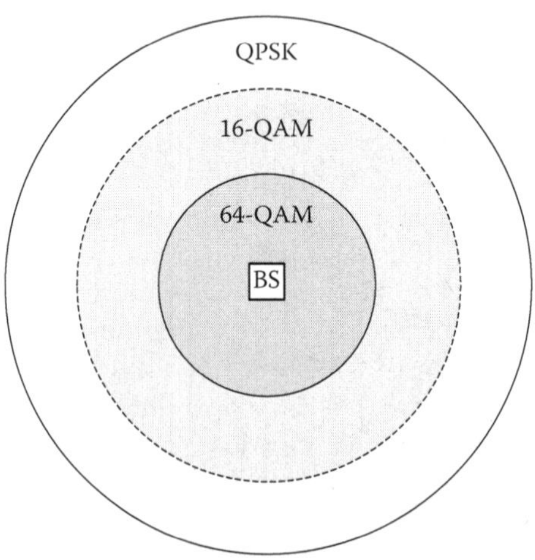

Figure 1.12 Example of the type of demodulation used inside a cell for transmission of a 64-QAM hierarchical constellation.

constellations while [Pursley and Shea 1999] studies the use of M-PSK hierarchical constellations in multimedia transmissions. Hierarchical 16-QAM and 64-QAM constellations have been incorporated into DVB-T standards [ETSI 2004].

1.2.5 Multiresolution Transmission Schemes

The introduction of multiresolution in a broadcast cellular system deals with source coding and the transmission of the output data streams. In a broadcast cellular system there is a heterogeneous network with different terminal capabilities and connection speeds. For the particular case of video, a common strategy presented in the literature to adapt its content within a heterogeneous communications environment is scalable video [Li 2001; Liu et al. 2003; Vetro et al. 2003; Dogan et al. 2004; Holma and Toskala 2007].

A flexible common channel, suitable for PMP transmissions, is already available in UMTS networks, namely the forward access channel (FACH), which is mapped onto the secondary common control physical channel (S-CCPCH).

If we do not have macrodiversity in Ref. [3GPP 2005b], about 40% of the sector total power has to be allocated to a single 64 Kbps MBMS if full cell coverage is required. This makes MBMS too expensive since the overall system capacity is limited by the power resource. To make MBMS affordable for the cellular system, its power consumption has to be reduced. Inner-loop power control (also called fast closed-loop power control) is not implemented in S-CCPCH channels because multiresolution schemes and/or macrodiversity allow using less transmit power. It is worth recalling that inner-loop power control in the downlink consists of the ability of the BS to adjust its output power in accordance with one or more commands received from the UE, in order to keep the received SNR at a given threshold.

Assuming that macrodiversity combining is not used, extra power budget has to be allocated to compensate for the receiving power fluctuations. A common approach consists of considering MBMS video streaming as scalable, with one basic layer to encode the basic quality and consecutive enhancement layers for higher quality. Only the most important stream (basic layer) is sent to all users in the cell to provide the basic service. The least important streams (enhancement layers) are sent with less amount of power or coding protection and only users who have better channel conditions are able to receive that additional information to enhance the video quality. Using this methodology, the transmission power for the most important MBMS stream can be reduced because the data rate is reduced, and the transmission power for the less important streams can also be reduced because the coverage requirement is relaxed.

Scalable video, depicted in Figure 1.13, provides a base layer for minimum requirements [Li 2001] and one or more enhancement layers to offer improved quality at increasing bit/frame rates and resolutions. Therefore, method significantly decreases the storage costs of the content provider.

Besides being a potential solution for content adaptation, scalable video schemes may also allow an efficient usage of power resources in MBMS, as suggested in Ref. [Cover 1972]. This is depicted in Figure 1.13, where two separate physical channels are provided for one MBMS service (e.g., at 256 Kbps): one for the base layer, at half bit rate of the

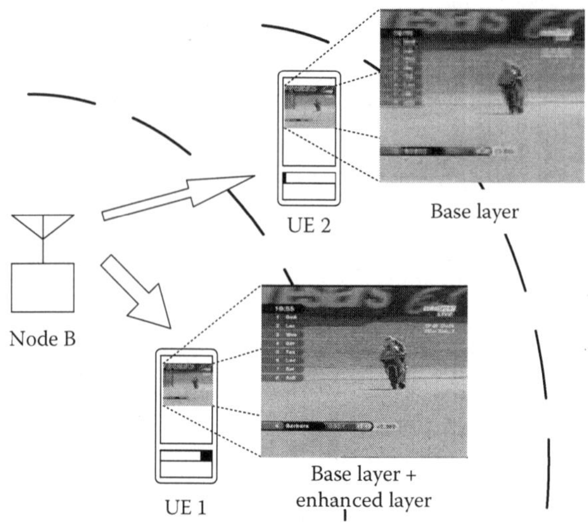

Figure 1.13 Scalable video transmission.

total bit rate (128 Kbps), and with a power allocation that can cover the whole cell range, and one for the enhanced layer, also at half bit rate of the total bit rate (128 Kbps), but with less power allocation than that of the base layer.

The system illustrated in Figure 1.13 consists of two QoS regions, where the first region receives all the information whereas the second region receives the most important data. The QoS regions are associated with the geometry factor that reflects the distance of the UE from the BS antenna.

Scalable video transmission can also be implemented using different techniques. One example consists of the use of hierarchical constellations. In case the channel conditions are not above a certain threshold, since the modulation considered is QPSK (instead of 16-QAM) the received bit rate is reduced to half. In addition, hierarchical constellations may be combined with different channel coding rates, which corresponds to the concept of AMC [Souto et al. 2007b].

Another possibility of implementing scalable video transmission is the use of the SM MIMO technique, where each transmit antenna sends a different data stream. The first data stream (most powerful) may include the base layer, whereas the enhanced layer may be sent by a second antenna (less powerful data stream). Depending on the power and channel conditions, a certain UE may receive successfully either the two streams or only the base layer.

1.2.6 Macrodiversity

There are two types of networks to be considered: the multifrequency networks (MFNs) and the single-frequency networks (SFNs). Macrodiversity refers to the transmission of the same information by different Node Bs to the UE in the downlink. The Node Bs, to which a terminal is linked to, are referred to as the active set. Macrodiversity aims at supplying additional diversity in situations where the terminal is far from the Node Bs, in order to compensate the path loss affecting the transmission to a UE located at the edge of the cell, and to reduce the amount of transmit power needed to reach a distant receiver, thus increasing network capacity.

It is worth noting that in broadcast, the global channel impulse response (CIR) is longer due to the distance between the transmitter (BS) and the several different receivers (UEs). Nevertheless, if the cyclic extension is long enough, the global CIR will be the sum of the independent CIRs, which enables the SFN concept, allowing the macrodiversity effect. This can be seen from Figure 1.14.

Therefore, macrodiversity is used during soft handover to ensure smooth transitions between two cells or two sectors of the same cell, thus reducing the risk to drop the call.

Note that the performance gain brought by macrodiversity depends on the diversity order of the channel, that is, a two-path channel benefits more from macrodiversity than a six-path channel because the latter exhibits already a high multipath diversity order. In opposition to the dedicated physical channel, macrodiversity for MBMS does not consume network resources, as MBMS is broadcasted simultaneously in several cells. In the case of MFN, the UE is required to estimate the carrier of each Node B it is linked to. This increases its power consumption. Moreover, the signals received from different Node Bs (especially far ones) may be significantly delayed with regard to those received from

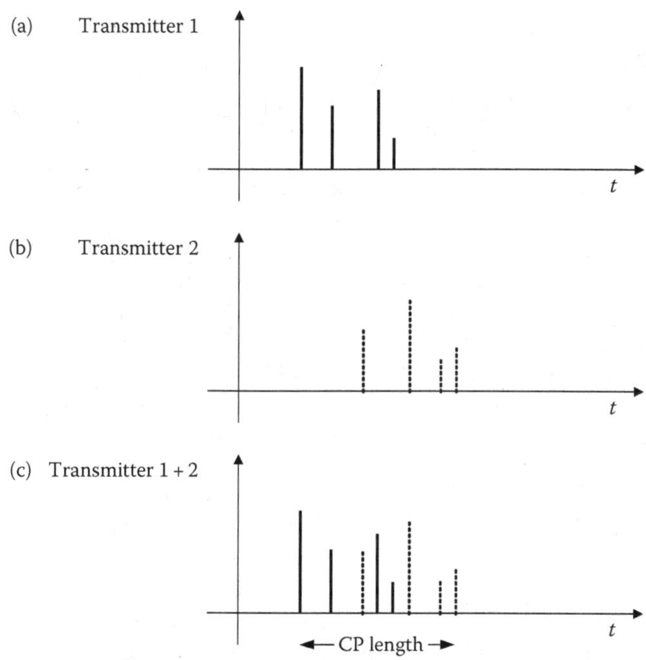

Figure 1.14 Global CIR (c) is composed of the sum of the CIR of the several transmitters (a + b).

near Node Bs. This requires extra memory at the terminal in order to store the received signals for further combining or an additional synchronization procedure among the Node Bs transmitters. In the downlink, the combining takes place at the mobile, which has to demodulate and then combine the signals received from the different Node Bs in the active set. The extra complexity added by macrodiversity then depends on the receiver type. In the case of an equalizer, one has to be set up and operated for each Node B the UE is linked to. Moreover, the UE must estimate one transmission channel per Node B. In the special case of OFDM, two main cases for macrodiversity can be distinguished:

Node Bs are synchronized, at least to allow UE's receiving signals from two or more Node Bs with a time difference smaller than the CP.
Node Bs are not synchronized.

In the first case, Node Bs can transmit identical signals to the terminal on the same time–frequency resource. This is possible because the signals will superpose within the CP: no ISI occurs as long as the sum of the time differences plus the maximum delay of the CIRs is shorter than the CP. In this case, the terminal can employ a single receiver to demodulate the superimposed signals. This means that it will perform a unique DFT. In this case, macrodiversity behaves just like transmit diversity (from a unique transmitter with multiple spaced antennas). When different Node Bs transmit the same data over the same subcarriers, the resulting propagation channel is equivalent to the allegorical sum of all propagation channels, which increase the diversity gain. When Node Bs send the same data over different subcarriers, maximum diversity can be achieved since each data symbol benefits from the summation of the propagation channel powers. It could also be possible to form a MIMO scheme but, in that case, the data and distinct pilot signals have to be sent on orthogonal time–frequency resources and several propagation channels have to be estimated. If the Node Bs cannot be assumed to be synchronized, the terminal will need separated receiver chains to demodulate the signals from the distinct Node Bs. Moreover, to avoid interference, orthogonal time–frequency resources have to be allocated to the different Node Bs. This is still very complex to fulfill; thus in the general case, interference will occur.

Fast cell selection is one option for macrodiversity for unicast data. Intra-Node-B selection should be able to operate on a subframe basis. An alternative Intra-Node-B macrodiversity scheme for unicast is a simultaneous multicell transmission with soft combining. The basic idea of multicell transmission is that, instead of avoiding interference at the cell border by means of intercell interference coordination, both cells are used for transmission of the same information to a UE, thus reducing intercell interference as well as improving the overall available transmit power. Another possibility of intra-Node-B multicell transmission is to explore the diversity gain between cells with space–time processing (e.g., by employing STBC through two cells). Assuming Node-B-controlled scheduling and that fast/tight coordination between different Node Bs is not feasible, multicell transmission should be limited to cells belonging to the same Node B. For multicell broadcast, soft combining of radio links should be supported, assuming a sufficient degree of inter-Node-B synchronization, at least among a subset of Node Bs.

1.2.7 Multihop Relays

One key to improve the coverage and capacity for high-quality multimedia broadcast and multicast transmissions in mobile networks is to increase the data rates for mobile stations (MSs) at the cell edge. MSs at the cell edge suffer from high propagation loss and high intercell interference from neighbor cells. Other MSs reside in areas that suffer from strong shadowing effects. Thus the overall goal of this activity is to bring more power to the cell edge and into shadowed areas while inducing minimal additional interference for neighbor cells. There are several measures to implement multihop relays [Sydir and Taori 2009]. The next paragraphs describe some of these methods.

Adaptive relaying: As stated before, MSs at the cell edge suffer from high propagation loss and high intercell interference from neighbor cells. Other MSs reside in areas that suffer from strong shadowing effects. Thus, the overall goal of adaptive relaying is to bring more power to the cell edge and into shadowed areas while inducing minimal additional interference for neighbor cells. The obvious solution for this would be to decrease cell sizes by installing additional BSs which would of course mean to increase the network infrastructure costs.

In opposition to conventional repeaters working with the amplify-and-forward strategy, adaptive relays are understood to work in a decode-and-forward style. By doing this, relays amplify and retransmit only the wanted component of the signal they receive and suppress the unwanted portions (they do not amplify unwanted signals). The disadvantages of relays compared to simple repeaters are the additional delay that they introduce into the transmission path between BS and UE and, depending on the algorithms, a possible signaling overhead.

The gain is based on the fact that the transmission path is split up into smaller parts that can reduce propagation loss.

Fixed RSs positioned at a specified distance from the BS (Figure 1.15) could help to increase the probability that an MS receives enough power from several BSs. This

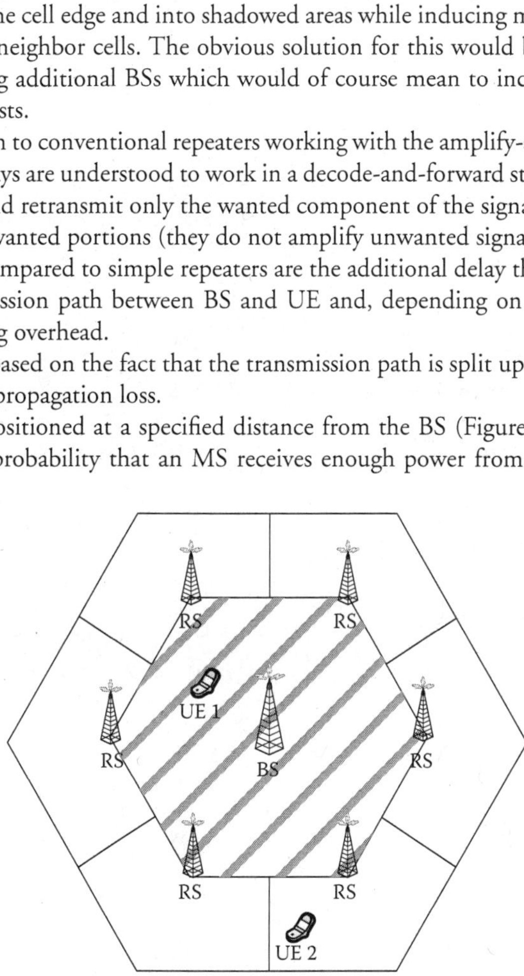

Figure 1.15 Two-hop relaying architecture.

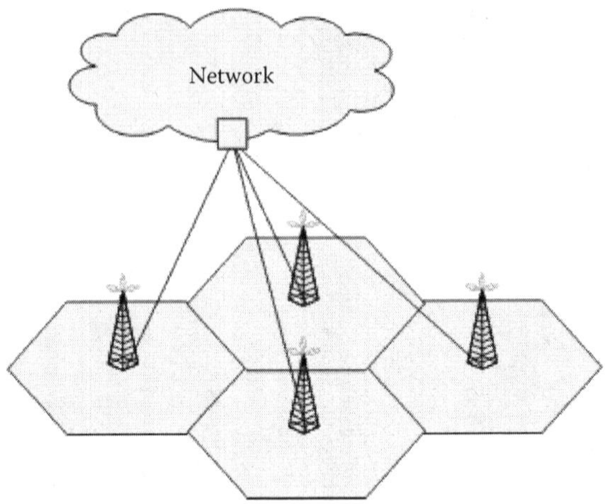

Figure 1.16 Present cellular network.

deployment concept would sectorize the cell in an inner region where the UEs (e.g., UE 1 in the figure) can receive their signal from the BS plus some RSs and an outer region where only the signal from the RSs is strong enough (e.g., UE 2 in the figure).

Configurable virtual cell sizes: Kudoh and Adachi [Kudoh and Adachi 2003] proposed a wireless multihop virtual cellular network. It consists of a so-called central port that corresponds to the BS that acts as a gateway to the core network. The so-called wireless ports correspond to the RSs that communicate with the MSs and relay the signal from and to the central port. The wireless ports that are communicating directly with the MSs are called end wireless ports. The wireless ports are stationary and can act together with the central port as one virtual BS. The central port and the end wireless ports introduce additional diversity into the cell, so that the transmit power may be reduced, which means also a reduced interference for other virtual cells. The differences between present cellular networks and virtual cellular networks are illustrated in Figures 1.16 and 1.17.

From the perspective of multimedia broadcasting and multicasting, configurable virtual cell sizes could be used to adapt the cell size to the user distribution and their service needs that can be very different in spatial or temporal dimensions.

1.3 Current and Emergent Wireless Systems

1.3.1 HSPA Improvements

3GPP wireless systems based on WCDMA radio access technology have been globally deployed. With all areas of the wireless industry developing rapidly, it has become necessary for the UMTS industry and its constituent members to provide their vision for the medium- and long-term evolution of UMTS networks and services. 3GPP has been

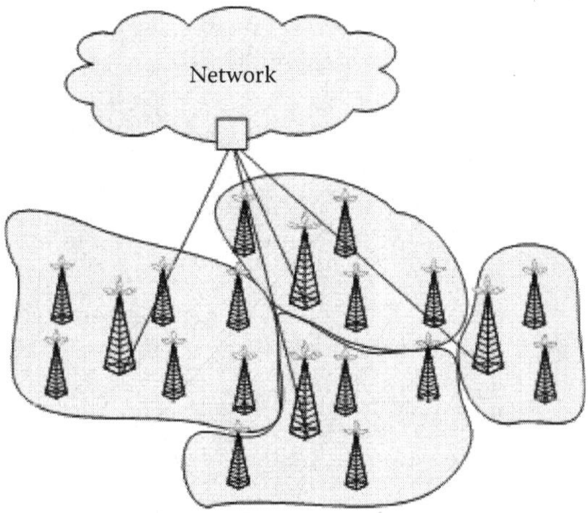

Figure 1.17 Virtual cellular network.

active to provide a vision of this evolution through its activities. Important milestones were achieved in 3GPP Releases 5 and 6 with IP multimedia subsystem (IMS), namely the high-speed downlink packet access (HSDPA) (enhancements enabling data transmission speeds of up to 14.4 Mbps per user), the high-speed uplink packet access (HSUPA), the incorporation of wireless local area network (WLAN) concept, and the integration of the IP Protocol as a transport protocol, in addition to circuit switching. High-speed packet access (HSPA) is a generic term adopted by the UMTS forum to refer to improvements in the UMTS Radio Interface in Releases 5 and 6 of the 3GPP standards. HSPA refers to both the improvements made in the UMTS downlink, often referred to as HSDPA, and the improvements made in the uplink, often referred to as HSUPA, but also referred to as enhanced dedicated channel (E-DCH). HSPA defines a series of straightforward upgrades to UMTS Release 99 (R99) networks, which offers improvements by a factor of ten in the speed of service delivery, improvements by a factor of five in network capacity, and a significant improvement in service latency. The next table summarizes the differences between the several different 3GPP releases.

New terminals are required to support these capabilities and the last terminals for HSPA were made available at the end of 2008; they are category 6 terminals, capable of supporting downloads up to 14.4 Mbps and uploads up to 5.76 Mbps.

Both HSDPA and HSUPA were implemented in the standard 5 MHz carrier of UMTS networks (Release 6), coexisting with the 1G of UMTS networks based on the 3GPP R99 standard. As HSPA standards refer uniquely to the radio access network (RAN), no change of the packet switching core network beyond handling higher traffic was required.

HSPA offered a cost-effective wide-area broadband mobility and played a significant role in stimulating the demand for data services, whether they are consumer multimedia and gaming or corporate e-mail and mobile access. By comparing HSPA with WiMAX (described below) it is clear that the higher penetration rates of the former are related to the latter. For the time being, HSPA connections are around 150 million globally, while WiMAX connections do not go beyond 3.5 million.

1.3.2 LTE Specifications

Emergent services are demanding more and more from wireless infrastructures. 3GPP is deeply involved in specifying and defining the architecture of the 3G evolution. A few years ago, 3GPP launched the study item evolved-UTRAN (E-UTRAN), which aims at defining the new air interface of the 3GPP LTE [Holma and Toskala 2007]. 3GPP LTE will be the natural evolution of the UMTS standard, in order to face the latest demands for voice and data services. In 2008, the increase in data rate was up to a factor of five [Bogineni et al. 2009].

The LTE air interface has been added to the specification in Release 8 and its initial deployment is expected to begin in 2010. Goals include improving spectral efficiency by a factor of 2–4 compared to HSPA Releases 6 and 7, making use of new spectrum and reframed spectrum opportunities, as well as better integration with other open standards [3GPP 2005a; NGMN 2006; NGMN 2007]. The LTE air interface is a completely new system based on OFDMA in the downlink and SC-FDE (also referred to as SC-FDMA by many authors) in the uplink. SC-FDE is the selected transmission technique for the uplink due to its lower peak to average power ratio, as compared to OFDM. In addition, LTE comprises MIMO as an option in order to achieve peak data rates in the downlink exceeding 300 Mbps and uplink peak data rates of approximately 75 Mbps [Dahlman et al. 2008]. Four different types of MIMO have been considered in 3GPP Release 8. Each one can be selected depending on the objective, namely STBC for improved performance, SM for increased data rate, SDMA for increased cell capacity, and beamforming for increased coverage [Astély et al. 2009]. The modulation comprises QPSK, 16-QAM, or 64-QAM. To handle occasional retransmission, LTE considers hybrid-automatic repeat request (H-ARQ) in the medium access control (MAC) sublayer.

Another important modification is that LTE will be based on all-IP architecture (i.e., all services in the future will be carried out on top of IP), instead of the existing UMTS circuit plus packet switching combined network. In fact, it is for use over any IP network, including WiMAX and Wi-Fi, and even wired networks [Dahlman et al. 2008]. Therefore, an important focus of this evolution is on enhancements for packet-based services.

The 3GPP community has been working on LTE and various contributions were made to implement MBMS in LTE [Astély et al. 2009]. In the 3GPP LTE project, two types of transmission scenarios exist.

Multicell transmission: Multimedia broadcast over an SFN (MBSFN) on a dedicated frequency layer or on a shared frequency layer.

Single-cell transmission: Single cell–point-to-multipoint (SC–PMP) on a shared frequency layer.

Multicell transmission in an SFN area is a way of improving the spectral efficiency: because all MBMS cells transmit the same MBMS session data, the signals can be combined for a UE located at a cell boundary. Furthermore, multicell transmission may be provided over a cell group that comprises cells that transmit the same service. In contrast, single-cell transmission covers only one cell.

The concept of a dynamic MBSFN area is introduced where MBMS transmission is switched off in some cells of the MBSFN area when a certain MBMS is not required. In some cases, the released resource can be reused for other MBMS or unicast services.

By considering MBSFN, LTE will allow delivering services such as mobile TV. It is worth noting that LTE standardization has already reached a state where specification changes are limited to corrections and bug fixes.

OFDM is an attractive choice to meet requirements for high data rates, with correspondingly large transmission BWs and flexible spectrum allocation [Bingham 1990; Liu and Li 2005]. OFDM also allows for a smooth migration from earlier radio access technologies and is known for high performance in frequency-selective channels. It enables further frequency-domain adaptation, provides benefits in broadcast scenarios, and is well suited for MIMO processing.

The possibility of operating in vastly different spectrum allocations is essential. In addition, different BWs are realized by varying the number of subcarriers used for transmission, while the subcarrier spacing remains unchanged. In this way, operation in spectrum allocations of 1.4, 2.5, 5, 10, 15, and 20 MHz can be supported, where the latter is required to provide the highest LTE data rates [Astély et al. 2006; Astély et al. 2009]. The draft document [3GPP 2006a], edited by RAN1 Working Group within 3GPP, defines and describes the potential physical layer evolution under consideration for E-UTRAN and UTRAN. Due to the fine frequency granularity offered by OFDM, a smooth migration of, for example, the 2G spectrum is possible. A 2G GSM operator can in principle migrate on a 200 kHz GSM carrier-by-carrier basis by using only a fraction of the available OFDM subcarriers. Frequency division duplexing (FDD), time division duplexing (TDD), and combined FDD/TDD are supported to allow for operation in the paired as well as unpaired spectrum.

By allocating the proper TSs and carrier frequencies, LTE provides intracell orthogonality between users in both the uplink and downlink. Nevertheless, intercell interference is a major problem, as compared to WCDMA/HSPA, especially for users at the edge of the cell. Several measures can be implemented in order to minimize this problem, such as power control and intercell interference coordination or advanced interference cancelation schemes [Marques da Silva and Correia 2000a; Marques da Silva and Correia 2000b; Marques da Silva and Correia 2003d; Marques da Silva and Correia 2003e; Marques da Silva et al. 2004b; Marques da Silva and Correia 2004a; Astély et al. 2009].

1.3.3 Expectations for 4G

Future mobile cellular communication systems beyond 3G aim at allowing subscribers ubiquitous access to a vitrtually unlimited amount of multimedia content with a guaranteed QoS. This constitutes extensive requirements for improvements to present mobile cellular communication systems to provide a greater throughput to subscribers. As described in Table 1.1, the expected data rates for the fourth-generation (4G) cellular system are in the range of 100 Mbps for vehicular mobility to 1 Gbps for nomadic access (in both indoor and outdoor environments).

Table 1.1 Comparison between Several Different 3GPP Releases

FDD *TDD*	*WCDMA* *TD-SCDMA*	*HSPA* *TD-HSDPA*	*HSPA+* *TD-HSUPA*	*LTE* *TD-LTE*	*LTE/IMT Advanced*
3GPP Release	99	5/6	7	8	10
Downlink data rate	384 Kbps	14.4 Mbps[a]	28 Mbps[a]	>160 Mbps[b]	100 Mbps mobility, 1 Gbps nomadic
Uplink data rate	128 Kbps	5.76 Mbps[a]	11 Mbps[a]	>60 Mbps[b]	
Switching	Circuit and packet switching	Circuit and packet switching	Circuit and packet switching	IP based (packet switching)	IP based (packet switching)
Transmission technique	WCDMA/TD-SCDMA	WCDMA/TD-SCDMA	WCDMA/TD-SCDMA	Downlink: OFDMA; Uplink: SC-FDMA	Downlink: OFDMA; Uplink: SC-FDMA
MIMO	No	No	Yes	Yes	Yes
Multihop relay	No	No	No	No	Yes
AMC	No	Yes	Yes	Yes	Yes
Deployment	2003	2006/8	2008/9	2010	

[a] Peak data rates.
[b] Assuming 20 MHz BW and 2×2 MIMO.

4G is also called International Mobile Telecommunications-Advanced (IMT-Advanced), as defined by the International Telecommunications Union–Radio communications (ITU-R) [ITU-R 2003]. Going back to IMT2000 technology, it is an umbrella term covering several standards, namely IEEE 802.16e/WiMAX, CDMA2000, WCDMA, and so on. In some bibliography 4G is also referred to as IMT-Advanced. In this sense, both LTE-Advanced (to be standardized by 3GPP) and IEEE 802.16m (standard from IEEE) are candidates for IMT-Advanced (i.e., 4G).

It is expected that, in future mobile radio networks, multihop relaying will be introduced [Sydir and Taori 2009]. Therefore, new topological approaches such as multihop or distributed antennas solutions and relaying allow an increased coverage of high rate data transmission as well as improved system performance and capabilities [3GPP 2008a]. Standardization is expected to be finalized in 3GPP Release 10. Nevertheless, for the time being, work on LTE-Advanced within 3GPP is still in an early phase, with several discussions taking place on the technologies to use in order to achieve the requirements.

Within 4G, voice, data, and streamed multimedia will be delivered to the user based on an all over IP packet switched platform, using IP version 6 (IPv6) [3GPP 2008a]. The goal is to reach the necessary QoS and data rates in order to accommodate emergent services such as mobile TV, HDTV, DVB, multimedia messaging service (MMS), video chat, and so on [3GPP 2008a]. All these services must be delivered in the concept of "anywhere" and "anytime."

Potential technologies for the air interface of 4G include several possibilities [Astély et al. 2009; Boudreau et al. 2009]: carrier aggregation composed of, for example, multiple components of 20 MHz in order to support transmission BWs of up to 100°MHz; advanced antenna systems (AASs) increasing the number of downlink transmission layers to eight and the number of uplink layers to four; multihop relay (adaptive relay, fixed RSs, configurable cell sizes, hierarchical cell structures, etc.) in order to improve coverage and data rates; advanced intercell interference cancelation schemes [Boudreau et al. 2009; Marques da Silva et al. 2008a]; multiresolution schemes, and so on. A candidate for IMT-Advanced is the IEEE 802.16m standard, which is expected to be composed of several advances to be implemented over the previous WiMAX version (entitled IEEE 802.16j, which already includes multihop relay and advanced MIMO schemes), in order to meet all the IMT-Advanced requirements [ITU-R M.2134], as defined by ITU-R. Another solution is to use 3GPP LTE (Release 8) as a basis for additional advances, in order to implement 4G.

Due to improvements in terms of space address with the 128 bits made available by IPv6, multicast and broadcast applications will be easily improved as well as security, reliability, intersystem mobility, and interoperability capabilities. In addition, as the 4G concept consists of a pool of wireless standards, this can be efficiently implemented using the software-defined radio (SDR) platform, being currently an interesting R&D area for many industries worldwide.

1.3.4 Wi-Fi Overview

Wi-Fi, often known as Wireless Fidelity, is a trademark of Wi-Fi Alliance for certified products based on the IEEE 802.11 standard. It consists of a technology dedicated to interconnect and allow Internet connection of wireless devices in a LAN environment, using the IEEE 802.11 standard.

This standard specifies the physical layer and the MAC sublayer, offering services to a common 802.2 Logical Link Control (LLC) [IEEE 1999].

Originally developed for cable replacement in companies, it quickly became very popular in providing IP connectivity in environments such as offices, restaurants, airports, residential campuses, and so on, covering typically ranges of the order of 100 m outdoors and 30 m indoors [IEEE 1999].

The initial version of Wi-Fi invented in 1991 by NCR Corporation/AT&T in The Netherlands was standardized as IEEE 802.11 supporting 1 or 2 Mbps in the 2.4 GHz band, using either the frequency hopping spread spectrum (FHSS) or the direct sequence spread spectrum (DSSS) [Geier 2002]. This version was upgraded by the following newer versions.

IEEE 802.11a, as a standard that consists of an extension to IEEE 802.11 that allows up to 54 Mbps in the 5 GHz band using the OFDM transmission technique, instead of FHSS or DSSS.

IEEE 802.11b, allowing a data rate of 11 Mbps in the 2.4 GHz band, using DSSS due to its relative immunity to interference.

IEEE 802.11g, as an extension to IEEE 802.11b that allows up to 54 Mbps in the 2.4 GHz band, using either OFDM or DSSS transmission techniques.

IEEE 802.11n, as an upgrade in order to allow over 100 Mbps in the 5 GHz band, by using both the OFDM transmission technique and the multistreaming MIMO scheme.

Wi-Fi uses the carrier sense multiple access/collision avoidance (CSMA/CA) algorithm as a MAC sublayer protocol (similar to the Ethernet considered by most fixed LANs), which is connectionless [Ohrtman and Roeder 2003]. Just as in an Ethernet LAN, having more users results in a reduction of throughput (within the coverage area). Therefore, its efficiency is limited to a reduced number of users and/or reduced traffic.

Wi-Fi-based products require at least 20 MHz for each channel (22 MHz in the 2.4 GHz band for IEEE 802.11b), and have specified only the license exempt bands 2.4 GHz ISM (Industrial, Scientific, Medical), 5 GHz ISM, and 5 GHz Unlicensed National Information Infrastructure (UNII) for operation [Ferro and Potorti 2005].

With the IEEE 802.11 family of standards, a wireless access point (WAP) connects a group of wireless devices into a single cable device (normally a router). A WAP is similar to a network cable switch, performing frame switching based on the MAC address.

Wi-Fi also allows ad hoc networks (peer-to-peer interconnection), which means that wireless devices can connect directly, without using an IEEE 802.11 WAP. In addition, by using two wireless bridges, an IEEE 802.11 link can be established to interconnect two cable LANs, as long as the two bridges are within Wi-Fi wireless range.

1.3.5 WiMAX Overview

WiMAX, standardized by the Institute of Electrical and Electronics Engineers as IEEE 802.16, was initially created in 2001 and updated by several newer versions. It consists of a technology that implements a wireless metropolitan area network (WMAN) using the IP [Eklund et al. 2002; Andrews et al. 2007; Peters and Heath 2009]. WiMAX means

Worldwide Interoperability for Microwave Access and allows fixed and mobile access. The basic idea is to provide wireless Internet access to the last mile, with a range of up to 50 km. Therefore, it can be seen as a complement or competitive to the existing asynchronous digital subscriber line (ADSL) or cable modem, providing the service with the minimum effort in terms of required infrastructures. On the other hand, fixed WiMAX can also be seen as backhaul for Wi-Fi, cellular BS, or mobile WiMAX. As the standard only defines the physical layer and the MAC sublayer, it can be used by either IP version 4 (IPv4) or IPv6 [or even asynchronous transfer mode (ATM)].

In order to allow the operation of WiMAX in different regulatory spectrum constraints faced by operators in different geographies, this standard specifies channel sizes ranging from 1.75 MHz up to 20 MHz, using either TDD or FDD, with many options in between [Yarali and Rahman 2008].

The initial version of WiMAX was upgraded by several newer versions:

IEEE 802.16-2004, also referred to as IEEE 802.16d. This version only specified the fixed interface of WiMAX, without providing any support for mobility [IEEE 2004]. This version of the standard was adopted by ETSI as a base for the High-Performance Metropolitan Area Network (HiperMAN).

IEEE 802.16-2005, also referred to as IEEE 802.16e. It consists of an amendment to the previous version. It introduced support for mobility, handover, and roaming, among other new capabilities [IEEE 2005]. In addition, MIMO schemes were improved in order to achieve better performances.

Relay specifications are included in the IEEE 802.16j amendment. The incorporation of multihop relay capability in the foundation of mobile IEEE 802.16-2005 is a way of increasing available throughput by a factor of 3–10 and/or coverage (and higher channel reuse factor), or even to fill the "coverage hole" of indoor coverage [Oyman et al. 2007; IEEE 2008; Peters and Heath 2009]. It is worth noting that, as previously mentioned, multihop relay capability is being considered as an option for inclusion in IMT-Advanced [Astély et al. 2009]. For backward compatibility, 802.16e Subscriber Station (SS) is able to benefit from this capability without awareness of the presence of the RS. In addition, a distributed MIMO system can be established using both the BS and RS transmit antennas on one side and the receive antennas of the MS on the other side. Alternatively, various adjacent BS and RS antennas can cooperatively perform space–time coding to send data packets.

In addition to these versions, requirements for the next version of mobile WiMAX entitled IEEE 802.16m [IEEE 2009] were completed by the end of 2007. End users will be able to use this new version beginning 2011. The goal of the IEEE 802.16m version is to reach all the IMT-Advanced requirements as proposed by ITU-R in Ref. [ITU-R M.2134 2008], making this standard a strong candidate for 4G. Advances in IEEE 802.16m include wider BWs (up to 100 MHz, shared between the uplink and downlink), adaptive and advanced TDMA/OFDMA access schemes, advanced relaying techniques (already incorporated in IEEE 802.16j), advanced multiple-antenna systems, adaptive modulation schemes such as hierarchical constellations and AMC, frequency adaptive scheduling, and so on [Ahmadi 2009]. In order to preserve investments, IEEE 802.16m will maintain backward compatibility with previous WiMAX versions.

The WiMAX forum is also making some adaptations to IEEE 802.16e in order to allow its use in mobile satellite services coupled with an ancillary terrestrial component, creating the so-called S-WiMAX [Ansari et al. 2009].

The original version of the standard specified a physical layer operating in the range of 10–66 GHz, based on OFDM and TDMA technology. IEEE 802.16-2004 added specifications for the 2–11 GHz range (licensed and unlicensed), whereas IEEE 802.16-2005 introduced scalable OFDMA (SOFDMA) with MIMO (STC based, SM based, or beamforming) or AASs [IEEE 2005] instead of the simple OFDM with 256 subcarriers considered by the previous version.

The introduction of SOFDMA, that is, scaling the DFT to the channel BW, is very effective, as it keeps the carrier spacing constant across different channel BWs. This results in higher spectral efficiency and improved resistance to multipath interference (MPI) in a wide range of delay spreads. In addition, with IEEE 802.16-2005 the number of subcarriers considered may vary from 128 up to 2048, instead of the rigid number of 256 subcarriers considered by IEEE 802.16-2004. Finally, IEEE 802.16-2005 introduced the hybrid automatic repeat request (HARQ) in between the SS and the BS, the AMC technique (BPSK, QPSK, 16-QAM, or 64-QAM, with modulation and coding rate chosen dynamically), as well as turbo coding and low-density parity check (LDPC), greatly enhancing performance and allowing a new capability for non line-of-sight (NLOS) and for mobility using TDD [IEEE 2005; Yarali and Rahman 2008]. On the contrary, the previous version only supported a fixed profile in line-of-sight (LOS) mode, using either TDD or FDD. Finally, the additional QoS made available with the enhanced real-time polling service brought by the 2005 version allowed the WiMAX standard to become much more appropriate for MBMS applications [IEEE 2005].

The Wi-Fi MAC algorithm based on CSMA/CA is not effective for MAN coverage, because of the high probability of distant SS being interrupted by closer SS, reducing their throughputs. This would degrade the QoSs as voice over internet protocol (VoIP) and Internet protocol-based television (IPTV). Therefore, the WiMAX working group has decided to use TDMA (2004 and 2005 versions) and/or OFDMA (2005 version), which is connection-oriented MAC. With such an access technique, the SS needs to compete only once, after which it is allocated a slot (time or frequency subcarriers) by the BS. In addition, based on the throughput requirements, the slot can enlarge or contract, in order to provide the required QoS. WiMAX runs on a connection-oriented MAC and is contention based, which allows providing QoS for a higher number of simultaneous users than Wi-Fi. WiMAX has been designed to scale from one up to hundreds of users within one RF channel.

In terms of throughputs and coverage, these two parameters are subject to a trade-off [IEEE 2005]: typically, mobile WiMAX provides up to 10 Mbps per channel (symmetric), over a range of 10 km in rural areas (LOS environment) or over a range of 2 km in urban areas (NLOS environment) [Ohrtman 2008]. With the fixed WiMAX this range can normally be extended. The mobile version considers an omnidirectional antenna, while the fixed WiMAX uses a high-gain antenna. Throughput and ranges may always change. Nevertheless, by enlarging one parameter the other has to reduce; otherwise BER will experience degradation. In the limit, WiMAX may deliver up to 70 Mbps per channel (in LOS, short distance and fixed access), and may cover up to 50 km (in LOS for fixed access), with a high-gain antenna [Ohrtman 2008], but not both parameters simultaneously and, in this case, a single SS may be supported by the BS. In opposition to UMTS where handover is detailed and specified, mobile WiMAX has three possibilities but only the first one is mandatory: hard handover (HHO), fast base station switching (FBSS), and macrodiversity handover (MDHO). FBSS and MDHO are optional, as it is up to the manufacturers to decide about their implementation. Therefore, there is the risk that handover is not possible between two BSs from different manufacturers. Another drawback of the use of WiMAX is still the speed allowed in mobility, which is limited to 60 km/h. For higher speeds, the user experiences a high degradation in performance.

In terms of physical layer characteristics, when comparing WiMAX against UMTS, it is clear that the former standard uses more advanced transmission techniques (e.g., OFDMA versus WCDMA), which is expected to be incorporated only in the next generation of cellular systems (LTE). The WiMAX version currently available (IEEE 802.16-2005) already incorporates most of the techniques not considered by UMTS but specified for LTE (from 3GPP). Such examples of techniques are OFDMA, MIMO, advanced turbo coding, all over IP architecture, and so on. In addition, the inclusion of multihop relay capabilities (IEEE 802.16j), which aims to improve the speed of service delivery and coverage by a factor of 3–10, will make the WiMAX standard closer to what is expected for IMT-Advanced. Recall that this latter standard foresees a target of 100 Mbps mobile and 1 Gbps nomadic. Therefore, the future of WiMAX may be a potential candidate for IMT-Advanced 4G. Although most of the techniques foreseen for IMT-Advanced are already integrated in the previous WiMAX versions, IEEE 802.16m will integrate and incorporate several advancements in these transmission techniques in order to meet all the IMT-Advanced requirements as defined by ITU-R [Osseiran et al. 2009].

Chapter 2

Transmission Techniques for Broadband Systems

This chapter describes the transmission techniques employed in different systems. We will consider DS-CDMA schemes, more specifically WCDMA adopted in UMTS, as well as block transmission schemes combined with frequency-domain receiver implementations such as OFDM, selected for DVB-T, Wi-Fi, WiMAX, and 3GPP LTE, as well as SC-FDE, also considered for 3GPP LTE.

This chapter is organized as follows: Section 2.1 describes spread spectrum communication systems, with emphasis on DS-CDMA techniques, both broadband and wideband. We also present DS-CDMA receivers suitable for multipath channels. Section 2.2 presents the basic principles behind multicarrier (MC) modulations in general and OFDM schemes in particular, as well as their relation with conventional SC modulations. Section 2.3 describes SC-FDE schemes and their relation with OFDM schemes. We include linear and iterative FDE receivers. Section 2.4 presents some performance comparisons between OFDM and SC-FDE.

2.1 Spread Spectrum Communications

As introduced in Section 1.2.1, digital communications employing spread spectrum signals are characterized by using a BW, B_{wd}, much greater than the information bit rate. This means that for a spread spectrum signal we have $\mathrm{SF} = B_{\mathrm{wd}}/R_{\mathrm{b}} \gg 1$, where R_{b} is the information bit rate and SF is the BW expansion factor (called spreading factor or processing gain). This BW expansion can be accomplished through different types of spread spectrum techniques [Glisic and Vucetic 1997; Proakis 2001]:

- *Direct sequence* (*DS*): Figure 2.1 shows the general scheme for a DS spread spectrum system. Information symbols are encoded and then modulated in combination with a pseudorandom sequence. In the receiver, the demodulator removes the pseudorandom impression from the received signal. DS techniques can be performed in two distinct ways. In the most common techniques transmitted bits are encoded using a code with a given coding rate (say 1/2 or 1/3) and then multiplication by the pseudorandom sequence increases the transmitted signal bit rate. This can be seen as applying a repetition code to the encoded bits and then adding the pseudorandom signature to the signal [Proakis 2001]. An alternative way of implementing DS techniques corresponds to performing all the spreading of the signal using a low rate code and then adding the pseudorandom signature to the signal. This approach is usually referred to as code spread (CS).
- *Frequency hopping* (*FH*): In this case the available BW is divided into several contiguous subband frequency slots. A pseudorandom sequence is used for selecting the frequency slot for transmission in each signaling interval.
- *Time hopping* (*TH*): In this method a time interval is divided into several time slots and the coded information symbols are transmitted in a time slot selected according to a pseudorandom sequence. The coded symbols are transmitted in the selected time slot as blocks of one or more code words.
- *Hybrid techniques*: DS, FH, and TH can be combined to obtain other types of spread spectrum signals. For example, a system can use a combination of DS and FH, where the transmitted signal is associated with two code sequences. One of the sequences is multiplied by the signal to be transmitted while the second is used for selecting the frequency slot for transmission in each signaling interval.

In the following, we will consider only DS spreading techniques. One common application of DS spread spectrum signals is CDMA communications where several users share the same channel BW for transmitting information simultaneously. If the spreading of each user is the joint effect of the channel encoder and the spreading code, then the transmission is designated as DS-CDMA. If BW expansion is performed by the channel encoder alone, then the transmission is referred to as CS-CDMA. In Ref. [Viterbi 1990] it is shown that this technique can achieve maximum theoretical performance. In fact, DS-CDMA can be regarded as a special case in CS-CDMA. All the users can transmit in the same frequency band and may be distinguished from each other by using a different

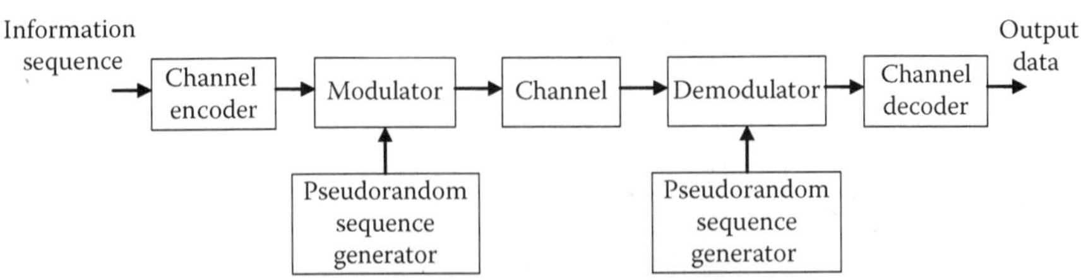

Figure 2.1 DS spread spectrum system general scheme.

pseudorandom sequence. If all the signals are transmitted with the symbols synchronized between them and we have a flat fading channel, then for each symbol of a user there will be only one interfering symbol from each of the other users. This is called synchronous CDMA transmissions and is usually employed in the downlink connection, between a BS and the users in a cell. If each transmitted symbol interferes with two symbols of any other user, then the transmission is asynchronous. This is the common method in the uplink connection.

2.1.1 General Model

To model a DS-CDMA system we will first consider single-user transmission. Figure 2.2 represents the basic communication model.

According to this model, a sequence of symbols, s_i, representing the information to be transmitted, enters the modulator, which outputs a sequence of ideal pulses $\delta(t - \mathrm{j}T_\mathrm{c})$ modulated by the spread symbols $\sqrt{E_\mathrm{c}} \cdot s_{\lfloor j/\mathrm{SF} \rfloor} \cdot c_j$ (T_chip is the chip duration, c_j is the jth chip of the spreading sequence, and E_c is the average chip energy). This sequence passes through a shaping filter with impulse response $h_\mathrm{T}(t)$ and frequency response $H_\mathrm{T}(f)$, resulting in a transmitted signal that can be written in the following equivalent low-pass form:

$$x(t) = \left(\sqrt{E_\mathrm{c}} \sum_{j=0}^{\infty} s_{\lfloor j/\mathrm{SF} \rfloor} \cdot c_j \cdot \delta\left(t - \mathrm{j}T_\mathrm{chip}\right) \right) * h_\mathrm{T}(t)$$
$$= \sqrt{E_\mathrm{c}} \sum_{j=0}^{\infty} s_{\lfloor j/\mathrm{SF} \rfloor} \cdot c_j \cdot h_\mathrm{T}\left(t - \mathrm{j}T_\mathrm{chip}\right). \quad (2.1)$$

The signal is then transmitted through the channel that is modeled by a time-invariant linear system with impulse response $h_\mathrm{C}(t)$ [frequency response $H_\mathrm{C}(f)$], using, for example, the model defined in Appendix A. At the receiver, the signal passes through a filter with impulse response $h_\mathrm{R}(t)$ [frequency response $H_\mathrm{R}(f)$]. The resulting signal is then obtained as the convolution of the transmitted signal with $h_\mathrm{C}(t)$ and $h_\mathrm{R}(t)$, that is,

$$r(t) = x(t) * h_\mathrm{C}(t) * h_\mathrm{R}(t) + n(t), \quad (2.2)$$

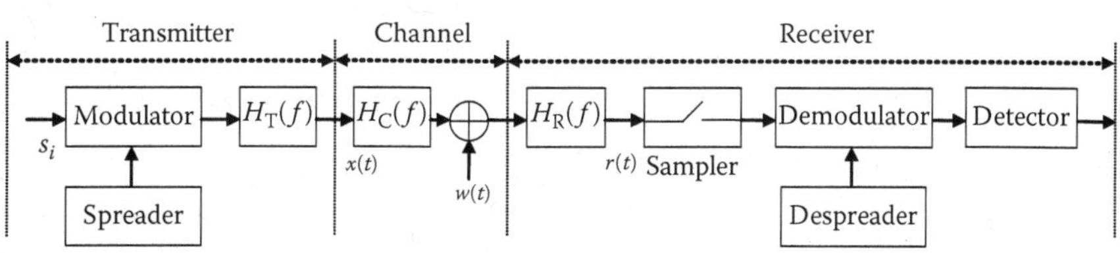

Figure 2.2 Basic DS-CDMA communication link.

where $n(t)$ is the noise component at the output of the receiver filter, that is, it is given by

$$n(t) = w(t) * h_R(t), \tag{2.3}$$

where $w(t)$ denotes the white noise at the receiver input. To design the shaping filter, $H_T(f)$, and the reception filter, $H_R(f)$, we will consider an ideal channel, that is, $h_C(t) = \delta(t)(H_C(f) = 1)$. Therefore, the received signal can be written as

$$r(t) = \sqrt{E_c}\sum_{j=0}^{\infty} s_{\lfloor j/\mathrm{SF}\rfloor} \cdot c_j \cdot p(t - jT_{\mathrm{chip}}) + n(t), \tag{2.4}$$

where $p(t)$ corresponds to the time response of the cascade of the transmitter and receiver filters:

$$p(t) = h_T(t) * h_R(t). \tag{2.5}$$

After the filter, the signal is sampled at rate f_{chip} $\left(f_{\mathrm{chip}} = 1/T_{\mathrm{chip}}\right)$. The sequence of samples r_k can be represented as (considering no delay in the transmission)

$$r_k \equiv r(t = kT_{\mathrm{chip}}) = \sqrt{E_c}\sum_{j=0}^{\infty} s_{\lfloor j/\mathrm{SF}\rfloor} \cdot c_j \cdot p(kT_{\mathrm{chip}} - jT_{\mathrm{chip}}) + n(kT_{\mathrm{chip}}) \tag{2.6}$$

or

$$\begin{aligned} r_k &= \sqrt{E_c}\sum_{j=0}^{\infty} s_{\lfloor j/\mathrm{SF}\rfloor} \cdot c_j \cdot p_{k-j} + n_k \\ &= \sqrt{E_c} s_{\lfloor k/\mathrm{SF}\rfloor} \cdot c_k \cdot p_0 + \sum_{\substack{j=0 \\ j\neq k}}^{\infty} s_{\lfloor j/\mathrm{SF}\rfloor} \cdot c_j \cdot p_{k-j} + n_k. \end{aligned} \tag{2.7}$$

In the last passage we admitted that $p_0 = 1$. The second term in the last expression represents the ISI that is not desired. The condition for no ISI is

$$p(kT_{\mathrm{chip}}) = p_k = \begin{cases} 1, & k = 0, \\ 0, & k \neq 0. \end{cases} \tag{2.8}$$

According to the *Nyquist pulse shaping criterion*, the necessary and sufficient condition for $p(t)$ to obey the above condition is that its Fourier transform $P(f)$ satisfies [Proakis 2001]

$$P_{\mathrm{eq}}(f) = \sum_{l=-\infty}^{\infty} P\left(f + \frac{l}{T_{\mathrm{chip}}}\right) = T_{\mathrm{chip}}, \tag{2.9}$$

where

$$P(f) = H_T(f)H_R(f). \tag{2.10}$$

There are several functions satisfying the Nyquist criterion. One of the most common functions used for $P(f)$ is the family of raised cosine functions, which can be expressed as

$$P(f) = \begin{cases} T_{\mathrm{chip}}, & |f| \leq \dfrac{1-\beta}{2T_{\mathrm{chip}}}, \\ \dfrac{T_{\mathrm{chip}}}{2}\left\{1+\cos\left[\dfrac{\pi T_{\mathrm{chip}}}{\beta}\left(|f|-\dfrac{1-\beta}{2T_{\mathrm{chip}}}\right)\right]\right\}, & \dfrac{1-\beta}{2T_{\mathrm{chip}}} < |f| \leq \dfrac{1+\beta}{2T_{\mathrm{chip}}}, \\ 0, & |f| > \dfrac{1+\beta}{2T_{\mathrm{chip}}}, \end{cases} \tag{2.11}$$

where β is called the roll-off factor with $0 \leq \beta \leq 1$. This factor represents the fraction of excess BW beyond the Nyquist frequency $1/(2T_{\mathrm{chip}})$, and hence $P(f)$ has a BW of

$$B_{\mathrm{wd}} = \frac{\beta+1}{2T_{\mathrm{chip}}}. \tag{2.12}$$

The corresponding impulse response in the time domain of the raised cosine function $P(f)$ is

$$p(t) = \sin\mathrm{c}\left(\frac{\pi t}{T_{\mathrm{chip}}}\right)\frac{\cos\left(\pi\beta t/T_{\mathrm{chip}}\right)}{1-\left(2\pi t/T_{\mathrm{chip}}\right)^2}. \tag{2.13}$$

The receiver filter can be matched to the transmitter filter, that is, $h_{\mathrm{R}}(t) = h_{\mathrm{T}}^*(-t)$, and thus $H_{\mathrm{T}}(f)$ and $H_{\mathrm{R}}(f)$ will both have root raised cosine responses, that is,

$$|H_{\mathrm{T}}(f)| = |H_{\mathrm{R}}(f)| = \sqrt{|P(f)|}. \tag{2.14}$$

As an example, in UMTS the shaping filter $h_{\mathrm{T}}(t)$ is a root raised cosine with a roll-off factor $\beta = 0.22$ [3GPP 2004c].

Therefore, if $p(t)$ satisfies the *Nyquist pulse shaping criterion*, Equation 2.7 reduces to

$$r_k = \sqrt{E_{\mathrm{c}}}s_{\lfloor k/\mathrm{SF}\rfloor} \cdot c_k + n_k. \tag{2.15}$$

The sequence of samples r_k is then despread by multiplying it by the conjugate of the spreading sequence and averaging over sets of SF chips. The resulting decision variables, z_i, for each information symbol, i, can be expressed as

$$z_i = \frac{1}{\mathrm{SF}}\sum_{k=i\cdot\mathrm{SF}}^{(i+1)\cdot\mathrm{SF}-1} c_k^* \cdot r_{0,k}. \tag{2.16}$$

Although the channel effect was considered ideal for obtaining Equation 2.14, in a real system the receiver has to compensate its effect. This can be accomplished either by including $h_{\mathrm{C}}(t)$ in the function $p(t)$ [resulting $p(t) = h_{\mathrm{T}}(t) * h_{\mathrm{R}}(t) * h_{\mathrm{C}}(t)$], and then computing the receiver filter as $H_{\mathrm{R}}(f) = P(f)/\left[H_{\mathrm{T}}(f)H_{\mathrm{C}}(f)\right]$ (this corresponds to channel equalization) where $P(f)$ satisfies the Nyquist criterion (using Equation 2.11, for example) or using some other channel compensation processing technique, as will be shown later.

2.1.2 Narrowband CDMA

CDMA systems can be considered as narrowband or wideband, depending on the mobile propagation conditions according to the definitions presented above. An example of a cellular system that employs the narrowband CDMA technique is the IS-95 standard. If the transmitted signal BW is lower than the coherence BW of the channel for the environments for which the system was projected, then there will be only one distinguishable received replica of the signal. In this case, the system is narrowband CDMA.

Let us consider now a multiuser environment where $N_u + 1$ users are transmitting simultaneously, as shown in Figure 2.3.

In this case, the signal transmitted by each user u can be expressed in a form similar to Equation 2.1:

$$x_u(t) = \sqrt{E_c}\sum_{j=0}^{\infty} s_{u,\lfloor j/\mathrm{SF}\rfloor} \cdot c_{u,j} \cdot h_\mathrm{T}\left(t - \mathrm{j}T_\mathrm{chip}\right), \tag{2.17}$$

where $s_{u,\lfloor j/\mathrm{SF}\rfloor}$ are the information symbols and $c_{u,j}$ are the symbols of the spreading sequence for user u.

First we will admit that the CDMA system is narrowband and thus the CIR of each user can be modeled as

$$h_{\mathrm{C},u}(t) = \alpha_u \delta\left(t - \tau_u\right), \tag{2.18}$$

where α_u is a complex attenuation. [α_u is a random process dependent on the time, but since we are admitting that $f_\mathrm{D} T_\mathrm{chip} \ll 1$, with f_D denoting the Doppler frequency ($f_\mathrm{D} = f_\mathrm{c} v/c$, with f_c denoting the carrier frequency, v the terminal speed, and c the speed of light), we can assume that it is approximately constant during a symbol period.] In Figure 2.4 a basic receiver for user 0 is shown. First the received signal passes through the matched filter (MF) $h_\mathrm{R}(t)$. From Equation 2.2, this signal can be written as

$$r(t) = \sqrt{E_c}\sum_{u=0}^{N_u}\sum_{j=0}^{\infty} s_{u,\lfloor j/\mathrm{SF}\rfloor} \cdot c_{u,j} \cdot \alpha_u \cdot p\left(t - \mathrm{j}T_\mathrm{chip} - \tau_u\right) + n(t). \tag{2.19}$$

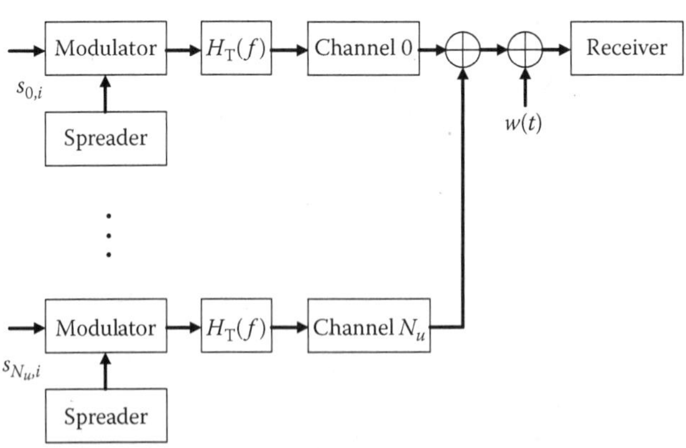

Figure 2.3 CDMA scheme in a multiuser environment.

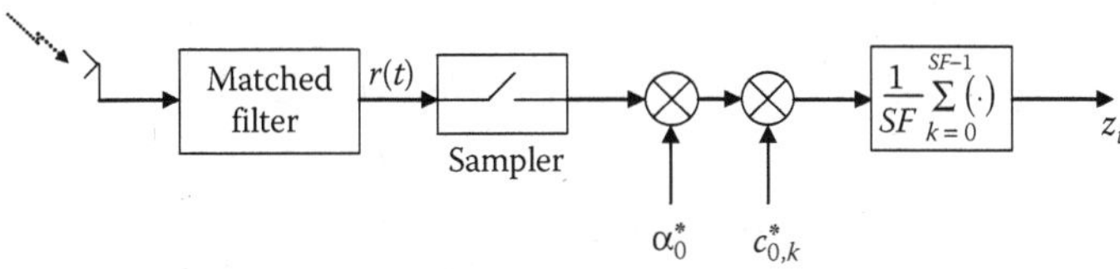

Figure 2.4 DS-CDMA receiver scheme for user 0.

The filter output is then sampled at times $t = kT_{\text{chip}} + \tau_0$ for extracting user 0 according to

$$\begin{aligned} r_{0,k} &\equiv r\left(t = kT_{\text{chip}} + \tau_0\right) = \sqrt{E_{\text{c}}}\sum_{u=0}^{N_u}\sum_{j=0}^{\infty} s_{u,\lfloor j/\text{SF}\rfloor}\cdot c_{u,j}\cdot\alpha_u \\ &\quad\cdot p\left(kT_{\text{chip}} + \tau_0 - \text{j}T_{\text{chip}} - \tau_u\right) + n\left(kT_{\text{chip}} + \tau_0\right) \\ &= \sqrt{E_{\text{c}}}\cdot s_{0,\lfloor k/\text{SF}\rfloor}\cdot c_{0,k}\cdot\alpha_0 + \sqrt{E_{\text{c}}}\sum_{u=1}^{N_u}\sum_{j=0}^{\infty} s_{u,\lfloor j/\text{SF}\rfloor}\cdot c_{u,j}\cdot\alpha_u \\ &\quad\cdot p\left(kT_{\text{chip}} + \tau_0 - \text{j}T_{\text{chip}} - \tau_u\right) + n_{0,k}. \end{aligned} \tag{2.20}$$

In this passage it was taken into account that $p(t)$ satisfies the *Nyquist pulse shaping criterion* and thus Equation 2.8 is valid. In Equation 2.20 the second term represents the interference component, from the other users, which, for an asynchronous transmission, has contributions from all the transmitted chips of those users since generally $\tau_0 - \tau_u \neq aT_{\text{chip}}(\forall a \in \mathbb{Z})$.

The channel is compensated by multiplying the sampled sequence by the complex conjugate of α_u (considering perfect channel knowledge). The resulting sequence is then despread by multiplying it by the complex conjugate of the respective spreading sequence, $c_{0,j}$, and each set of SF samples belonging to the same information symbol is summed. The decision variable $z_{0,i}$ for the ith information symbol can be expressed as

$$\begin{aligned} z_{0,i} &= \frac{1}{\text{SF}}\sum_{k=i\cdot\text{SF}}^{(i+1)\cdot\text{SF}-1} \alpha_0^*\cdot c_{0,k}^*\cdot r_{0,k} \\ &= \frac{\sqrt{E_{\text{c}}}\cdot|\alpha_0|^2}{\text{SF}}\sum_{k=i\cdot\text{SF}}^{(i+1)\cdot\text{SF}-1} s_{0,\lfloor k/\text{SF}\rfloor}\cdot|c_{0,k}|^2 \\ &\quad+\frac{\sqrt{E_{\text{c}}}}{\text{SF}}\sum_{k=i\cdot\text{SF}}^{(i+1)\cdot\text{SF}-1}\sum_{u=1}^{N_u}\sum_{j=0}^{\infty}\alpha_0^*\cdot c_{0,k}^*\cdot s_{u,\lfloor j/\text{SF}\rfloor}\cdot c_{u,j}\cdot\alpha_u \\ &\quad\times p\left(kT_{\text{chip}} + \tau_0 - \text{j}T_{\text{chip}} - \tau_u\right) + \frac{1}{\text{SF}}\sum_{k=i\cdot\text{SF}}^{(i+1)\cdot\text{SF}-1}\alpha_0^*\cdot c_{0,k}^*\cdot n_{0,k} \end{aligned}$$

$$= \sqrt{E_c} \cdot |\alpha_0|^2 \cdot s_{0,i} + \frac{\sqrt{E_c}}{\mathrm{SF}} \sum_{u=1}^{N_u} \sum_{j=0}^{\infty} \alpha_0^* \cdot s_{u,\lfloor j/\mathrm{SF} \rfloor} \cdot \alpha_u \cdot \sum_{k=i\cdot \mathrm{SF}}^{(i+1)\cdot \mathrm{SF}-1} c_{0,k}^* \cdot c_{u,j}$$

$$\times\, p\left(kT_{\mathrm{chip}} + \tau_0 - \mathrm{j}T_{\mathrm{chip}} - \tau_u\right) + \frac{1}{\mathrm{SF}} \sum_{k=i\cdot \mathrm{SF}}^{(i+1)\cdot \mathrm{SF}-1} \alpha_0^* \cdot c_{0,k}^* \cdot n_{0,k}. \qquad (2.21)$$

In this expression it was assumed that $|c_{0,k}|^2 = 1$. In Equation 2.21 the second term represents the interference from other users MAI, which depends on the cross-correlation between the desired spreading code and the spreading code of interferer u weighted by the pulse shaping function $p(t)$, which depends on the relative delays. In a downlink synchronous transmission, the fading coefficients and the delays are all the same, $\alpha_0 = \alpha_u$ and $\tau_0 = \tau_u (u = 1 \ldots N_u - 1)$, and Equation 2.21 simplifies to

$$z_{0,i} = \sqrt{E_c} \cdot |\alpha|^2 \cdot s_{0,i} + \frac{\sqrt{E_c}}{\mathrm{SF}} \cdot |\alpha|^2 \cdot \sum_{u=1}^{N_u} s_{u,i} \cdot \sum_{k=i\cdot \mathrm{SF}}^{(i+1)\cdot \mathrm{SF}-1} c_{0,k}^* \cdot c_{u,k}$$

$$+ \frac{\alpha^*}{\mathrm{SF}} \sum_{k=i\cdot \mathrm{SF}}^{(i+1)\cdot \mathrm{SF}-1} c_{0,k}^* \cdot n_{0,k}. \qquad (2.22)$$

In this case, to minimize multiuser interference it is only necessary to employ spreading sequences with low cross-correlation values. For some values of SF (e.g., if SF is a power of 2) it is possible to design sequences where

$$\sum_{k=0}^{\mathrm{SF}-1} c_{0,k}^* \cdot c_{u,k} = 0, \qquad (2.23)$$

that is, orthogonal sequences, in which case there will be no multiuser interference (in a synchronous and for the single-path propagation channel). This is what is done in the downlink connection of UMTS [3GPP 2004b].

2.1.3 Wideband CDMA

If the transmitted signal BW is greater than the coherence BW of the channel, the CDMA system will be possible to resolve several multipath components and it will correspond to a WCDMA system. The UMTS system, which employs DS-CDMA transmission techniques, is a wideband system. The CIR for WCDMA systems can be described using a tapped delay line (TDL) model [Silva et al. 2003] where the L multipaths are considered to be discrete. Considering that the channel is time invariant, the CIR for each user can be written as

$$h_{c,u}(t) = \sum_{l=1}^{L} \alpha_{u,l} \cdot \delta(t - \tau_{u,l}), \qquad (2.24)$$

where $\alpha_{u,l}$ is the fading coefficient affecting the transmitted signal of user u for the lth propagation path and $\tau_{u,l}$ is the respective time delay with $\tau_{u,l} = \tau_{u,1} + m_l T_{\mathrm{chip}}$, $m_l \in \mathrm{Z}$ (the relative path delays for the same user are integer multiples of T_{chip}).

Since in WCDMA systems the multipath replicas arriving with relative delays higher than the chip duration carry information about the transmitted signal, there is the possibility that when a replica is severely attenuated due to fading the others may be received in more favorable conditions. Because of the good autocorrelation properties of the spreading codes usually used in CDMA systems, it is possible to distinguish and extract the strongest replicas (with relative delays higher than the chip duration) present in the received signal and thus obtain diversity. These replicas can be extracted and combined using a RAKE receiver [Price and Green 1958] to help recover the transmitted signal, as shown in Figure 2.5. To maximize the resulting SNR at the receiver, the extracted replicas are weighted using the complex conjugate of the respective fading coefficient and then added. This technique is designated as maximal ratio combining (MRC) [Proakis 2001].

According to the scheme shown in Figure 2.5, first the received signal goes through an MF similar to what was done in the narrowband CDMA receiver. From Equations 2.2, 2.17, and 2.24, the filtered signal $r(t)$ can be expressed as

$$\begin{aligned} r(t) &= \left[\sum_{u=0}^{N_u-1} x_u(t) * h_{C,u}(t)\right] * h_R(t) + n(t) \\ &= \sqrt{E_c}\sum_{u=0}^{N_u}\sum_{l=1}^{L}\sum_{j=0}^{\infty} \alpha_{u,l} \cdot s_{u,\lfloor j/SF \rfloor} \cdot c_{u,j} \cdot p\left(t - jT_{chip} - \tau_{u,l}\right) + n(t). \end{aligned} \quad (2.25)$$

The rest of the RAKE is composed of up to L parallel branches denoted as fingers. Each finger extracts one of the received replicas for the target user. The processing steps inside each finger are similar to those performed in the receiver for the narrowband

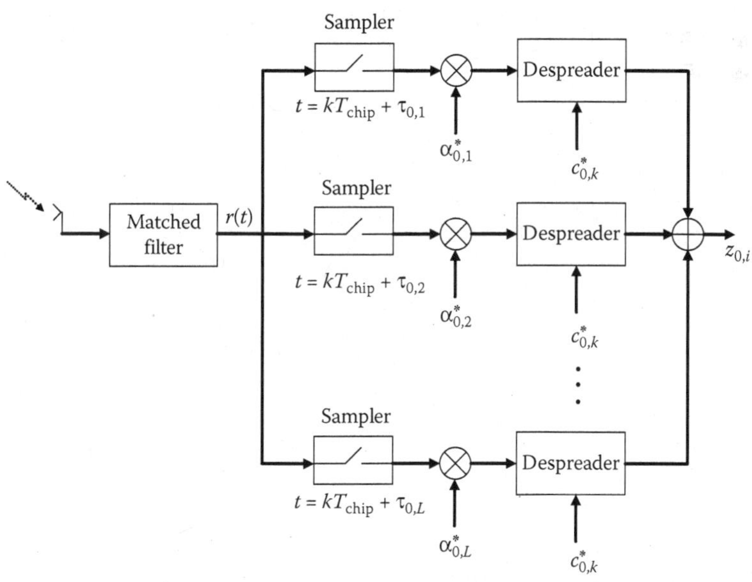

Figure 2.5 RAKE receiver scheme for user 0.

CDMA system. So, first the filter output is sampled at times $t = kT_{\text{chip}} + \tau_{0,f}$ for extracting user 0 in finger f. This can be expressed as

$$\begin{aligned}
r_{0,k,f} &\equiv r\left(t = kT_{\text{chip}} + \tau_{0,f}\right) = \sqrt{E_{\text{c}}} \sum_{u=0}^{N_u} \sum_{l=1}^{L} \sum_{j=0}^{\infty} \alpha_{u,l} \cdot s_{u,\lfloor j/\text{SF} \rfloor} \cdot c_{u,j} \cdot p(kT_{\text{chip}} \\
&\quad + \tau_{0,f} - \mathrm{j}T_{\text{chip}} - \tau_{u,l}) + n\left(kT_{\text{chip}} + \tau_{0,f}\right) \\
&= \sqrt{E_{\text{c}}} \cdot s_{0,\lfloor k/\text{SF} \rfloor} \cdot c_{0,k} \cdot \alpha_{0,f} + \sqrt{E_{\text{c}}} \sum_{\substack{l=1 \\ l \neq f}}^{L} \sum_{j=0}^{\infty} \alpha_{0,l} \cdot s_{0,\lfloor j/\text{SF} \rfloor} \cdot c_{0,j} \cdot p\left(kT_{\text{chip}}\right. \\
&\quad \left. + \tau_{0,f} - \mathrm{j}T_{\text{chip}} - \tau_{0,l}\right) + \sqrt{E_{\text{c}}} \sum_{u=1}^{N_u} \sum_{l=1}^{L} \sum_{j=0}^{\infty} \alpha_{u,l} \cdot s_{u,\lfloor j/\text{SF} \rfloor} \cdot c_{u,j} \cdot p\left(kT_{\text{chip}}\right. \\
&\quad \left. + \tau_{0,f} - \mathrm{j}T_{\text{chip}} - \tau_{u,l}\right) + n_{0,k,f} \\
&= \sqrt{E_{\text{c}}} \cdot s_{0,\lfloor k/\text{SF} \rfloor} \cdot c_{0,k} \cdot \alpha_{0,f} + \sqrt{E_{\text{c}}} \sum_{\substack{l=1 \\ l \neq f}}^{L} \alpha_{0,l} \cdot s_{0,\lfloor (k+(\tau_{0,f}-\tau_{0,l})/T_{\text{chip}})/\text{SF} \rfloor} \\
&\quad \times c_{0,k+(\tau_{0,f}-\tau_{0,l})/T_{\text{chip}}} + \sqrt{E_{\text{c}}} \sum_{u=1}^{N_u} \sum_{l=1}^{L} \sum_{j=0}^{\infty} \alpha_{u,l} \cdot s_{u,\lfloor j/\text{SF} \rfloor} \cdot c_{u,j} \cdot p(kT_{\text{chip}} \\
&\quad + \tau_{0,f} - \mathrm{j}T_{\text{chip}} - \tau_{u,l}) + n_{0,k,f}.
\end{aligned} \tag{2.26}$$

To obtain this expression it was assumed that $p(t)$ satisfies the *Nyquist pulse shaping criterion* and thus Equation 2.8 is valid.

The channel is compensated by multiplying the sampled sequence by the complex conjugate of $\alpha_{u,f}$ (assuming perfect channel knowledge). The resulting sequence is then despread and the result is summed with the outputs of the other fingers. The decision variable $z_{0,i}$ for the ith information symbol can be expressed as

$$\begin{aligned}
z_{0,i} &= \frac{1}{\text{SF}} \sum_{f=1}^{L} \sum_{k=i\cdot\text{SF}}^{(i+1)\cdot\text{SF}-1} \alpha_{0,f}^{*} \cdot c_{0,k}^{*} \cdot r_{0,k,f} \\
&= \sqrt{E_{\text{c}}} \cdot s_{0,i} \cdot \sum_{f=1}^{L} \left|\alpha_{0,f}\right|^2 + \frac{\sqrt{E_{\text{c}}}}{\text{SF}} \sum_{f=1}^{L} \sum_{k=i\cdot\text{SF}}^{(i+1)\cdot\text{SF}-1} \sum_{\substack{l=1 \\ l \neq f}}^{L} \alpha_{0,f}^{*} \cdot c_{0,k}^{*} \cdot \alpha_{0,l} \\
&\quad \times s_{0,\lfloor (k+(\tau_{0,f}-\tau_{0,l})/T_{\text{chip}})/\text{SF} \rfloor} \cdot c_{0,k+(\tau_{0,f}-\tau_{0,l})/T_{\text{chip}}} \\
&\quad + \frac{\sqrt{E_{\text{c}}}}{\text{SF}} \sum_{u=1}^{N_u} \sum_{f=1}^{L} \sum_{k=i\cdot\text{SF}}^{(i+1)\cdot\text{SF}-1} \sum_{l=1}^{L} \sum_{j=0}^{\infty} \alpha_{0,f}^{*} \cdot c_{0,k}^{*} \cdot \alpha_{u,l} \cdot s_{u,\lfloor j/\text{SF} \rfloor} \cdot c_{u,j} \cdot p(kT_{\text{chip}} \\
&\quad + \tau_{0,f} - \mathrm{j}T_{\text{chip}} - \tau_{u,l}) + \frac{1}{\text{SF}} \sum_{f=1}^{L} \sum_{k=i\cdot\text{SF}}^{(i+1)\cdot\text{SF}-1} \alpha_{0,f}^{*} \cdot c_{0,k}^{*} \cdot n_{0,k,f}.
\end{aligned} \tag{2.27}$$

In this expression the second term represents interference caused by the user's own signal due to the delayed multipath replicas and the third term represents the interference component from all the multipath replicas of the other users. These interference components can be reduced through the use of MUD schemes.

Much research has been undertaken in the area of MUDs [Marques da Silva and Correia 2003a; Marques da Silva and Correia 2000a,b] for WCDMA technology. An optimal MUD, usually known as a maximum likelihood sequence detector, is too complex for practical application. Its complexity increases exponentially with an increase in the number of users. On the other hand, a suboptimal MUD has a complexity that increases linearly with an increase in the number of users. The latter can take two different forms: linear MUDs (e.g., decorrelating [Marques da Silva et al. 2005a], MMSE [Glisic and Vucetic 1997]) and subtractive MUDs (e.g., successive interference cancelation and parallel interference cancelation (PIC) [Marques da Silva and Correia 2003a; Glisic and Vucetic 1997]).

Linear suboptimal detectors apply a linear transformation to the bank output of conventional detectors, in order to decrease the level of MAI seen by each user.

There are two main types of implementations for subtractive MUDs: the (PIC) scheme [Varanasi and Aazhang 1990] and the successive interference cancelation (SIC) scheme [Patel and Holtzman 1994; Johansson and Svensson 1999]. The main difference between both schemes relies on how interference subtraction is performed on the received signal. While the SIC detector relies on removing the interfering signals from the received signal, one at a time as they are detected, the PIC removes all the interfering signals simultaneously after they are detected.

Nevertheless, a MUD is normally employed at the BS (uplink) where there is enough power processing capability and where it is easier to know/estimate the uplink CIR and the spreading and scrambling sequences of interfering users. Such power processing capability is normally not available at the mobile station (MS) side; for this reason it is important to employ alternative schemes to improve the performance.

In Ref. [Souto et al. 2004] we evaluated the performance of UMTS uplink transmissions using PIC and SIC schemes for several transmission rates. Different interferers' patterns and UMTS environments (Indoor A, Pedestrian A and Vehicular A) were simulated and it was seen that these interference cancelation schemes achieved good performances.

If the transmission is synchronous and comes from a BS (downlink), the channel coefficients and delays do not depend on the user, $\alpha_{0,l} = \alpha_{u,l}$ and $\tau_{0,l} = \tau_{u,l} (u = 1 \ldots N_u - 1, l = 1 \ldots L)$, and this expression simplifies to

$$\begin{aligned} z_{0,i} &= \sqrt{E_c} \cdot s_{0,i} \cdot \sum_{f=1}^{L} |\alpha_f|^2 + \frac{\sqrt{E_c}}{\mathrm{SF}} \sum_{u=1}^{N_u} s_{u,i} \cdot \sum_{f=1}^{L} |\alpha_f|^2 \cdot \sum_{k=i\cdot\mathrm{SF}}^{(i+1)\cdot\mathrm{SF}-1} c_{0,k}^* \cdot c_{u,k} \\ &\quad + \frac{\sqrt{E_c}}{\mathrm{SF}} \sum_{u=0}^{N_u} \sum_{f=1}^{L} \sum_{k=i\cdot\mathrm{SF}}^{(i+1)\cdot\mathrm{SF}-1} \sum_{\substack{l=1 \\ l\neq f}}^{L} \alpha_f^* \cdot \alpha_l \cdot c_{0,k}^* \cdot c_{u,k+(\tau_f-\tau_l)/T_{\mathrm{chip}}} \\ &\quad \times s_{u,\lfloor (k+(\tau_f-\tau_l)/T_{\mathrm{chip}})/\mathrm{SF} \rfloor} + \frac{1}{\mathrm{SF}} \sum_{f=1}^{L} \sum_{k=i\cdot\mathrm{SF}}^{(i+1)\cdot\mathrm{SF}-1} \alpha_f^* \cdot c_{0,k}^* \cdot n_{0,k,f}. \end{aligned} \tag{2.28}$$

In this expression interference is grouped in a different form. The second term represents interference from the signals of other users time aligned with the desired signal, while the third term represents interference caused by the multipath replicas of all signals. The second term can be canceled if orthogonal spreading sequences are employed. In this case Equation 2.23 is valid and Equation 2.28 simplifies to

$$z_{0,i} = \sqrt{E_{\mathrm{c}}} \cdot s_{0,i} \cdot \sum_{f=1}^{L} |\alpha_f|^2 + \frac{\sqrt{E_{\mathrm{c}}}}{\mathrm{SF}} \sum_{u=0}^{N_u} \sum_{f=1}^{L} \sum_{k=i\cdot \mathrm{SF}}^{(i+1)\cdot \mathrm{SF}-1} \sum_{\substack{l=1 \\ l\neq f}}^{L} \alpha_f^* \cdot \alpha_l \cdot c_{0,k}^*$$
$$\times\, c_{u,k+(\tau_f-\tau_l)/T_{\mathrm{chip}}} \cdot s_{u,\lfloor (k+(\tau_f-\tau_l)/T_{\mathrm{chip}})/\mathrm{SF} \rfloor} + \frac{1}{\mathrm{SF}} \sum_{f=1}^{L} \sum_{k=i\cdot \mathrm{SF}}^{(i+1)\cdot \mathrm{SF}-1} \alpha_f^* \cdot c_{0,k}^* \cdot n_{0,k,f}. \tag{2.29}$$

Nevertheless, the type of spreading sequences to use must be carefully selected because orthogonal spreading sequences may give rise to high levels of MAI caused by the multipath environment, even in a synchronous environment (downlink).

2.2 OFDM Schemes

2.2.1 Signal Characterization

A brief description of the OFDM transmission technique was introduced in Section 1.2.2. In order to describe the OFDM signal in detail, let us first consider conventional SC modulation, that is, modulation where the energy associated with each symbol is spread out along the total transmission band. The complex envelope of an N-symbol burst (N even) can be written as

$$s(t) = \sum_{n=-N/2}^{N/2-1} s_n r(t - nT_{\mathrm{s}}), \tag{2.30}$$

where $r(t)$ represents the transmitted impulse, T_{s} is the symbol duration, and s_n is a complex coefficient representing the nth symbol resulting from a direct mapping rule of the original data bits into a selected signal constellation [e.g., a phase shift keying (PSK) or a quadrature amplitude modulation (QAM) constellation]. By applying the FT to both sides of Equation 2.30, we obtain the following equivalent frequency-domain expression:

$$S(f) = F\{s(t)\} = \sum_{k=-N/2}^{N/2-1} s_k R(f) \exp(-\mathrm{j}2\pi fkT_{\mathrm{s}}), \tag{2.31}$$

where $R(f)$ represents the FT of $r(t)$. Clearly, the transmission band associated with each symbol s_n is the same as the band occupied by $R(f)$.

Let us consider now multicarrier modulation. In this case, symbols are sequentially transmitted in the frequency domain, each one on a different subcarrier during the time interval T. The complex envelope of a multicarrier burst can then be written as

$$s(t) = \sum_{k=-N/2}^{N/2-1} S_k r(t) \exp(j2\pi tkF), \tag{2.32}$$

where $F = 1/T$ is the subcarriers' separation, N is the number of subcarriers, and S_k is the kth frequency-domain symbol. Applying the FT to both sides of Equation 2.32 gives

$$S(f) = \mathcal{F}\{s(t)\} = \sum_{k=-N/2}^{N/2-1} S_k R(f - kF), \tag{2.33}$$

and the power spectral density (PSD) of the transmitted signals is proportional to

$$\sum_{k=-N/2}^{N/2-1} E\left[|S_k|^2\right] \left|R(f - kF)\right|^2. \tag{2.34}$$

The simplest case of multicarrier modulation is a conventional frequency division multiplexing (FDM) scheme where the spectrum associated with different symbols does not overlap. If we assume that the BW associated with $R(f)$ is smaller than $F/2$,* then each symbol S_k occupies a fraction $1/N$ of the total transmission band, as shown in Figure 2.6.

Comparing Equation 2.30 with 2.33 and Equation 2.31 with 2.32, it is clear that multicarrier modulation can be regarded as a dual version of conventional SC modulation, with the time and frequency domains exchanged.

As already mentioned in Section 2.1.1, for SC modulations, the orthogonality condition between impulses $r(t)$ associated with different symbols that ensures an ISI-free transmission at the receiver's MF output is given by

$$\int_{-\infty}^{\infty} r(t - kT_s) r^*(t - k'T_s)\, dt = 0, \quad k \neq k'. \tag{2.35}$$

From the duality relation just referred to, it is easy to see that the orthogonality condition between subcarriers for multicarrier modulation is given by

$$\int_{-\infty}^{\infty} R(f - kF) R^*(f - k'F)\, df = 0, \quad k \neq k', \tag{2.36}$$

---

* Clearly, F is the bilateral BW and $F/2$ is the unilateral BW.

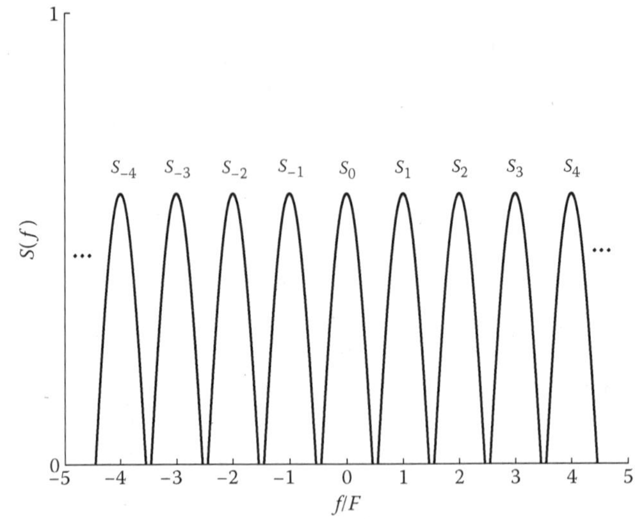

Figure 2.6 Conventional FDM.

which, from Parseval's theorem, is equivalent to

$$\int_{-\infty}^{\infty} |r(t)|^2 \exp(-2j\pi(k-k')Ft)\,dt = 0, \quad k \neq k'. \tag{2.37}$$

It is known that with linear SC modulations impulses $\{r(t-nT), n = \ldots, -1, 0, 1, \ldots\}$ can be orthogonal even if they overlap in time. For example,

$$r(t) = \sin c\left(\frac{t}{T_s}\right), \tag{2.38}$$

with $\sin c(x) = \sin(\pi x)/(\pi x)$, satisfies the orthogonality condition 2.35. Similarly, for multicarrier modulations it is possible to verify the orthogonality between subcarriers, expressed by Equation 2.36 (or Equation 2.37), even when $\{R(f-kF), k = 0, 1, \ldots, N-1\}$ overlap in the frequency domain. This means that we do not need to restrict ourselves to the conventional FDM case of Figure 2.6, where the spectra associated with different frequency channels do not overlap. In fact, if we have

$$R(f) = \sin c\left(\frac{f}{F}\right), \tag{2.39}$$

then the corresponding time-domain impulse $r(t)$ is rectangular with duration $T = 1/F$ (for instance, a rectangular impulse from t_0 to $t_0 + T$). In this case,

Equation 2.37 reduces to

$$\int_{t_0}^{t_0+T} \exp(-2j\pi(k-k')Ft)\,dt = 0, \quad k \neq k', \tag{2.40}$$

and, for a subcarrier separation F, the N subcarriers are orthogonal when $T = 1/F$ (see Figure 2.7).

For conventional OFDM modulations, the adopted impulses $r(t)$ are rectangular:

$$r(t) = \begin{cases} 1, & [-T_G, T], \\ 0 & \text{elsewhere}, \end{cases} \tag{2.41}$$

where $T = 1/F$ and $T_G \geq 0$ represents the guard period. It will be shown in the following that this guard period can be used to cope with time-dispersive channels. Although Equation 2.37 is not verified with impulses given by Equation 2.41, we can say that the different subcarriers are orthogonal for the interval $[0, T]$, which will be the effective

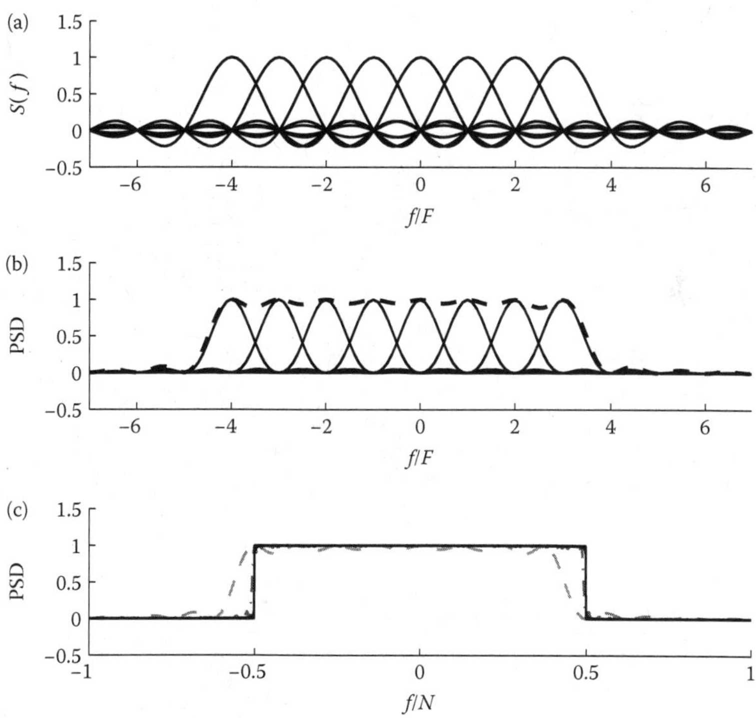

Figure 2.7 Spectrum of each OFDM subcarrier: (a) PSD of an OFDM signal with $N = 8$ (orthogonal) subcarriers (–) and (b) PSD of each subcarrier (—) and (c) PSD of an OFDM signal with $N = 8$ (- - -), $N = 64$ (- · · -), and $N = 512$ (—) subcarriers.

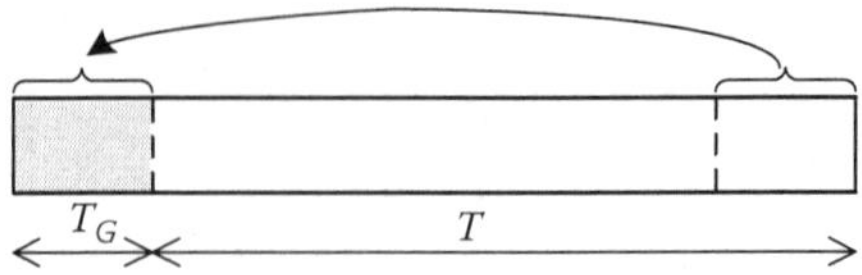

Figure 2.8 Repetitions of the MC burst's final part in the guard interval.

detection interval. In fact,

$$\int_0^T |r(t)|^2 \exp(-j2\pi(k-k')Ft)\,dt = \int_0^T \exp(-j2\pi(k-k')Ft)\,dt = 0, \quad k \neq k'. \tag{2.42}$$

In this case, since

$$s^{(P)}(t) = \sum_{k=-N/2}^{N/2-1} S_k \exp\left(j2\pi \frac{kt}{T}\right) \tag{2.43}$$

is a time-domain periodic function with period T, the complex envelope associated with the guard period is a repetition of the MC burst's final part as illustrated in Figure 2.8, that is,

$$s(t) = s(t+T), -T_G \leq t \leq 0. \tag{2.44}$$

Contrary to classical FDM systems where the total frequency band is subdivided into N nonoverlapping frequency subchannels, each one modulated with a separate symbol and subsequently frequency multiplexed, OFDM modulations are multicarrier modulations that make a much more efficient use of BW because they verify the orthogonality conditions between different subcarriers although the spectra of individual subchannels overlap.*

2.2.2 Transmitter Structure

The complex envelope of OFDM signals can be described as a sum of bursts of duration $T_B \geq T$ (i.e., the bursts are transmitted at a rate $1/T_B \leq F$), with the separation between subcarriers denoted by F and the duration of the useful part $T = 1/F$, that is,

$$s^{Tx}(t) = \sum_m s^{(m)}(t - mT_B). \tag{2.45}$$

* In practice, an efficient use of the spectrum is still possible with nonoverlapping FDM channels provided that $R(f)$ satisfies the 1st Nyquist criterion. However, the transmitter and receiver complexities are much higher.

Given Equation 2.32, the mth OFDM burst can be written as

$$s^{(m)}(t) = \sum_{k=-N/2}^{N/2-1} S_k^{(m)} r(t) \exp(\mathrm{j}2\pi kFt) = s^{(m,P)}(t)r(t), \quad (2.46)$$

where, according to Equation 2.43,

$$s^{(m,P)}(t) = \sum_{k=-N/2}^{N/2-1} S_k^{(m)} \exp\left(\mathrm{j}2\pi \frac{kt}{T}\right). \quad (2.47)$$

The coefficient $S_k^{(m)}$ denotes the kth symbol of the mth burst and $r(t)$, a rectangular impulse, whose duration should be greater than $1/F(T_\mathrm{B} = T + T_\mathrm{G} \geq T = 1/F)$ and appropriately adjusted so as to deal with the time-dispersive conditions of the channels, as described in the following.

The characterization of the complex envelope of OFDM signals given by Equations 2.45 and 2.46 suggests a conceptual FDM transmitter structure with a bank of N parallel SC modulators with frequencies $f_k = f_c + kF$, $k = 0, 1, \ldots, N-1$, with f_c denoting the frequency of the first subcarrier, as shown in Figure 2.9. For a small number of subcarriers this conceptual OFDM transmitter structure could be easily employed. However, for large N this is not a practical structure because hundreds or even thousands of local oscillators and multipliers would be necessary.

However, a single orthogonal modulator can be employed for OFDM schemes. The basic OFDM transmitter structure is shown in Figure 2.10, where the "in-phase" and "quadrature" components of each OFDM burst are obtained from a sequence of samples corresponding to the IDFT of the block to be transmitted.

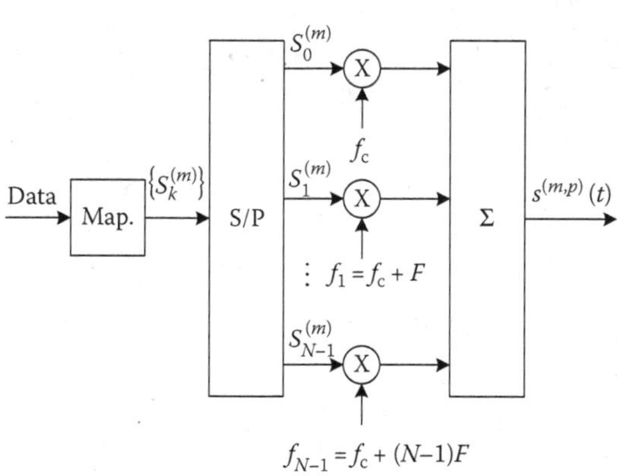

Figure 2.9 Conceptual OFDM transmitter structure.

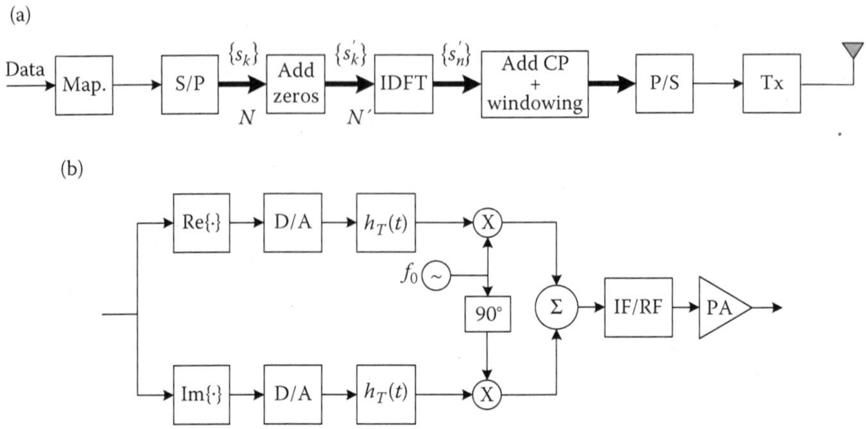

Figure 2.10 OFDM transmitter structure (a) and details of the "Tx" block (b).

The motivation behind this transmitter implementation is the following. Let us consider the signal

$$s^{(P)}(t) = \sum_{k=0}^{N-1} S_k \exp\left(j2\pi \frac{kt}{T}\right). \tag{2.48}$$

By applying the FT to both sides of Equation 2.48, we obtain the spectrum of $s^{(P)}(t)$:

$$S^{(P)}(f) = F\{s^{(P)}(t)\} = \sum_{k=0}^{N-1} S_k \delta\left(f - \frac{k}{T}\right). \tag{2.49}$$

From Equation 2.49, it is clear that the band occupied by $s^{(P)}(t)$ is $N/T = NF$. According to the sampling theorem, $s^{(P)}(t)$ can be completely recovered from its samples taken in the interval $[0, T]$ with a sample rate $1/T_s = N/T$, that is,*

$$\begin{aligned} s_n^{(P)} \triangleq s^{(P)}\left(\frac{nT}{N}\right) &= \sum_{k=0}^{N-1} S_k \exp\left(j2\pi \frac{kn}{N}\right) \\ &= N\left(\frac{1}{N} \sum_{k=0}^{N-1} S_k \exp\left(j2\pi \frac{kn}{N}\right)\right) = N s_n. \end{aligned} \tag{2.50}$$

This means that the block of time-domain samples $\{s_n; n = 0, 1, \ldots, N-1\}$ is the IDFT of $\{S_k; k = 0, 1, \ldots, N-1\}$, where the IDFT is defined as $\{x_n; n = 0, 1, \ldots,$

* With our DFT definition, the first sample corresponds to instant (or frequency) "0".

$N-1\} = \mathrm{IDFT}\{X_k; k = 0, 1, \ldots, N-1\}$ with

$$x_n = \frac{1}{N}\sum_{k=0}^{N-1} X_k \exp\left(\mathrm{j}2\pi\frac{kn}{N}\right). \tag{2.51}$$

This means that, apart from a scalar factor N, the sampled version of $s^{(P)}(t)$ in the interval $[0, T]$ corresponds to the IDFT of the block $\{S_k; k = 0, \ldots, N-1\}$ in the frequency domain,* which can be efficiently implemented with the well-known FFT algorithm [Cooley and Tukey 1965].

Although the power spectrum of an OFDM signal with bursts given by Equation 2.46 has an approximately rectangular shape with BW $N/T = NF$, the required sample rate for an exact representation of that signal can be much higher than NF due to high "out-of-band radiation," especially for small values of N (see Figure 2.7). Fortunately, an exact representation of Equation 2.46 is not necessary and therefore a sample rate of N/T, or slightly higher, is enough in practice.

After generating the samples of $s^{(P)}(t)$, the wave shape associated with a given burst is obtained by multiplying those samples with the samples of the "time window" $r(t)$ whose duration is higher than T (see Equation 2.46). This means that OFDM burst samples are given by $s_n r_n$, with $r_n \triangleq r(nT/N)$ (note that, contrary to samples s_n, samples $s_n r_n$ are not periodic).

Finally, the analog signal associated with a given OFDM burst is generated with the samples $s_n r_n$ by digital-to-analog (D/A) conversion followed by reconstruction filtering (see Figure 2.10b). The complex envelope can be written as

$$\begin{aligned} s(t) &= \left(\left(\sum_{n=-\infty}^{+\infty} s_n\delta\left(t - n\frac{T}{N}\right)\right)\cdot r(t)\right) * h_{\mathrm{T}}(t) \\ &= \left(\sum_{n=-\infty}^{+\infty} s_n r_n\delta\left(t - n\frac{T}{N}\right)\right) * h_{\mathrm{T}}(t) = \sum_{n=-\infty}^{+\infty} s_n r_n h_{\mathrm{T}}\left(t - n\frac{T}{N}\right), \end{aligned} \tag{2.52}$$

where $h_{\mathrm{T}}(t)$ is the impulsive response of the reconstruction filter.

To simplify the reconstruction filter $h_{\mathrm{T}}(t)$, usually samples of the OFDM burst given by Equation 2.46 are taken with a sample rate $M_{Tx}N/T > N/T$, that is, with an oversampling factor $M_{Tx} > 1$, not necessarily integer. Usually, the original block $\{S_k; k = -N/2, -N/2+1, \ldots, N/2-1\}$ already includes $2N_{\mathrm{I}}$ "idle" subcarriers (i.e., with $S_k = 0$), half of them at the beginning and the other half at the end of the burst. This is in fact equivalent to oversampling an OFDM burst by a factor

$$M_{Tx} = \frac{N}{N - 2N_{\mathrm{I}}} \tag{2.53}$$

with $N - 2N_{\mathrm{I}}$ useful subcarriers.

---

* We also assumed that the time domain and frequency domain inherent to the DFT are periodic with period N.

When we have an oversampling factor M_{Tx} for a reference burst with N subcarriers, samples of $s^{(P)}(t)$ in the interval $[0, T]$ are given by

$$s_n^{(M_{Tx})} \triangleq s^{(P)}\left(n\frac{T}{N'}\right) = \sum_{k=-N/2}^{N/2-1} S_k \exp\left(j2\pi\frac{nk}{N'}\right), \quad n = 0, 1, \ldots, N' - 1, \quad (2.54)$$

where $N' = NM_{Tx}$. In this case,

$$s_n^{(M_{Tx})} = N'\left(\frac{1}{N'}\sum_{k=0}^{N-1} S'_k \exp\left(j2\pi\frac{nk}{N'}\right)\right) = N's'_n, \quad (2.55)$$

$$n = 0, 1, \ldots, N' - 1, \text{ with}$$

$$s'_n = \frac{1}{N'}\sum_{k=0}^{N'-1} S'_k \exp\left(j2\pi\frac{nk}{N'}\right). \quad (2.56)$$

Clearly, the time-domain block $\{s'_n; n = 0, 1, \ldots, N'-1\}$ is the IDFT of the extended block $\{S'_k; k = 0, 1, \ldots, N' - 1\}$ obtained by adding $N' - N$ zeros to the original block in the frequency domain, $\{S_k; k = -N/2, -N/2+1, \ldots, N/2 - 1\}$, in the following way:

$$S'_k = \begin{cases} S_k, & 0 \le k \le \frac{N}{2} - 1, \\ 0, & \frac{N}{2} \le k \le N' - \frac{N}{2} - 1, \\ S_{k-N'}, & N' - \frac{N}{2} \le k \le N' - 1. \end{cases} \quad (2.57)$$

Once again, apart from a scalar factor N', the sampled version of Equation 2.48 with an oversampling factor M_{Tx} corresponds to the IDFT of the block $\{S'_k; k = 0, \ldots, N' - 1\}$.

The complex envelope of the analog signal associated with a given OFDM burst with an oversampling factor M_{Tx} is given by

$$s^{(M_{Tx})}(t) = \sum_{n=-\infty}^{+\infty} s'_n r_n h_T\left(t - n\frac{T}{N'}\right). \quad (2.58)$$

Although signal $s^{(M_{Tx})}(t)$ does not exactly equal the reference representation of the OFDM burst given by Equation 2.46, the difference is small, especially for a large number of subcarriers and/or when the oversampling factor is high, with differences manifesting mainly at the extreme vicinities of the interval occupied by $r(t)$; obviously, if $M_{Tx} \to +\infty$, the signal given by Equation 2.58 converges to the OFDM reference burst.

As for the "time window" $r(t)$, it is common to employ a square-root raised-cosine window instead of a rectangular shape window to reduce the out-of-band radiation levels

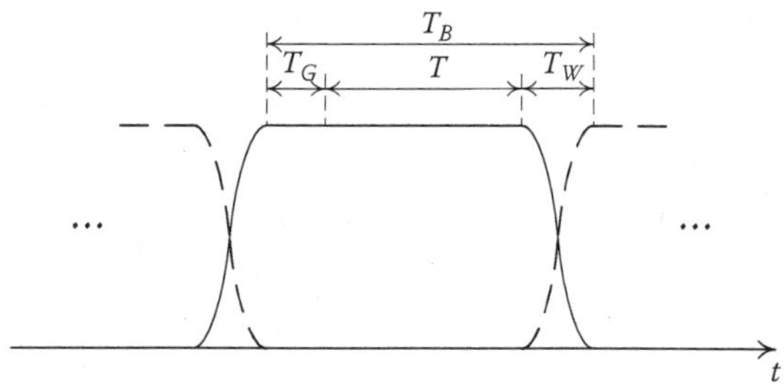

Figure 2.11 Raised-cosine time window.

on the spectrum of OFDM bursts. In practice, the signal associated with a given burst still has a complex envelope given by Equation 2.58, but with

$$r(t) = r'(t) * h_W(t), \tag{2.59}$$

where

$$h_W(t) = \frac{\pi}{2T_W} \cos\left(\frac{\pi t}{T_W}\right) \text{rect}\left(\frac{t}{T_W}\right) \tag{2.60}$$

and $r'(t)$ is a rectangular impulse with duration $T_B = T + T_G + T_W$. As shown in Figure 2.11, the duration of the time window $r(t)$ is $T_B + T_W = T + T_G + 2T_W$, resulting in an overlap of T_W between adjacent bursts. This means that this raised-cosine window has a roll-off factor of $T_W/(T + T_G + T_W)$. In the following we assume that $T_W = 0$.

2.2.3 Receiver Structure

Figure 2.12a shows the OFDM receiver structure and Figure 2.12b shows the input section equivalent scheme of the receiver, corresponding to radio frequency (RF) and intermediate frequency (IF) stages and the signal down-conversion and filtering circuitries to the orthogonal demodulator. Signal sampling is performed with a sample rate

$$\frac{1}{T_s} = \frac{N'}{T}, \tag{2.61}$$

that is, the reception oversampling factor can be the same as the transmission oversampling factor.

Because of multipath propagation, the received burst will overlap, as illustrated in Figure 2.13a. Moreover, some interference between the different subcarriers of the same burst will occur. However, because the detection of OFDM signals operates on signal samples associated with a useful period of duration T, this means that the use of CPs with duration T_G, longer than the overall CIR (which includes the impact of the transmission

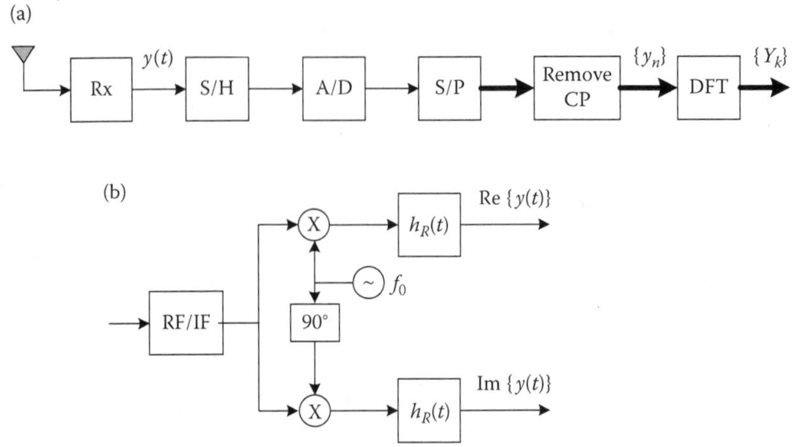

Figure 2.12 OFDM receiver structure (a) and details of "Rx" block (b).

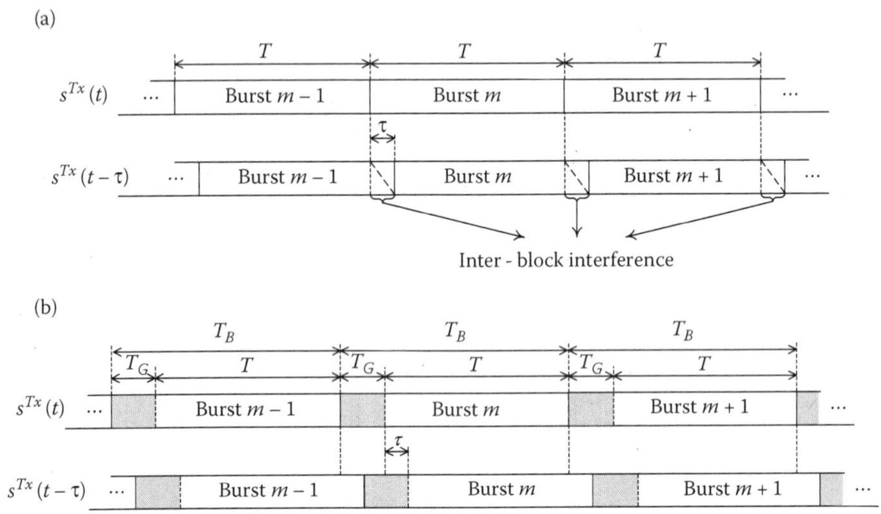

Figure 2.13 Multipath channel impact leading to IBI (a) and elimination of IBI through guard periods (b).

and detection filters as well as the radio channel itself), prevents the effects of the burst's overlapping in the received samples associated with the useful interval (see Figure 2.13b); this is usually referred to as the "absence of ISI".*

* In this context, an OFDM block is usually referred as an "OFDM symbol." The term IBI (Interblock interference) would probably be more adequate.

Let us consider the received time-domain samples y_n given by the discrete convolution

$$y_n = \sum_{l=0}^{N_h-1} s_{n-l} h_l + \nu_n, \tag{2.62}$$

where $\{s_n; n = 0, \ldots, N-1\}$ is the block of samples associated with a transmitted burst, $\{h_n; n = 0, \ldots, N_h - 1\}$ is the CIR, with $N_h < N$ denoting the channel length, and $\{\nu_n; n = 0, \ldots, N-1\}$ are the corresponding Gaussian channel noise samples, assumed to be independent and identically distributed (i.i.d.) in each received burst. Rewriting Equation 2.62 in matrix notation, it is easy to see that it is equivalent to

$$\mathbf{y} = \mathbf{h}\mathbf{s} + \boldsymbol{\nu} \quad \Leftrightarrow \quad \begin{bmatrix} y_0 \\ y_1 \\ \vdots \\ y_{N-1} \\ \hline y_N \\ \vdots \\ y_{N+N_h-2} \end{bmatrix}$$

$$= \left[\begin{array}{cccccccccc}
h_0 & 0 & 0 & \cdots & & & & & & 0 \\
h_1 & h_0 & 0 & 0 & \cdots & & & & & 0 \\
 & \ddots & \ddots & & & & & & & \\
h_{N_h-1} & h_{N_h-2} & \cdots & h_1 & h_0 & 0 & 0 & \cdots & & 0 \\
0 & h_{N_h-1} & h_{N_h-2} & \cdots & h_1 & h_0 & 0 & 0 & \cdots & 0 \\
 & & & & & \ddots & \ddots & & & \\
0 & & \cdots & & 0 & h_{N_h-1} & h_{N_h-2} & \cdots & h_1 & h_0 \\
\hline
0 & & & \cdots & & 0 & h_{N_h-1} & h_{N_h-2} & \cdots & h_1 \\
 & & & & & & & \ddots & \ddots & \\
0 & \cdots & & & & & & 0 & h_{N_h-1} & h_{N_h-2} \\
0 & \cdots & & & & & & & 0 & h_{N_h-1}
\end{array}\right]$$

$$\times \begin{bmatrix} s_0 \\ s_1 \\ \vdots \\ s_{N-1} \\ \hline s_N \\ \vdots \\ s_{N+N_h-2} \end{bmatrix} + \begin{bmatrix} \nu_0 \\ \nu_1 \\ \vdots \\ \nu_{N-1} \\ \hline \nu_N \\ \vdots \\ \nu_{N+N_h-2} \end{bmatrix}. \tag{2.63}$$

From Equation 2.63, we can observe that the received time-domain samples associated with the useful part of a given block are

$$\begin{bmatrix} y_0 \\ y_1 \\ \vdots \\ y_{N-1} \end{bmatrix} = \underbrace{\begin{bmatrix} h_0 & 0 & \cdots & & & & & 0 \\ h_1 & h_0 & 0 & \cdots & & & & 0 \\ & \ddots & \ddots & & & & & \\ 0 & \cdots & & 0 & h_{N_h-1} & \cdots & h_1 & h_0 & 0 \\ 0 & \cdots & & & 0 & h_{N_h-1} & \cdots & h_1 & h_0 \end{bmatrix}}_{h_{0[N\times N]}} \begin{bmatrix} s_0 \\ s_1 \\ \vdots \\ s_{N-1} \end{bmatrix} + \begin{bmatrix} v_0 \\ v_1 \\ \vdots \\ v_{N-1} \end{bmatrix} \tag{2.64}$$

and the received time-domain samples associated with the interference from the $(m-1)$th block are

$$\begin{bmatrix} y_N \\ y_{N+1} \\ \vdots \\ y_{2N-1} \end{bmatrix} = \underbrace{\begin{bmatrix} 0 & \cdots & 0 & h_{N_h-1} & \cdots & h_2 & h_1 \\ 0 & \cdots & & 0 & h_{N_h-1} & \cdots & h_2 \\ & & & & \ddots & \ddots & \\ 0 & \cdots & & & & 0 & h_{N_h-1} \\ 0 & \cdots & & & & & 0 \\ & & & \vdots & & & \\ 0 & \cdots & & & & & 0 \end{bmatrix}}_{h_{1[N\times N]}} \begin{bmatrix} s_N \\ s_{N+1} \\ \vdots \\ s_{2N-1} \end{bmatrix} + \begin{bmatrix} v_N \\ v_{N+1} \\ \vdots \\ v_{2N-1} \end{bmatrix}. \tag{2.65}$$

Therefore, the received samples associated with the mth block can be written as

$$y^{(m)} = h_0 s^{(m)} + h_1 s^{(m-1)} + v^{(m)}. \tag{2.66}$$

Since for cyclic extended bursts with duration $T_B = T_G + T$, at least the first N_h samples associated with the guard period T_G are a repetition of the L final burst samples, this means that $s^{(m)} = s^{(m-1)}$ for the guard period, leading to

$$y^{(m)} = (h_0 + h_1)\, s^{(m)} = h_{CP}\, s^{(m)} + v^{(m)}, \tag{2.67}$$

where h_{CP} is a size- $N \times N$ circulant matrix given by

$$h_{CP} = \begin{bmatrix} h_0 & 0 & \cdots & & 0 & h_{N_h-1} & \cdots & h_2 & h_1 \\ h_1 & h_0 & 0 & \cdots & & 0 & h_{N_h-1} & \cdots & h_2 \\ & \ddots & \ddots & & & & \ddots & \ddots & \\ 0 & \cdots & 0 & h_{N_h-1} & \cdots & h_0 & 0 & \cdots & h_{N_h-1} \\ & & & & & \ddots & \ddots & & \\ 0 & \cdots & & 0 & h_{N_h-1} & \cdots & h_1 & h_0 & 0 \\ 0 & \cdots & & & 0 & h_{N_h-1} & \cdots & h_1 & h_0 \end{bmatrix}, \tag{2.68}$$

whose elements verify $[h_{CP}]_{i,i'} = h_{(i-i') \bmod N}$ ($x \bmod y$ denotes the modulo operation, that is, the remainder of division of x by y).

It is well known that circulant matrices can be diagonalized using a Fourier matrix, that is,

$$\mathbf{h}_{CP} = \mathbf{F}^{-1}\boldsymbol{\Lambda}\mathbf{F}, \tag{2.69}$$

where the size- $N \times N$ matrix $\mathbf{F}$, given by

$$\mathbf{F} = \frac{1}{\sqrt{N}} \begin{bmatrix} 1 & 1 & 1 & \cdots & 1 \\ 1 & \omega & \omega^2 & \cdots & \omega^{N-1} \\ 1 & \omega^2 & \omega^4 & \cdots & \omega^{2(N-1)} \\ \vdots & \vdots & \vdots & & \vdots \\ 1 & \omega^{N-1} & \omega^{2(N-1)} & \vdots & \omega^{(N-1)(N-1)} \end{bmatrix}, \tag{2.70}$$

with $\omega = \exp(-j2\pi/N)$, is the unitary (i.e., $F^H = F^{-1}$) DFT matrix whose columns are the eigenvectors of h_{CP}, and $\boldsymbol{\Lambda}$ is a size- $N \times N$ diagonal matrix whose elements, the eigenvalues of h_{CP}, equal the DFT of the first column of h_{CP}, that is,

$$\boldsymbol{\Lambda} = \begin{bmatrix} \lambda_0 & & & 0 \\ & \lambda_1 & & \\ & & \ddots & \\ 0 & & & \lambda_{N-1} \end{bmatrix} = \begin{bmatrix} H_0 & & & 0 \\ & H_1 & & \\ & & \ddots & \\ 0 & & & H_{N-1} \end{bmatrix} = \mathbf{H}, \tag{2.71}$$

where

$$\lambda_k = \sum_{n=0}^{N-1} h_n \exp\left(-j2\pi\frac{kn}{N}\right) = \mathrm{DFT}\{h_n\} = H_k, \quad k = 0, 1, \ldots, N-1. \tag{2.72}$$

Using Equations 2.69 and 2.71 in Equation 2.67, it follows that (for the sake of notation simplicity, we will drop the superscript m)

$$\begin{aligned} \mathbf{y} &= \mathbf{F}^{-1}\mathbf{H}\mathbf{F}\mathbf{s} + \boldsymbol{\nu} \;\Leftrightarrow \\ \Leftrightarrow\; \mathbf{F}\mathbf{y} &= \mathbf{H}\mathbf{F}\mathbf{s} + \mathbf{F}\boldsymbol{\nu} \;\Leftrightarrow \\ \Leftrightarrow\; \mathbf{Y} &= \mathbf{H}\mathbf{S} + \mathbf{N}, \end{aligned} \tag{2.73}$$

with the vectors

$$\mathbf{Y} = \mathbf{F}\mathbf{y} = \begin{bmatrix} Y_0 \\ Y_1 \\ \vdots \\ Y_{N-1} \end{bmatrix}, \tag{2.74}$$

$$\mathbf{S} = \mathbf{F}\mathbf{s} = \begin{bmatrix} S_0 \\ S_1 \\ \vdots \\ S_{N-1} \end{bmatrix}, \tag{2.75}$$

and

$$\mathbf{N} = \mathbf{F}\boldsymbol{\nu} = \begin{bmatrix} N_0 \\ N_1 \\ \vdots \\ N_{N-1} \end{bmatrix} \tag{2.76}$$

denoting the DFTs of $\mathbf{y}$, $\mathbf{s}$, and $\boldsymbol{\nu}$, respectively. This means that the received sample at the kth subcarrier is given by

$$Y_k = H_k S_k + N_k, \tag{2.77}$$

$k = 0, 1, \ldots, N-1$, where H_k is the channel frequency response for the kth subcarrier and N_k is the channel noise component for that subcarrier. Therefore, the channel acts like a simple multiplicative factor for each subcarrier, preserving the orthogonality between subcarriers in the useful interval; this is usually referred to as the "absence of interchannel interference" (ICI). In fact, because CP corresponds to a cyclic extension of each burst, this means that, for each useful interval T, the transmitted signal corresponds, not as a sequence of OFDM bursts expressed by Equation 2.45 but as the periodic signal $s^{(P)}(t)$ given by Equation 2.43 (see Figure 2.14). Therefore, the linear convolution associated with the channel is formally equivalent to a circular convolution with respect to the useful part of the OFDM block.

It should be mentioned that the elimination of ISI can be achieved by extending each burst with any known fixed sequence, including an all-zero sequence [zero padding (ZP)] or a pseudonoise (PN) symbol sequence, denoted as PN extension or unique word (UW). ZP schemes can be a good alternative to CP-assisted schemes [Muquet

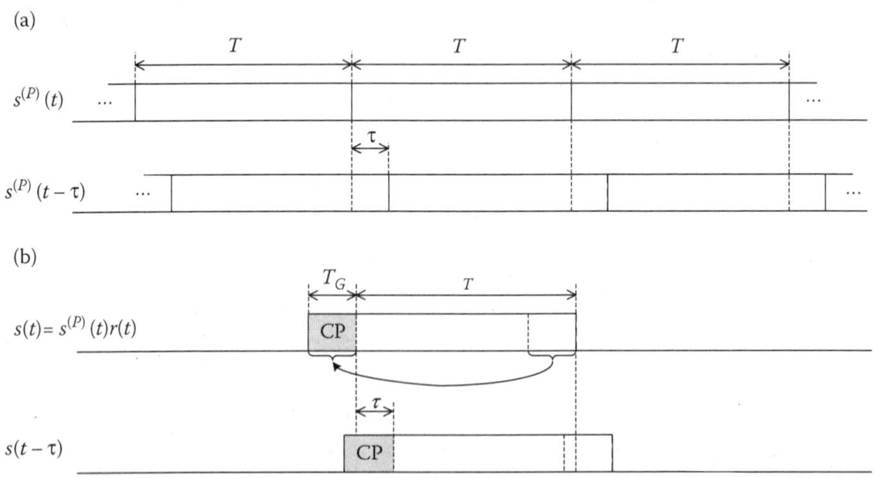

Figure 2.14 Multipath channel impact on $s^{(P)}$ (a) and on the corresponding CP extended OFDM burst (b).

et al. 2000a]. However, complex receiver structures must be employed, involving the inversion and/or the multiplication of matrixes whose dimensions can grow with block length, which are not suitable when large blocks are employed. By employing overlap-and-add techniques [Muquet et al. 2000b], receiver complexity becomes similar to that of conventional CP-assisted schemes, but performance is also identical. Efficient FFT-based receivers can also be designed for ZP OFDM [Araújo and Dinis 2006].

To avoid power and spectral degradation, the CP length should be a small fraction of the overall length of the blocks. In this work we will consider only CP-assisted block transmission.

Since CP-assisted transmission can be seen as an N parallel nonselective subchannel transmission expressed by Equation 2.77, channel distortion effects for an uncoded OFDM transmission can easily be compensated for using the receiver depicted in Figure 2.15a, where the equalized frequency-domain sample at the kth subcarrier $\tilde{S}_k$ is obtained by

$$\tilde{S}_k = F_k Y_k, \tag{2.78}$$

with the set of coefficients $\{F_k; k = 0, 1, \ldots, N-1\}$ given by

$$F_k = \frac{1}{H_k} = \frac{H_k^*}{|H_k|^2}, \tag{2.79}$$

corresponding to an FDE under the zero-forcing (ZF) criterion.

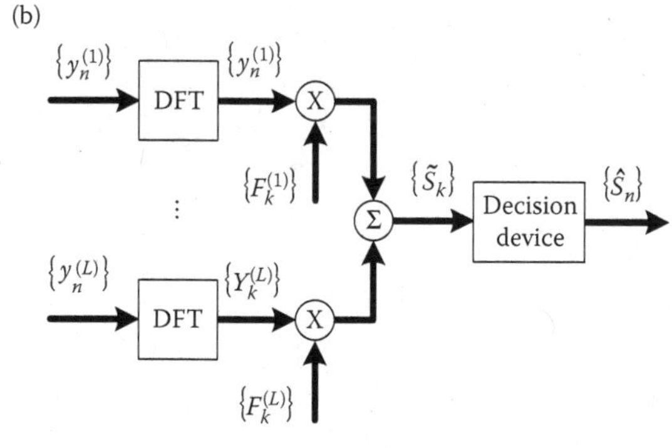

Figure 2.15 Channel distortion effects compensation for an uncoded OFDM transmission with no space diversity (a) and with an *L*-order space diversity receiver (b).

In the case where we have L-order space diversity, the received sample at the kth subcarrier and the lth diversity branch is given by

$$Y_k^{(l)} = S_k H_k^{(l)} + N_k^{(l)}, \quad l = 1, \ldots, L, \tag{2.80}$$

where $H_k^{(l)}$ and $N_k^{(l)}$ denote the channel frequency response and the noise term for the kth subcarrier and the lth diversity branch, respectively. The corresponding equalized sample is

$$\tilde{S}_k = \sum_{l=1}^{L} F_k^{(l)} Y_k^{(l)}, \tag{2.81}$$

where the set $\left\{F_k^{(l)}; k = 0, 1, \ldots, N-1\right\}$ $(l = 1, \ldots, L)$ denotes the FDE coefficients associated with the lth diversity branch, which can be set as

$$F_k^{(l)} = \frac{H_k^{(l)*}}{\sum_{l'=1}^{L} |H_k^{(l')}|^2}. \tag{2.82}$$

Therefore, the receiver can be the one depicted in Figure 2.15b and, from Equation 2.81,

$$\tilde{S}_k = S_k + \frac{\sum_{l=1}^{L} H_k^{(l)*}}{\sum_{l'=1}^{L} |H_k^{(l')}|^2} N_k^{(l)}. \tag{2.83}$$

A decision on the symbol transmitted through subcarrier k can be made based on $\tilde{S}_k$. Clearly, the receiver structure of Figure 2.15b implements a maximal-ratio combining (MRC) diversity scheme for each subcarrier k.

2.2.4 Envelope Fluctuations

One important drawback of conventional OFDM transmission schemes is their strong envelope fluctuation and high peak-to-mean envelope power ratio (PMEPR), as depicted in Figure 2.16, which lead to amplification difficulties. In fact, to avoid the out-of-band radiation levels that are inherent in nonlinear distortion, power amplifiers for OFDM transmission are required to have linear characteristics and/or a significant input backoff has to be adopted. For this reason several methods have been proposed to reduce these amplification difficulties by means of digital signal processing [Jones and Wilkinson 1996; Müller and Huber 1997; Cimini Jr. and Sollenberger 2000]. One of the most promising techniques to reduce the envelope fluctuations of OFDM signals employ a digital clipping operation followed by a frequency-domain filtering operation [Dinis and Gusmão 2004]. In this case, the OFDM transmitted could be the one depicted in Figure 2.17 where the block of modified time-domain samples to be transmitted is generated as follows. The augmented frequency-domain block $\{S'_k; k = 0, 1, \ldots, N'-1\}$, with S'_k given by Equation 2.57, is passed to the time domain by an IDFT operation, leading to the block

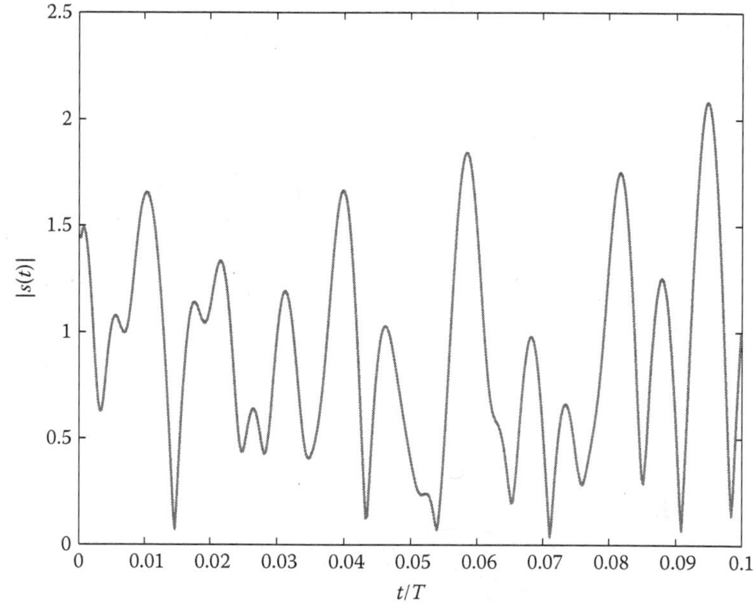

Figure 2.16 Evolution of the envelope of an OFDM signal.

$\{s'_n; n = 0, 1, \ldots, N' - 1\}$, with s'_n given by Equation 2.56. These time-domain samples are submitted to a clipping device, leading to the samples

$$s_n^{\mathrm{C}} = \begin{cases} s'_n, & \left|s'_n\right| < s_M, \\ s_M \exp(\mathrm{j} \arg(s'_n)), & \left|s'_n\right| \geq s_M. \end{cases} \tag{2.84}$$

The block $\{s_n^{\mathrm{C}}; n = 0, 1, \ldots, N' - 1\}$ is passed to the frequency domain by a DFT operation, leading to the block $\{S_k^{\mathrm{C}}; k = 0, 1, \ldots, N' - 1\}$, where a filtering operation is performed to reduce/eliminate the out-of-band radiation. The final frequency-domain block to be transmitted is $\{S_k^{\mathrm{CF}} = S_k^{\mathrm{C}} G_k; k = 0, 1, \ldots, N' - 1\}$, with $G_k = 1$ for the in-band subcarriers and 0 for the out-of-band ones. The corresponding time-domain block to be transmitted is $\{s_n^{\mathrm{CF}}; k = 0, 1, \ldots, N' - 1\} = \mathrm{IDFT}\{S_k^{\mathrm{CF}}; k = 0, 1, \ldots, N' - 1\}$.

These techniques allow significant reductions on the envelope fluctuations of the transmitted signals while maintaining the spectral occupation of conventional OFDM

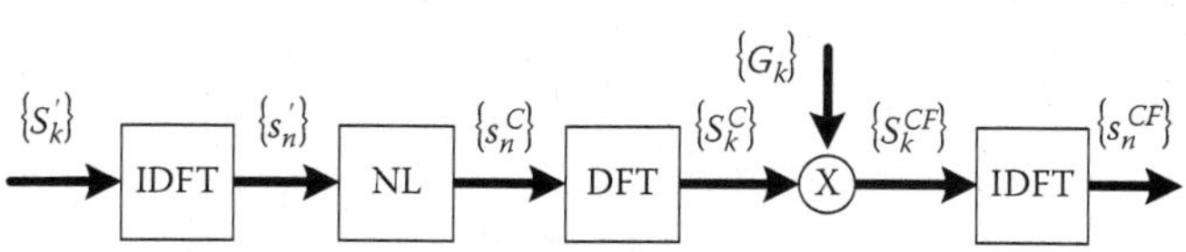

Figure 2.17 Clipping and filtering technique.

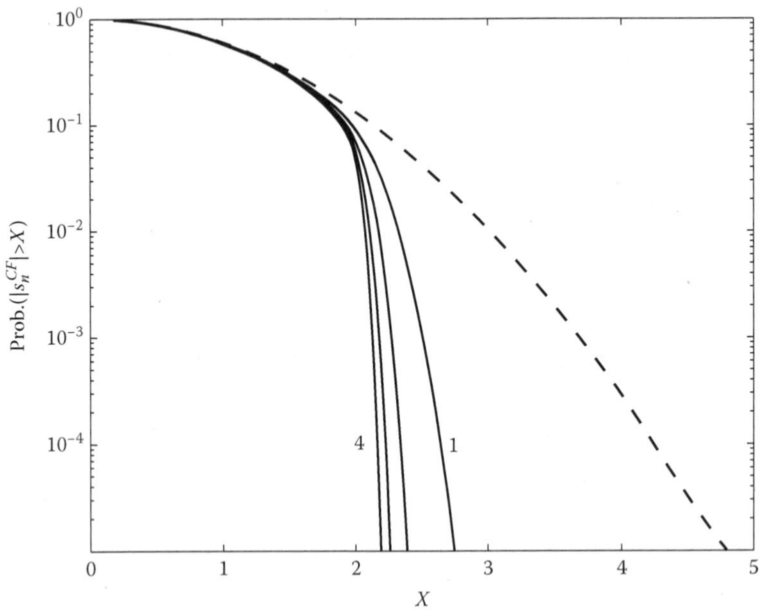

Figure 2.18 Envelope distribution for iterations 1 to 4 (solid lines), together with the envelope distribution for conventional OFDM (dashed line).

signals. Since the filtering operation leads to some regrowth on the envelope fluctuations of transmitted signals, the clipping and filtering operations can be repeated several times allowing additional reduction on the envelope fluctuations of transmitted signals [Armstrong 2001; Dinis and Gusmão 2004] (see Figure 2.18). However, the use of repeated iterative clipping and filtering operations leads to an increase in the nonlinear distortion levels as well as to an increase in the signal processing requirements. Moreover, even for the most sophisticated techniques, the transmitted signals still have PMEPRs higher than the PMEPRs for SC signals based on similar constellations. This is the motivation behind SC-FDE schemes [Sari et al. 1994].

2.3 SC with FDE

2.3.1 Linear FDE

To understand SC-FDE schemes, as well as SC-FDMA schemes adopted for the uplink of 3GPP LTE [3GPP 2005a], let us consider an SC-based block transmission with N useful modulation symbols per block $\{s_n; n = 0, 1, \ldots, N-1\}$ resulting from a direct mapping of the original data into a selected signal constellation, plus a suitable CP that has essentially the same function as with OFDM schemes. The receiver can

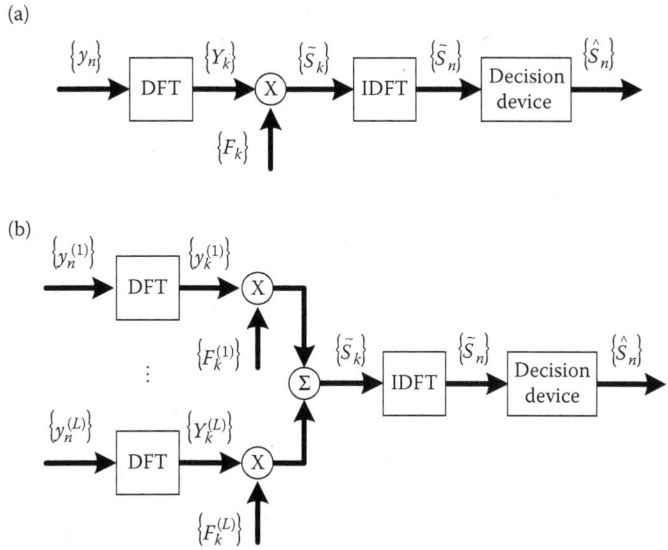

Figure 2.19 SC-FDE receiver structure with no space diversity (a) and with an *L*-order space diversity receiver (b).

have the structure represented in Figure 2.19a, where it is assumed that after being down-converted and filtered, the signal is sampled and A/D converted. The resulting signal is S/P converted and the CP samples are removed, leading to the received time-domain samples $\{y_n; n = 0, 1, \ldots, N-1\}$. These samples are passed to the frequency domain by an N-point DFT, leading to the corresponding frequency-domain samples $\{Y_k; k = 0, 1, \ldots, N-1\}$, where $Y_k = H_k S_k + N_k$ (as with OFDM schemes).

For an FDE optimized under the ZF criterion, the equalized frequency-domain samples, given by Equation 2.78, can be obtained with the set of coefficients F_k given by Equation 2.69. However, for a typical frequency-selective channel, deep notches in the channel frequency response lead to significant noise enhancement effects when the ZF criterion is employed. To minimize the combined effect of ISI and channel noise, the equalized samples $\{\widetilde{S}_k; k = 0, 1, \ldots, N-1\}$ are obtained with the coefficients $\{F_k; k = 0, 1, \ldots, N-1\}$ usually optimized under the MMSE criterion. The mean-squared error (MSE) after returning to the time domain can be expressed as

$$\Theta(k) = \frac{1}{N^2} \sum_{k=0}^{N-1} \Theta_k, \tag{2.85}$$

where

$$\Theta_k = E\left[\left|\widetilde{S}_k - S_k\right|^2\right] = E\left[|Y_k F_k - S_k|^2\right]. \tag{2.86}$$

To minimize the MSE, we can minimize Θ_k for each k separately, that is,

$$\min_{F_k} \left(E\left[|Y_k F_k - S_k|^2\right]\right), \quad k = 0, 1, \ldots, N-1, \tag{2.87}$$

which leads to the set of optimized FDE coefficients

$$F_k = \frac{H_k^*}{\beta + |H_k|^2}, \quad k = 0, 1, \ldots, N-1, \tag{2.88}$$

where β is the inverse of the SNR, given by

$$\beta = \frac{\sigma_N^2}{\sigma_S^2}, \tag{2.89}$$

with

$$\sigma_N^2 = \frac{E\left[|N_k|^2\right]}{2} \tag{2.90}$$

and

$$\sigma_S^2 = \frac{E\left[|S_k|^2\right]}{2} \tag{2.91}$$

denoting the variances of the real and imaginary parts of the channel noise components $\{N_k; k = 0, 1, \ldots, N-1\}$ and the data samples $\{S_k; k = 0, 1, \ldots, N-1\}$, respectively.

Since for SC modulations the data contents of a given block are transmitted in the time domain, the equalized samples $\{\tilde{S}_k; k = 0, \ldots, N-1\}$ are converted back to the time domain by an IDFT operation leading to the block of time-domain equalized samples $\{\tilde{s}_n; n = 0, \ldots, N-1\}$. These equalized samples will then be used to make decisions on the transmitted symbols.

The SC-FDE receiver can be easily extended to an L-branch diversity scenario, as depicted in Figure 2.19b. In this case, the frequency-domain samples at the FDE's output are given by Equation 2.67, where the set $\left\{F_k^{(l)}; k = 0, 1, \ldots, N-1\right\}$ $(l = 1, \ldots, L)$ can be selected under the ZF or MMSE criterion. Under "equal noise level" conditions [i.e., $\sigma_n^{(1)} = \ldots = \sigma_n^{(L)} \triangleq \sigma_n$, with $\left(\sigma_n^{(l)}\right)^2$ denoting the variance of the input noise samples at the lth branch] and an MMSE criterion, it can be easily shown that the optimized FDE coefficients are given by [Gusmão et al. 2003]

$$F_k^{(l)} = \frac{H_k^{(l)*}}{\beta + \sum_{l'=1}^{L} \left|H_k^{(l')}\right|^2}, \quad l = 1, 2, \ldots, L. \tag{2.92}$$

Contrary to the OFDM schemes where ZF and MMSE criterions yield the same performance [Gusmão et al. 2003], for the SC case the performances with ZF and MMSE criterions are only identical when the channel frequency response across the transmission BW is practically constant (i.e., for $H_0^{(l)} = \ldots = H_{N-1}^{(l)} \triangleq H^{(l)}, l = 1, \ldots, L$). An

FDE optimized under the MMSE criterion does not attempt to fully invert the channel when we have a deep fade, reducing noise enhancement effects and allowing better performances.

Figure 2.20 shows the compensation of the channel distortion effect when the FDE coefficients $\{F_k; k = 0, \ldots, N-1\}$ are optimized under the ZF or MMSE criterions for a frequency-selective channel realization with deep in-band notches. Clearly, with the ZF criterion the channel is completely inverted, resulting in a perfect equalized channel after the FDE, while the MMSE criterion leads to imperfect channel equalization. However, as shown in Figure 2.21, in spite of perfectly equalizing the channel, the ZF criterion significantly enhances the channel noise at subchannels with local deep notches, while with the MMSE criterion, the noise-dependent term β in Equation 2.92 avoids noise enhancement effects when the channel frequency response is very low.

For comparison purposes, Figure 2.22 shows the block diagrams for the transmission chains of OFDM and SC-FDE options. From this figure it is clear that both schemes are closely related. Their overall signal processing efforts, measured in terms of DFT/IDFT

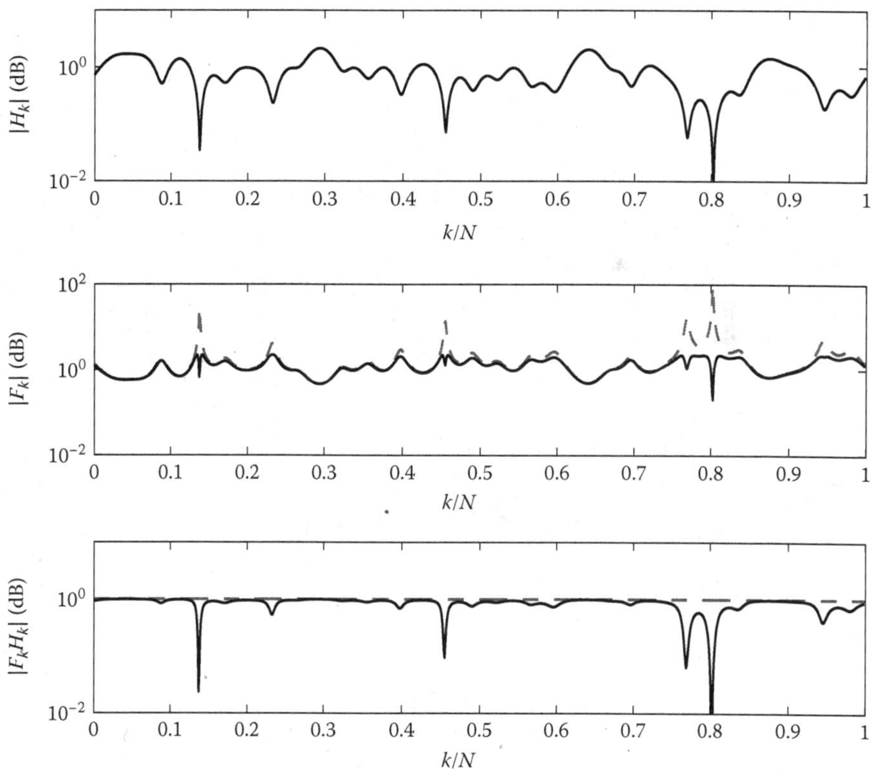

Figure 2.20 Channel frequency response (top), FDE coefficients optimized under the ZF (- - -) and MMSE criterions (—) (middle) and the corresponding channel distortion effects compensation (bottom).

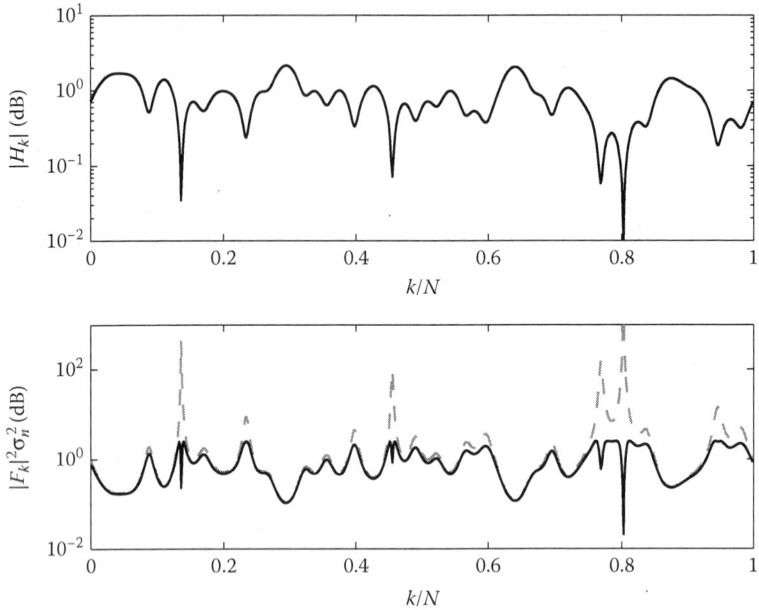

Figure 2.21 Channel frequency response (top) and noise power at the output of an FDE optimized under the ZF (- - -) and MMSE (—) criterions (bottom).

blocks, are similar, with the only difference being the shift of the IDFT block from the OFDM transmitter to the SC-FDE receiver. However, there are pros and cons to both sides of these two transmission techniques. One of the most important issues that must be taken into consideration is the already mentioned strong envelope fluctuations and

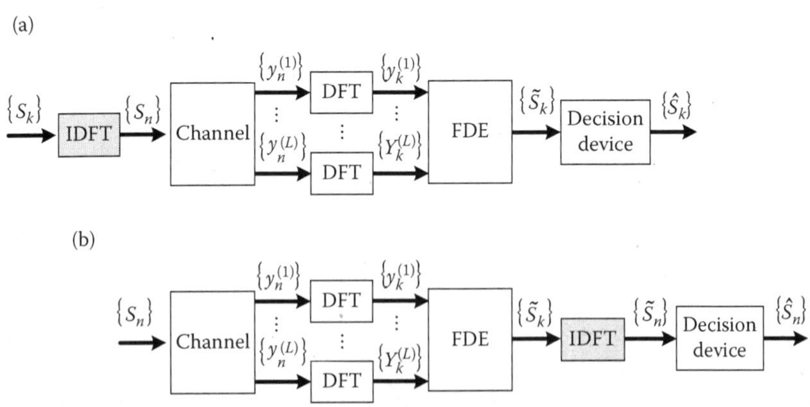

Figure 2.22 Generic transmission chain for OFDM (a) and SC-FDE (b).

high PMEPR of OFDM signals with a large number of subcarriers, leading to power amplification difficulties. This requires the use of highly linear power amplifiers at the transmitter and/or more power backoff than with comparable SC signals. This aspect is particularly important for uplink transmission since low-cost and low-consumption power amplifiers are desirable at the MT/UE. In fact, even when suitable signal processing techniques are employed to reduce the envelope fluctuations of OFDM signals [Dinis and Gusmão 2004], the resulting envelope fluctuations are still higher than with the corresponding SC schemes.

Nevertheless, when the wireless network includes fixed terminals (e.g., within BSs and/or for broadcasting systems), OFDM schemes are good candidates. Having in mind the compatibility between SC-FDE and OFDM options we can choose an SC-FDE scheme, exhibiting low envelope fluctuations, for the uplink and an OFDM scheme for the downlink. This means an implementation advantage for the MTs/UEs, where simple SC transmissions and OFDM reception functions are then carried out. The implementation charge is concentrated in the BSs (where increased power consumption and cost are not so critical), concerning both signal processing effort and power amplification difficulties [Gusmão et al. 2000; Falconer et al. 2002].

2.3.2 IB-DFE Receivers

It is well known that decision feedback equalizers (DFEs) [Proakis 2001] can significantly outperform linear equalizers (in fact, DFEs include as a special case linear equalizers). Time-domain DFEs have good performance/complexity trade-offs, provided that the CIR is not too long. However, if the CIR expand over a large number of symbols (such as in the case of severely time-dispersive channels), conventional time-domain DFEs are too complex. For this reason, a hybrid time–frequency SC-DFE was proposed in Ref. [Benvenuto and Tomasin 2002], employing a frequency-domain feedforward filter and a time-domain feedback filter. This hybrid time–frequency-domain DFE has a better performance than a linear FDE. However, as with conventional, time-domain DFEs, it can suffer from error propagation, especially when the feedback filters have a large number of taps. A promising iterative block-DFE (IB-DFE) approach for SC transmission was proposed in Ref. [Benvenuto and Tomasin 2002] and extended to transmit/diversity scenarios in Ref. [Dinis et al. 2003]. Within these IB-DFE schemes, both the feedforward and the feedback parts are implemented in the frequency domain, as depicted in Figure 2.23.

Let us consider an L-order space diversity IB-DFE. For a given ith iteration, the output samples are given by

$$\tilde{S}_k^{(i)} = \sum_{l=1}^{L} F_k^{(l,i)} Y_k^{(l)} - B_k^{(i)} \hat{S}_k^{(i-1)}, \tag{2.93}$$

where $\left\{F_k^{(l,i)}; k = 0, 1, \ldots, N-1\right\}$ $(l = 1, \ldots, L)$ and $\left\{B_k^{(i)}; k = 0, 1, \ldots, N-1\right\}$ denote the feedforward and feedback equalizer coefficients, respectively, and $\left\{\hat{S}_k^{(i-1)}; k = 0, 1, \ldots, N-1\right\}$ is the DFT of the hard-decision block $\{\hat{s}_n^{(i-1)};$

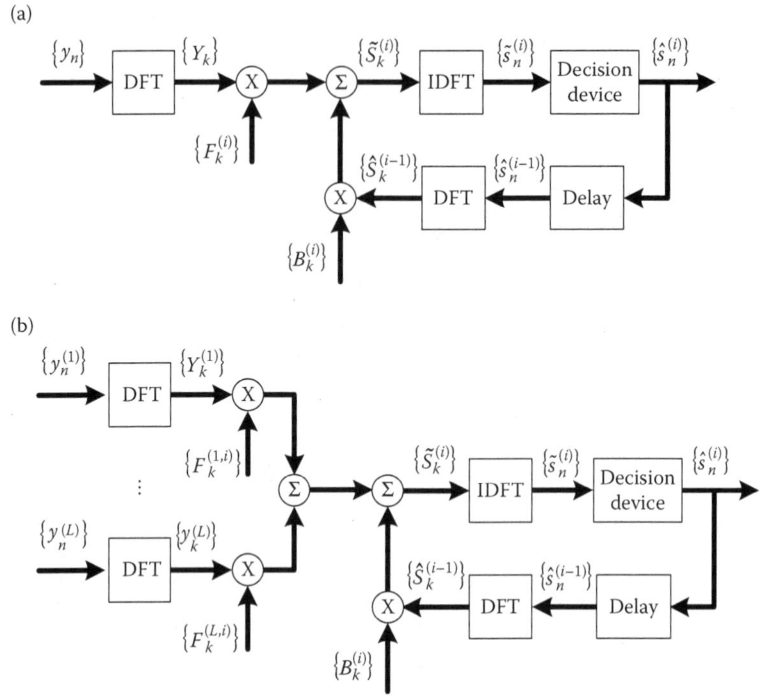

Figure 2.23 IB-DFE receiver structure with no diversity (a) and with an *L*-branch space diversity (b).

$n = 0, 1, \ldots, N-1\}$, of the $(i-1)$th iteration, associated with the transmitted time-domain block $\{s_n; n = 0, 1, \ldots, N-1\}$.

The forward and backward IB-DFE coefficients $\left\{F_k^{(l,i)}; k = 0, 1, \ldots, N-1\right\}$ $(l = 1, \ldots, L)$ and $\left\{B_k^{(i)}; k = 0, 1, \ldots, N-1\right\}$, respectively, are chosen so as to maximize the "signal-to-interference plus noise ratio" (SINR). In Ref. [Dinis et al. 2003] it is shown that the optimum feedforward and feedback coefficients are given by

$$F_k^{(l,i)} = \frac{\kappa_F^{(i)} H_k^{(l)*}}{\beta + \left[1 - \left(\rho^{(i-1)}\right)^2\right] \sum_{l'=1}^{L} \left|H_k^{(l')}\right|^2}, \quad l = 1, 2, \ldots, L \tag{2.94}$$

and

$$B_k^{(i)} = \rho^{(i-1)} \left(\sum_{l'=1}^{L} F_k^{(l',i)} H_k^{(l')} - 1 \right), \tag{2.95}$$

respectively, where $\kappa_F^{(i)}$ is selected to ensure that $\gamma^{(i)} = 1$, β is given by Equation 2.75, and

$$\rho^{(i)} = \frac{E\left[s_n^* \hat{s}_n^{(i)}\right]}{E\left[|s_n|^2\right]} \tag{2.96}$$

is a measure of the reliability of the decisions used in the feedback loop. Since the IB-DFE coefficients take into account the overall block reliability, the error propagation problem is significantly reduced. Consequently, the IB-DFE techniques offer much better performances than the noniterative methods. In fact, the IB-DFE schemes can be regarded as low-complexity turbo equalizers [Tüchler and Hagenauer 2000; Tüchler and Hagenauer 2001; Tüchler et al. 2002] since the feedback loop uses the equalizer outputs instead of the channel decoder outputs. For the first iteration we do not have any information about s_n, which means that $\rho = 0$, $B_k^{(0)} = 0$, and $F_k^{(l,0)}$ are given by Equation 2.92. Therefore, the IB-DFE reduces to a linear FDE.

Clearly, for the first iteration ($i = 0$), no information exists about S_k and the correlation coefficient in Equation 2.96 is zero. This means that $B_k^{(0)} = 0$ and

$$F_k^{(l,0)} = \frac{\kappa_F^{(0)} H_k^{(l)*}}{\beta + \sum_{l'=1}^{L} \left|H_k^{(l')}\right|^2}, \quad l = 1, 2, \ldots, L, \tag{2.97}$$

corresponding to the optimum frequency-domain equalizer coefficients under the MMSE criterion. After that first iteration, and if the residual BER is not too high, we can use the feedback coefficients to eliminate a significant part of the residual interference. When $\rho \approx 1$ (after several iterations and/or moderate-to-high SNR), we have an almost full cancelation of the residual ISI through these coefficients, while the feedforward coefficients perform an approximate matched filtering.

Clearly, Equation 2.93 could be written as

$$\tilde{S}_k^{(i)} = \sum_{l=1}^{L} F_k^{(l,i)} Y_k^{(l)} - B_k^{(i)} \overline{S}_k^{(i-1)}, \tag{2.98}$$

with

$$\overline{S}_k^{(i-1)} = \rho^{(i-1)} \hat{S}_k^{(i-1)}. \tag{2.99}$$

Since $\rho^{(i-1)}$ can be regarded as the blockwise reliability of the estimates $\hat{S}_k^{(i-1)}$, $\overline{S}_k^{(i-1)}$ is the overall block average of $S_k^{(i-1)}$ at the FDE output. To improve the performances, we could replace the "blockwise averages" by "symbol averages," which can be done as described in the following.

If we assume that the transmitted symbols are selected from a QPSK constellation under a Gray mapping rule (the generalization to other cases is straightforward), that is, $s_n = \pm 1 \pm j = s_n^{\mathrm{I}} + j s_n^{\mathrm{Q}}$, with $s_n^{\mathrm{I}} = \mathrm{Re}\{s_n\}$ and $s_n^{\mathrm{Q}} = \mathrm{Im}\{s_n\}$ (and similar definitions for $\tilde{s}_n$, $\bar{s}_n$, and $\hat{s}_n$), then it can be shown that the log likelihood ratios (LLRs) of the "in-phase bit" and the "quadrature bit," associated with s_n^{I} and s_n^{Q}, respectively, are given by

$$L_n^{\mathrm{I}} = \frac{2\tilde{s}_n^{\mathrm{I}}}{\sigma_p^2} \tag{2.100}$$

and

$$L_n^{\mathrm{Q}} = \frac{2\tilde{s}_n^{\mathrm{Q}}}{\sigma_p^2}, \tag{2.101}$$

respectively, where

$$\sigma_p^2 = \frac{1}{2}E\left[|s_n - \tilde{s}_n|^2\right] \approx \frac{1}{2N}\sum_{n=0}^{N-1} E\left[|\hat{s}_n - \widetilde{S}_n|^2\right]. \tag{2.102}$$

Under a Gaussian assumption, it can be shown that the mean value of s_n conditioned to the FDE output $\widetilde{S}_n$ is

$$\begin{aligned}\bar{s}_n &= \tanh\left(\frac{L_n^{\mathrm{I}}}{2}\right) + \mathrm{j}\tanh\left(\frac{L_n^{\mathrm{Q}}}{2}\right)\\ &= \rho_n^{\mathrm{I}}\hat{s}_n^{\mathrm{I}} + \mathrm{j}\rho_n^{\mathrm{Q}}\hat{s}_n^{\mathrm{Q}},\end{aligned} \tag{2.103}$$

where the hard decisions $\hat{s}_n^{\mathrm{I}} = \pm 1$ and $\hat{s}_n^{\mathrm{Q}} = \pm 1$ are defined according to the signs of L_n^{I} and L_n^{Q}, respectively, and ρ_n^{I} and ρ_n^{Q} can be regarded as the reliabilities associated with the "in-phase" and "quadrature" bits of the nth symbol, given by

$$\rho_n^{\mathrm{I}} = \frac{E\left[s_n^{\mathrm{I}*}\hat{s}_n^{\mathrm{I}}\right]}{E\left[|s_n^{\mathrm{I}}|^2\right]} = \tanh\left(\frac{|L_n^{\mathrm{I}}|}{2}\right) \tag{2.104}$$

and

$$\rho_n^{\mathrm{Q}} = \frac{E\left[s_n^{\mathrm{Q}*}\hat{s}_n^{\mathrm{Q}}\right]}{E\left[\left|s_n^{\mathrm{Q}}\right|^2\right]} = \tanh\left(\frac{|L_n^{\mathrm{Q}}|}{2}\right) \tag{2.105}$$

(for the first iteration, $\rho_n^{\mathrm{I}} = \rho_n^{\mathrm{Q}} = 0$ and $\bar{s}_n = 0$).

The feedforward coefficients are still obtained from Equation 2.94, with the blockwise reliability given by

$$\rho^{(i)} = \frac{1}{2N}\sum_{n=0}^{N-1}\left(\rho_n^{\mathrm{I}(i)} + \rho_n^{\mathrm{Q}(i)}\right). \tag{2.106}$$

Therefore, the receiver with "blockwise reliabilities," denoted in the following as IB-DFE with hard decisions, and the receiver with "symbol reliabilities," denoted in the following as IB-DFE with soft decisions, employ the same feedforward coefficients;

however, in the first the feedback loop uses the "hard decisions" on each data block, weighted by a common reliability factor, while in the second the reliability factor changes from symbol to symbol (in fact, the reliability factor is different in the real and imaginary component of each symbol).

It is also possible to define turbo FDE receivers based on IB-DFE receivers that, as conventional turbo equalizers, employ the channel decoder outputs instead of the uncoded "soft decisions" in the feedback loop. The receiver structure is similar to the IB-DFE with soft decisions, but with a soft-in, soft-out (SISO) channel decoder employed in the feedback loop. The SISO block, which can be implemented as defined in Ref. [Vucetic and Yuan 2002], provides the LLRs of both the "information bits" and the "coded bits." The inputs of the SISO block are the LLRs of the "coded bits" at the FDE output. Once again, the feedforward coefficients are obtained from Equation 2.94, with the blockwise reliability given by Equation 2.106.

2.4 Performance Comparisons between OFDM and SC-FDE

To give some insight into the BER performance of OFDM and SC-FDE modulations, we now present a set of simulation results concerning an uncoded transmission over the strong frequency-selective channel. We consider perfect synchronization and channel estimation conditions, as well as linear power amplification. Blocks have length $N = 512$ useful modulation symbols that are selected from a QPSK constellation, under a Gray mapping rule. For both modulation choices a suitable CP is added to each time-domain block. We also consider an L-order space diversity receiver with uncorrelated receive antennas.

Figures 2.24 and 2.25 show the average uncoded BER performance (averaged over all blocks) for both OFDM and SC-FDE modulations (under the MMSE criterion) with $L = 1$, 2, and 4, respectively. BERs are expressed as a function of E_b/N_0, where E_b denotes the average bit energy and N_0 the one-sided power spectral density of the channel noise. For the sake of comparisons, we also include the corresponding AWGN channel performance with Lth order diversity

$$P_b = Q\left(\sqrt{\frac{2LE_b}{N_0}}\right), \tag{2.107}$$

the MF bound (MFB) performance, defined as

$$P_{b,\mathrm{MFB}} = E\left[Q\left(\sqrt{\frac{2E_b}{N_0}\frac{1}{N}\sum_{k=0}^{N-1}\sum_{l=1}^{L}\left|H_k^{(l)}\right|^2}\right)\right], \tag{2.108}$$

where the expectation is over the set of channel realizations (it is assumed that $E\left[\left|H_k^{(l)}\right|^2\right] = 1$, for any k), and the performance for a Rayleigh fading channel with

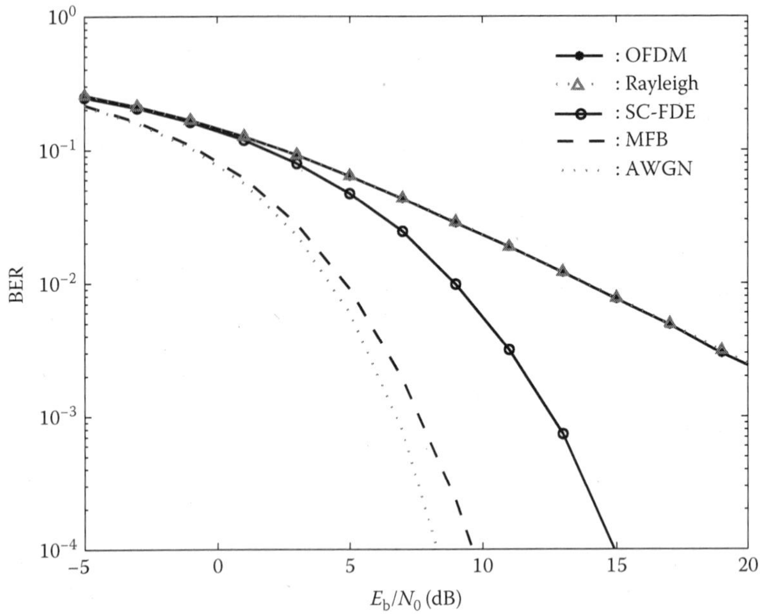

Figure 2.24 Uncoded BER performance for OFDM and SC-FDE without receiver diversity ($L = 1$) (for comparison purposes, the corresponding performance for an AWGN channel, for the MFB, and for a Rayleigh fading channel are also included).

L-order space diversity, given by [Proakis 2001]

$$P_{\mathrm{b,Ray}} = \left(\frac{1-\mu}{2}\right)^L \sum_{l=0}^{L-1} \binom{L-1+l}{l} \left(\frac{1+\mu}{2}\right)^l, \tag{2.109}$$

with

$$\mu = \sqrt{\frac{E_{\mathrm{b}}/N_0}{1+E_{\mathrm{b}}/N_0}}. \tag{2.110}$$

From these figures it is clear that SC-FDE schemes outperform OFDM schemes in all cases. As expected, the use of diversity considerably improves the performance of both modulation schemes. We can also observe that the performance of OFDM schemes is very close to the performance under Rayleigh fading conditions. Naturally, the poor performance of OFDM schemes results from the fact that the overall performance is conditioned by the subchannels in deep fades. To overcome this problem we can employ suitable channel coding schemes.

Let us consider now the impact of channel coding. We consider the rate-1/2 64-state convolutional code with generators $1+D^2+D^3+D^5+D^6$ and $1+D+D^2+D^3+D^6$. For the OFDM scheme an appropriate intrablock interleaving is employed. Figure 2.26

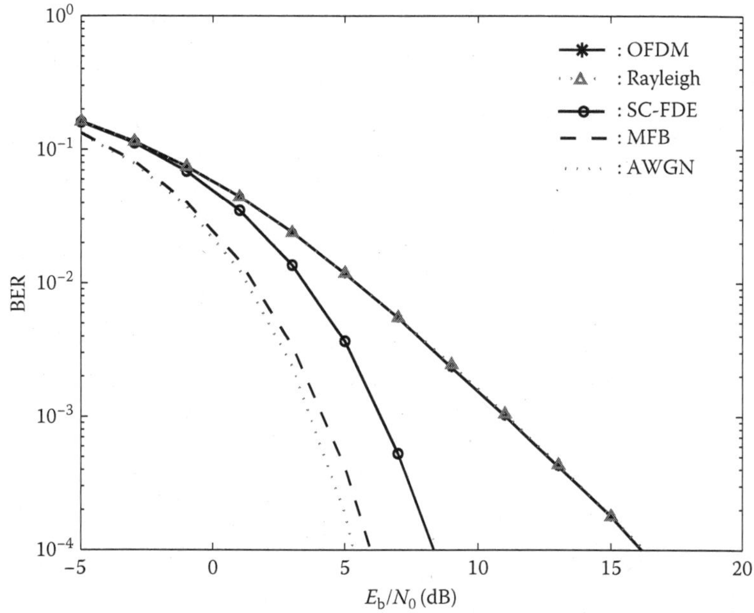

Figure 2.25 As in Figure 2.19 but for 2-branch receiver diversity.

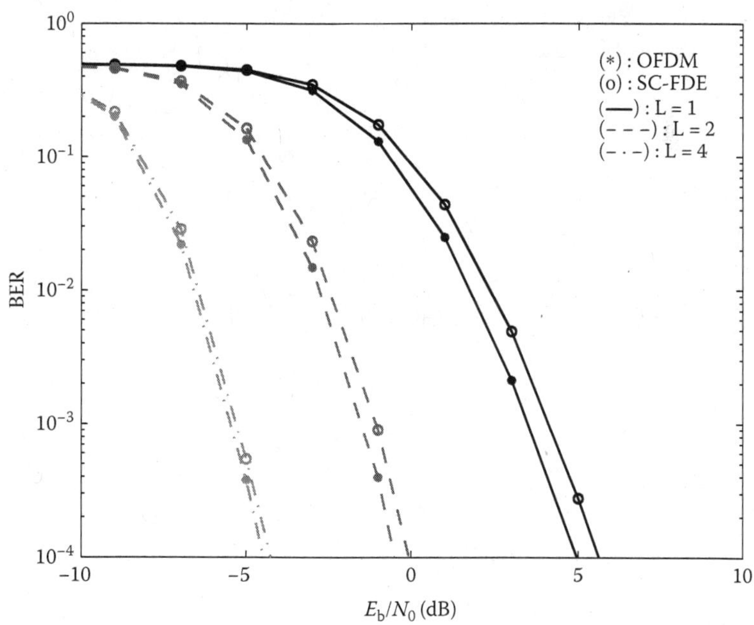

Figure 2.26 Coded BER performance for OFDM and SC-FDE with a rate-1/2 convolutional code, with $L = 1$, 2, and 4 branch receiver diversity.

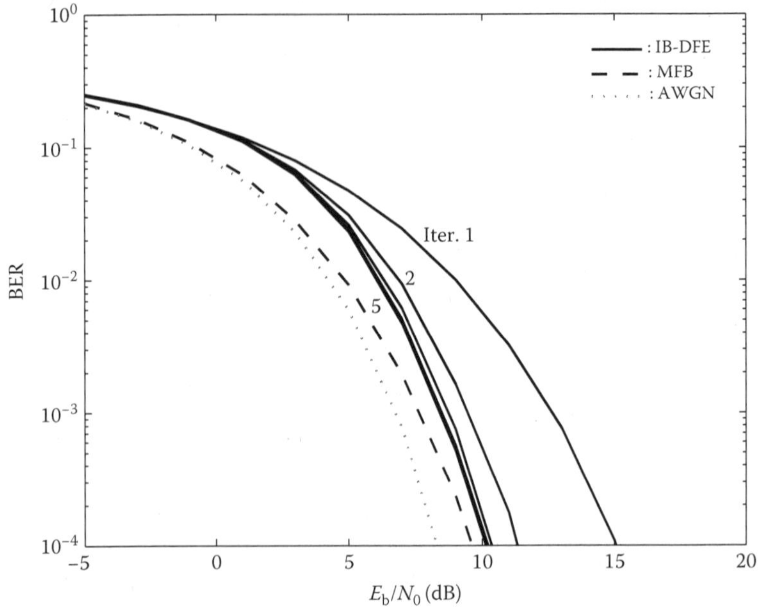

Figure 2.27 Uncoded BER performance for an IB-DFE with 1 to 5 iterations without receiver diversity ($L = 1$) (for comparison purposes, the performances for an AWGN channel and for the MFB are also included).

shows the average coded BER for both modulations, when $L = 1, 2$, and 4. From this figure, the benefits of channel coding become clear, especially for the OFDM choice, yielding similar performances for SC-FDE and OFDM. This means that channel coding can compensate the worse uncoded performance of OFDM modulations, as long as the code rate is low enough (as expected, for higher code rates SC-FDE can significantly outperform OFDM).

Finally, let us consider now an uncoded transmission with SC modulation where the linear FDE is replaced by an IB-DFE (once again, under the MMSE criterion). Figure 2.27 shows the average uncoded BER performance for iterations 1–5 with $L = 1$, as well as the corresponding performance of the MFB and AWGN channel. From this figure we can observe that the iterative receiver allows a significant improvement on BER performance: the required E_b/N_0 for BER $= 10^{-4}$ is about 15 dB for the first iteration (corresponding to the conventional linear SC-FDE), dropping to about 10.5 dB after just three iterations; iterations 4 and 5 provide less significant improvements. Moreover, the asymptotic BER performances become very close to the corresponding MFB after just a few iterations. These results show the potential of IB-DFE receivers.

Chapter 3

Channel Estimation

Noncoherent receivers are suitable for fast varying channels since they do require channel estimation. However, their performance is worse than that of coherent receivers, provided that accurate channel estimates are available at the receiver. For this reason, we usually consider coherent receivers in wireless systems. In fact, the receivers described in Chapter 2 for WCDMA and block transmission schemes such as OFDM and SC-FDE are coherent, assuming perfect channel knowledge.

Typically, the channel estimates are obtained with the help of training symbols that are multiplexed with the data symbols, either in the time domain or in the frequency domain [Lodge and Moher 1987; Cavers 1991; Hoher et al. 1997; Sanzi and Speidel 2000; Deneire et al. 2001]. Owing to this multiplexing of pilot/training and data symbols, some of the available BW has to be consumed for accomplishing the transmission of the pilot symbols. Since the CIR is usually very long, especially for block transmission schemes, the required channel estimation overheads can be high, particularly for fast varying scenarios. This leads to a reduction of the useful bit rate, decreasing the spectral efficiency of the systems.

A promising method of overcoming this problem is to employ implicit pilots, also known as pilot embedding or superimposed pilots, which are added to the data block instead of being multiplexed with it [Ho et al. 2001; Orozco-Lugo et al. 2004]. This means that we can increase significantly the pilots' density while keeping the system capacity unchanged. In fact, we can even have a pilot for each data symbol. However, the interference levels between the data symbols and pilots might be high. This means that the channel estimates are corrupted by the data signal, leading to irreducible noise floors (i.e., the channel estimates cannot be improved beyond a given level, even without channel noise). Moreover, there is also interference to the data symbols due to the pilots, leading to performance degradation. Therefore, this approach usually requires extra signal processing at the receiver to reduce the cross interference between pilots and data.

This chapter is concerned with channel estimation issues. Section 3.1 deals with the channel estimation requirements with or without receiver diversity. Section 3.2 describes conventional channel estimation techniques for block transmission schemes based on multiplexed pilots or training. Section 3.3 deals with channel estimation techniques using implicit pilots, once again for block transmission schemes. Finally, Section 3.4 deals with channel estimation issues for WCDMA systems.

3.1 Channel Estimation Requirements

3.1.1 *Performance in Flat Rayleigh Fading Channels without Diversity*

Let us consider the transmission of a quaternary phase shift keying (QPSK) symbol, s, in a flat fading environment. We can model the received signal sample as

$$r = \alpha \cdot s + n, \tag{3.1}$$

where α is the channel coefficient and n represents additive white thermal noise. Both α and n are modeled as complex Gaussian random variables with $E[\alpha] = 0, E\left[|\alpha|^2\right] = 2\sigma_\alpha^2 s$ (average fading power), $E[n] = 0$, and $E\left[|n|^2\right] = N_0$ ($N_0/2$ is the two-sided noise power spectrum density). As a consequence, and for a fixed transmitted symbol, r is also a complex Gaussian random variable with $E[r] = 0$ and

$$E\left[|r|^2 \, |s\right] = 2\sigma_r^2 = 2\,|s|^2\,\sigma_\alpha^2 + N_0. \tag{3.2}$$

Each received sample is multiplied by the conjugate of the channel coefficient estimate, $\hat{\alpha}$, leading to

$$z = r \cdot \hat{\alpha}^* s. \tag{3.3}$$

This complex decision variable can then be used for taking a decision on each of the two bits mapped to the QPSK symbol. Owing to the symmetry of the I and Q branches, the decision is independent of each of the two bits requiring only the real part or the imaginary part of Equation 3.3, that is, $z_{\rm re}$ and $z_{\rm im}$. Therefore, we can develop our study using the decision variable for only one of the axes, $z_{\rm re}$ or $z_{\rm im}$. In the following derivations, we will work with $z_{\rm re}$ only. Figure 3.1 shows the projection of a QPSK constellation onto the I-axis and the respective bit decision regions.

Assuming an unbiased linear channel estimator, channel estimate $\hat{\alpha}$ will be a complex Gaussian random variable with $E\left[\hat{\alpha}\right] = 0$ and $E\left[|\hat{\alpha}|^2\right] = 2\sigma_{\hat{\alpha}}^2$ ($\sigma_{\hat{\alpha}}^2$ depends on the channel estimation algorithm employed). Note that r and $\hat{\alpha}$ will be correlated with the joint moment expressed as

$$E\left[r \cdot \hat{\alpha}^* |s\right] = sE\left[\alpha \cdot \hat{\alpha}^* |s\right] + E\left[n \cdot \hat{\alpha}^* |s\right]. \tag{3.4}$$

The reliability of the channel estimate can depend on the symbol position in the transmitted frame. This can happen, for example, when pilot symbols are periodically

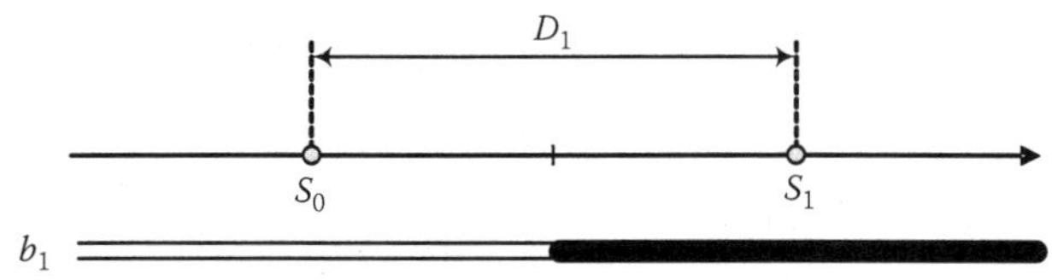

Figure 3.1 Decision regions for a QPSK constellation projected into the I branch. The dark area corresponds to a "1" and the white area to a "0".

inserted into the frame. In this situation, the channel will evolve between two pilot symbols and it is expected that data symbols closer to the pilot symbols will have more accurate channel estimates. As a result, the BER will also depend on the position in the frame, which means that it must be averaged over all possible locations t in a frame of size N_F as

$$P_b = \frac{1}{N_F} \sum_{t=1}^{N_F} P_e(t), \tag{3.5}$$

where $P_e(t)$ represents the probability of bit error in frame position t. Assuming equi-probable symbols and taking into account the symmetry of the constellation, we can rewrite Equation 3.5 as

$$\begin{aligned} \Psi_k(\upsilon i) &= F\left\{p\left(z_{re_k}|s\right)\right\} \\ &= -\frac{2H_k}{F_k\left(\upsilon i - G_k - H_k\right)\left(\upsilon i - G_k + H_k\right)}, \end{aligned} \tag{3.6}$$

where $\text{Prob}\{z_{re} < 0|s_0, t\}$ is the probability of the decision variable being lower than 0 when the transmitted symbol is s_0 and its position in the frame is t. To compute the average BER it is necessary to obtain an expression for this probability. In the following derivation, we will drop the index t(symbol position in the frame) for simplicity of notation, although we should bear in mind that the channel estimates can depend on it.

First, it is necessary to derive the probability density function (PDF) of z_{re} conditioned on s. We will follow a derivation similar to the one presented in Ref. [Webb and Hanzo 1994]. The multivariate complex Gaussian PDF of a complex random vector $\mathbf{x}$ with size n and mean $E(\mathbf{x})$ is equal to

$$p(\mathbf{x}) = \frac{1}{\pi^n \det\left(\mathbf{C_x}\right)} \exp\left[-\left(\mathbf{x} - E(\mathbf{x})\right)^{\mathrm{H}} \mathbf{C_x}^{-1} \left(\mathbf{x} - E(\mathbf{x})\right)\right], \tag{3.7}$$

where $\mathbf{C_x}$ is the covariance matrix defined as

$$\mathbf{C_x} = E\left[\left(\mathbf{x} - E(\mathbf{x})\right)\left(\mathbf{x} - E(\mathbf{x})\right)^{\mathrm{H}}\right]. \tag{3.8}$$

Here, H denotes the Hermitian matrix. On the basis of this definition, we can obtain the joint PDF of r and $\hat{\alpha}_s$, which are complex random Gaussian variables, by computing their covariance matrix as

$$\mathbf{C_x} = \begin{bmatrix} E\left(|r|^2 |s\right) & E\left(r^* \cdot \hat{\alpha}|s\right) \\ E\left(r \cdot \hat{\alpha}^*|s\right) & E\left(\left|\hat{\alpha}\right|^2 |s\right) \end{bmatrix} = \begin{bmatrix} 2\sigma_r^2 & 2\mu^*\sigma_r\sigma_{\hat{\alpha}} \\ 2\mu\sigma_r\sigma_{\hat{\alpha}} & 2\sigma_{\hat{\alpha}}^2 \end{bmatrix}, \tag{3.9}$$

where we have used the following equality:

$$E\left(r \cdot \hat{\alpha}^*|s\right) = \left(E\left(r^* \cdot \hat{\alpha}|s\right)\right)^* = \mu\sqrt{E\left[|r|^2 |s\right] E\left[\left|\hat{\alpha}\right|^2 |s\right]} = 2\mu\sigma_r\sigma_{\hat{\alpha}},$$

with μ being the cross-correlation coefficient between r and $\hat{\alpha}$, which is defined as

$$\mu = \frac{E\left[r\hat{\alpha}^*|s\right]}{\sqrt{E\left[|r|^2 |s\right] E\left[\left|\hat{\alpha}\right|^2 |s\right]}} == |\mu|\, e^{-\varepsilon j}. \tag{3.10}$$

Using this covariance matrix, the following joint PDF is obtained:

$$p\left(r, \hat{\alpha}|s\right) = \frac{1}{4\pi^2\sigma_{\hat{\alpha}}^2\sigma_r^2\left(1-|\mu|^2\right)} \exp\left[-\frac{|r|^2\sigma_{\hat{\alpha}}^2 + \left|\hat{\alpha}\right|^2\sigma_r^2 - 2\sigma_{\hat{\alpha}}\sigma_r \mathrm{Re}\left\{\mu r\hat{\alpha}^*\right\}}{2\sigma_{\hat{\alpha}}^2\sigma_r^2\left(1-|\mu|^2\right)}\right]. \tag{3.11}$$

From the definition of a conditional PDF, we have

$$p\left(r|s, \hat{\alpha}\right) = \frac{p\left(r, \hat{\alpha}|s\right)}{p\left(\hat{\alpha}\right)}. \tag{3.12}$$

Since channel estimate, $\hat{\alpha}$, is a complex random Gaussian variable, its PDF can be expressed as

$$p\left(\hat{\alpha}\right) = \frac{1}{2\pi\sigma_{\hat{\alpha}}^2} \exp\left[-\frac{\left|\hat{\alpha}\right|^2}{2\sigma_{\hat{\alpha}}^2}\right]. \tag{3.13}$$

Inserting Equation 3.13 into Equation 3.12, we get

$$p\left(r|s, \hat{\alpha}\right) = \frac{1}{2\pi\sigma_r^2\left(1-|\mu|^2\right)} \exp\left[-\frac{\left|r - \mu^*\hat{\alpha}(\sigma_r/\sigma_{\hat{\alpha}})\right|^2}{2\sigma_r^2\left(1-|\mu|^2\right)}\right]. \tag{3.14}$$

Performing the change of variables $z = r \cdot \hat{\alpha}^*$ in Equation 3.14, the PDF of z conditioned on s and $\hat{\alpha}$ is obtained as

$$p\left(z|s, \hat{\alpha}\right) = \frac{1}{2\pi\sigma_r^2\left(1-|\mu|^2\right)\left|\hat{\alpha}\right|^2} \exp\left[-\frac{\left|z - \mu^*\left|\hat{\alpha}\right|^2(\sigma_r/\sigma_{\hat{\alpha}})\right|^2}{2\sigma_r^2\left(1-|\mu|^2\right)\left|\hat{\alpha}\right|^2}\right]. \tag{3.15}$$

Integrating this expression multiplied by the PDF of $\hat{\alpha}$ over the region of the complex plane of possible values of $\hat{\alpha}, R_{\hat{\alpha}}$, results in the following expression for the PDF of z conditioned on s

$$
\begin{aligned}
p(z|s) &= \int_{R_{\hat{\alpha}}} p\left(z, \hat{\alpha}|s\right) \mathrm{d}\hat{\alpha} \\
&= \int_{R_{\hat{\alpha}}} p\left(z|s, \hat{\alpha}\right) p\left(\hat{\alpha}\right) \mathrm{d}\hat{\alpha} \\
&= \frac{1}{2\pi\sigma_{\hat{\alpha}}^2\sigma_r^2\left(1-|\mu|^2\right)} \times \exp\left[\frac{|\mu|\left(z_{\text{re}}\cos\varepsilon + z_{\text{im}}\sin\varepsilon\right)}{\sigma_r\sigma_{\hat{\alpha}}\left(1-|\mu|^2\right)}\right] \\
&\quad \times K_0\left[\frac{\sqrt{z_{\text{re}}^2+z_{\text{im}}^2}}{\sigma_r\sigma_{\hat{\alpha}}\left(1-|\mu|^2\right)}\right],
\end{aligned}
\tag{3.16}
$$

where $K_0(\cdot)$ denotes the modified Hankel function of order zero (also named the modified Bessel function of the second kind of order zero).

Integrating $p(z|s)$ over z_{im} yields the marginal PDF of the decision variable z_{re}

$$
\begin{aligned}
p\left(z_{\text{re}}|s\right) &= \int_{-\infty}^{+\infty} p(z|s)\, \mathrm{d}z_{\text{im}} \\
&= \frac{1}{2\pi\sigma_{\hat{\alpha}}^2\sigma_r^2\left(1-|\mu|^2\right)} \int_{-\infty}^{+\infty} \exp\left[\frac{|\mu|\left(z_{\text{re}}\cos\varepsilon + z_{\text{im}}\sin\varepsilon\right)}{\sigma_r\sigma_{\hat{\alpha}}\left(1-|\mu|^2\right)}\right] \\
&\quad \times K_0\left[\frac{\sqrt{z_{\text{re}}^2+z_{\text{im}}^2}}{\sigma_r\sigma_{\hat{\alpha}}\left(1-|\mu|^2\right)}\right] \mathrm{d}z_1,
\end{aligned}
\tag{3.17}
$$

A modified Hankel function of order v can be represented as

$$
K_v(x) = \frac{1}{2}\left(\frac{x}{2}\right)^v \int_0^{\infty} \frac{1}{t^{v+1}} \exp\left[-t-\frac{x^2}{4t}\right] \mathrm{d}t, \quad |\arg x| < \frac{\pi}{2}, \quad \operatorname{Re}\left\{x^2\right\} > 0. \tag{3.18}
$$

In Equation 3.17 the order of the Hankel function is zero, and since the argument is a real number, Equation 3.18 is always valid and becomes

$$
K_0(x) = \frac{1}{2}\int_0^{+\infty} \frac{1}{t} \exp\left[-t-\frac{x^2}{4t}\right] \mathrm{d}t. \tag{3.19}
$$

Introducing Equation 3.19 into Equation 3.17 results in

$$
\begin{aligned}
p\left(z_{\text{re}}|s\right) = {} & \frac{1}{2\pi\sigma_{\hat{\alpha}}^2\sigma_r^2\left(1-|\mu|^2\right)} \int_{-\infty}^{+\infty} \exp\left[\frac{|\mu|\left(z_{\text{re}}\cos\varepsilon + z_{\text{im}}\sin\varepsilon\right)}{\sigma_r\sigma_{\hat{\alpha}}\left(1-|\mu|^2\right)}\right] \\
& \times \frac{1}{2}\int_0^{+\infty} \frac{1}{t} \exp\left[-t-\frac{z_{\text{re}}^2+z_{\text{im}}^2}{4t\sigma_r^2\sigma_{\hat{\alpha}}^2\left(1-|\mu|^2\right)^2}\right] \mathrm{d}t\, \mathrm{d}z_{\text{im}}
\end{aligned}
$$

$$= \frac{\exp\left[\frac{|\mu| z_{\mathrm{re}} \cos\varepsilon}{\sigma_r \sigma_{\hat{\alpha}} (1-|\mu|^2)}\right]}{4\pi\sigma_{\hat{\alpha}}^2 \sigma_r^2 \left(1 - |\mu|^2\right)} \int_{-\infty}^{+\infty} \int_0^{+\infty} \exp\left[\frac{|\mu| z_{\mathrm{im}} \sin\varepsilon}{\sigma_r \sigma_{\hat{\alpha}} \left(1 - |\mu|^2\right)}\right.$$

$$\left. -t - \frac{z_{\mathrm{re}}^2 + z_{\mathrm{im}}^2}{4t\sigma_r^2 \sigma_{\hat{\alpha}}^2 \left(1 - |\mu|^2\right)^2}\right] \frac{1}{t} \mathrm{d}t\, \mathrm{d}z_{\mathrm{im}}$$

$$= \frac{\exp\left[\frac{|\mu| z_{\mathrm{re}} \cos\varepsilon}{\sigma_r \sigma_{\hat{\alpha}} (1-|\mu|^2)}\right]}{4\pi\sigma_{\hat{\alpha}}^2 \sigma_r^2 \left(1 - |\mu|^2\right)} \int_0^{+\infty} \frac{1}{t} \exp\left[-t - \frac{z_{\mathrm{re}}^2}{4t\sigma_r^2 \sigma_{\hat{\alpha}}^2 \left(1 - |\mu|^2\right)^2}\right]$$

$$\times \int_{-\infty}^{+\infty} \exp\left[\frac{4t|\mu|\sigma_r \sigma_{\hat{\alpha}} \left(1 - |\mu|^2\right) z_{\mathrm{im}} \sin\varepsilon - z_{\mathrm{im}}^2}{4t\sigma_r^2 \sigma_{\hat{\alpha}}^2 \left(1 - |\mu|^2\right)^2}\right] \mathrm{d}z_{\mathrm{im}}\, \mathrm{d}t\, s. \qquad (3.20)$$

It can be shown that

$$\int_{-\infty}^{+\infty} \exp\left[\frac{4t|\mu|\sigma_r \sigma_{\hat{\alpha}} \left(1 - |\mu|^2\right) z_{\mathrm{im}} \sin\varepsilon - z_{\mathrm{im}}^2}{4t\sigma_r^2 \sigma_{\hat{\alpha}}^2 \left(1 - |\mu|^2\right)^2}\right] \mathrm{d}z_i$$
$$= 2\sigma_r \sigma_{\hat{\alpha}} \left(1 - |\mu|^2\right) \sqrt{\pi t} \exp\left(|\mu|^2 (\sin\varepsilon)^2 t\right), \qquad (3.21)$$

which means that Equation 3.20 becomes

$$p\left(z_{\mathrm{re}} | s\right) = \frac{\sqrt{\pi} \exp\left[\frac{|\mu| z_{\mathrm{re}} \cos\varepsilon}{\sigma_r \sigma_{\hat{\alpha}} (1-|\mu|^2)}\right]}{2\pi\sigma_r \sigma_{\hat{\alpha}}}$$

$$\times \int_0^{+\infty} \frac{1}{\sqrt{t}} \exp\left[\left(|\mu|^2 (\sin\varepsilon)^2 - 1\right) t - \frac{z_{\mathrm{re}}^2}{4t\sigma_r^2 \sigma_{\hat{\alpha}}^2 \left(1 - |\mu|^2\right)^2}\right] \mathrm{d}t$$

$$= \frac{\sqrt{\pi} \exp\left[\frac{|\mu| z_{\mathrm{re}} \cos\varepsilon}{\sigma_r \sigma_{\hat{\alpha}} (1-|\mu|^2)}\right]}{2\pi\sigma_r \sigma_{\hat{\alpha}}} \sqrt{\frac{\pi}{1 - |\mu|^2 (\sin\varepsilon)^2}} \exp\left[-\sqrt{\frac{\left(1 - |\mu|^2 (\sin\varepsilon)^2\right) z_{\mathrm{re}}^2}{\sigma_r^2 \sigma_{\hat{\alpha}}^2 \left(1 - |\mu|^2\right)^2}}\right]$$

$$= \frac{1}{2\sigma_r \sigma_{\hat{\alpha}} \sqrt{1 - |\mu|^2 (\sin\varepsilon)^2}} \exp\left[\frac{|\mu| z_{\mathrm{re}} \cos\varepsilon}{\sigma_r \sigma_{\hat{\alpha}} \left(1 - |\mu|^2\right)}\right]$$

$$\times \exp\left[-\sqrt{1 - |\mu|^2 (\sin\varepsilon)^2} \frac{|z_{\mathrm{re}}|}{\sigma_r \sigma_{\hat{\alpha}} \left(1 - |\mu|^2\right)}\right]. \qquad (3.22)$$

The desired probability, Prob $(z_{re} < 0|s)$, can be obtained by simply integrating the PDF from $-\infty$ to 0, resulting in

$$
\begin{aligned}
&\text{Prob}\,(z_{re} < 0|\,s) \\
&= \int_{-\infty}^{0} p\,(z_{re}|s)\,\mathrm{d}z_{re} \\
&= \frac{1}{2\sigma_r\sigma_{\hat{\alpha}}\sqrt{1-|\mu|^2(\sin\varepsilon)^2}} \int_{-\infty}^{0} \exp\left[\frac{\left(|\mu|\cos\varepsilon+\sqrt{1-|\mu|^2(\sin\varepsilon)^2}\right)z_{re}}{\sigma_r\sigma_{\hat{\alpha}}\left(1-|\mu|^2\right)}\right]\mathrm{d}z_{re} \\
&= \frac{1-|\mu|^2}{2\left(1-|\mu|^2(\sin\varepsilon)^2+|\mu|\cos\varepsilon\sqrt{1-|\mu|^2(\sin\varepsilon)^2}\right)}.
\end{aligned}
\tag{3.23}
$$

Using the derived expressions, we present in Figure 3.2 several BER curves for QPSK with imperfect channel estimation. Channel estimates are modeled as $\hat{\alpha} = a \cdot \alpha + \xi$, where a is a complex value denoting the bias of the estimate (which will be assumed as $a = 1$) and ξ is a zero-mean complex Gaussian variable representing the channel estimation error. We will consider that the channel estimation error variance is the same for all bit positions in the frame. Curves for perfect channel estimation as well as for imperfect channel estimation with $E\left[|\xi|^2\right] = 0.1 \cdot \mathrm{N}_0$, $E\left[|\xi|^2\right] = \mathrm{N}_0$, and $E\left[|\xi|^2\right] = 0.1 \cdot \mathrm{N}_0 + 0.001 \cdot \Omega^2$ are presented. The latter case corresponds to channel estimates with irreducible errors and leads to performances with irreducible BER floors, as can be seen

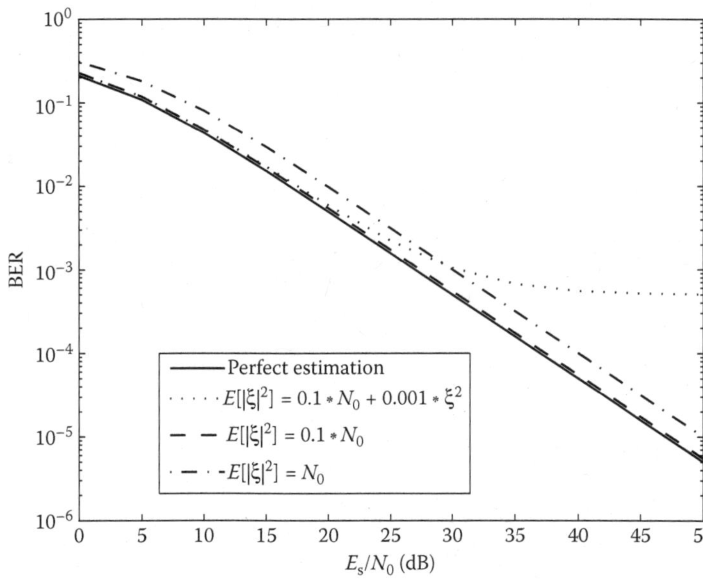

Figure 3.2 QPSK BER performance for several channel estimation error variances.

in the graph for high E_S/N_0 values. The other two cases correspond to channel estimates corrupted by thermal noise only and basically cause a shift in the curves relative to the perfect channel estimation one. Note that for low E_S/N_0 values, it is the presence of thermal noise in the channel that estimates the main source of performance degradation relative to the perfect channel estimation curve as can be seen by the nearly overlapping curves with $E\left[|\xi|^2\right] = 0.1 \cdot \mathrm{N}_0$ and $E\left[|\xi|^2\right] = 0.1 \cdot \mathrm{N}_0 + 0.001 \cdot \Omega^2$.

3.1.2 Performance in Flat Rayleigh Fading Channels with Diversity

We will now extend the BER expressions to the case of receiver diversity with L branches, where different diversity branches can have different fading powers. In this situation, each received signal sample can be written as

$$r_k = \alpha_k \cdot s + n_k, \quad k = 1 \ldots L. \tag{3.24}$$

Assuming independent diversity branches, we have $E\left[\alpha_k \cdot \alpha_{k'}^* |s\right] = 0$ and $E\left[n_k \cdot n_{k'}^* |s\right] = 0$ for $k \neq k'$. The second-order moment of r_k is obtained as

$$E\left[|r_k|^2 |s\right] = 2\sigma_{r_k}^2 = 2|s|^2 \sigma_{\alpha_k}^2 + N_0. \tag{3.25}$$

The receiver multiplies each of the received signals by the complex conjugate of the respective channel coefficient estimate, $\hat{\alpha}_k$, and adds the results of the L products. This corresponds to performing MRC of the received replicas and the decision variable can be expressed as

$$z_{\mathrm{re}} = \mathrm{Re}\left\{\sum_{k=1}^{L} r_k \cdot \hat{\alpha}_k^*\right\} = \sum_{k=1}^{L} \mathrm{Re}\left\{r_k \cdot \hat{\alpha}_k^*\right\} = \sum_{k=1}^{L} z_{\mathrm{re}_k}. \tag{3.26}$$

Assuming an unbiased linear channel estimator, channel estimates, $\hat{\alpha}_k$, are still complex Gaussian random variables with $E\left[\hat{\alpha}_k\right] = 0$ and $E\left[\left|\hat{\alpha}_k\right|^2\right] = 2\sigma_{\hat{\alpha}_k}^2$. Similarly to the case of no diversity, r_k and $\hat{\alpha}_k$ will be correlated with a joint moment $E\left[r_k\hat{\alpha}_k^*|s\right]$ s dependent on the channel estimation algorithm employed and a cross-correlation coefficient defined as

$$\mu_k = \frac{E\left[r_k\hat{\alpha}_k^*|s\right]}{\sqrt{E\left[|r_k|^2 |s\right] E\left[\left|\hat{\alpha}_k\right|^2 |s\right]}} = |\mu_k|\, e^{-\varepsilon_k \cdot j}. \tag{3.27}$$

In this type of environment, BER expression 3.6 is still valid, only the method by which $\mathrm{Prob}\{z_{\mathrm{re}} < 0|s_0, t\}$ is computed will be different. Therefore, we need to find an analytical expression for the PDF of z_{re} conditioned on s. To obtain this function, we will start with the PDF for an individual channel, $p\left(z_{\mathrm{re}_k}|s\right)$, corresponding to the expression for the no diversity case Equation 3.22, which we rewrite here as

$$p\left(z_{\mathrm{re}_k}|s\right) = \frac{1}{F_k} \exp\left[G_k \cdot z_{\mathrm{re}_k}\right] \exp\left[-H_k \cdot \left|z_{\mathrm{re}_k}\right|\right]. \tag{3.28}$$

with

$$\begin{aligned} F_k &= 2\sigma_{r_k}\sigma_{\hat{\alpha}_k}\sqrt{1-|\mu_k|^2(\sin\varepsilon_k)^2}, \\ G_k &= \frac{|\mu_k|\cos\varepsilon_k}{\sigma_{r_k}\sigma_{\hat{\alpha}_k}\left(1-|\mu_k|^2\right)}, \\ H_k &= \frac{\sqrt{1-|\mu_k|^2(\sin\varepsilon_k)^2}}{\sigma_{r_k}\sigma_{\hat{\alpha}_k}\left(1-|\mu_k|^2\right)}. \end{aligned} \tag{3.29}$$

The decision variable for the case of diversity corresponds to the sum of L variables with PDFs similar to Equation 3.28. The PDF of the sum of independent random variables can be computed either with the convolution of the individual PDFs or with the inverse FT of the product of the individual characteristic functions. The characteristic function of Equation 3.28 is obtained applying the FT (defined over variable υ) resulting in

$$\begin{aligned} \Psi_k(\upsilon i) &= F\left\{p\left(z_{\text{re}_k}|s\right)\right\} \\ &= -\frac{2H_k}{F_k(\upsilon i - G_k - H_k)(\upsilon i - G_k + H_k)}. \end{aligned} \tag{3.30}$$

The characteristic function of z_{re} is then given by the product of the L individual characteristic functions, that is,

$$\begin{aligned} \Psi(\upsilon i) &= \prod_{k=1}^{L} \Psi_k(\upsilon i) \\ &= \prod_{k=1}^{L} \frac{-2H_k}{F_k(\upsilon i - G_k - H_k)(\upsilon i - G_k + H_k)}. \end{aligned} \tag{3.31}$$

Although we assume the same distribution for the different channels, they can have different fading powers. Therefore, we will group the L diversity branches into L' different sets according to the fading powers so that each set is composed of θ_k $(k = 1, \ldots, L')$ diversity branches with the same fading powers. These sizes satisfy

$$L = \sum_{k=1}^{L'} \theta_k. \tag{3.32}$$

Using an auxiliary complex variable $\zeta = \upsilon i$ we can rewrite Equation 3.31 as

$$\begin{aligned} \Psi(\zeta) &= \prod_{k=1}^{L} \Psi_k(\zeta) \\ &= \prod_{k=1}^{L} \frac{-2H_k}{F_k(\zeta - G_k - H_k)(\zeta - G_k + H_k)} \\ &= \prod_{k=1}^{L'} \left(\frac{-2H_k}{F_k}\right)^{\theta_k} \Psi'(\zeta). \end{aligned} \tag{3.33}$$

where

$$\Psi'(\zeta) = \prod_{l=1}^{L'} \frac{1}{(\zeta - G_l - H_l)^{\theta_l}(\zeta - G_l + H_l)^{\theta_l}}. \tag{3.34}$$

Expression 3.34 can be written as a sum of simple fractions, that is, as

$$\Psi'(\zeta) = \sum_{l=1}^{L'}\sum_{i=1}^{\theta_l} \frac{A_{l,i}^1}{(\zeta - G_l - H_l)^i} + \sum_{l=1}^{L'}\sum_{i=1}^{\theta_l} \frac{A_{l,i}^2}{(\zeta - G_l + H_l)^i}, \tag{3.35}$$

with coefficients $A_{l,i}^1$ and $A_{l,i}^2$ computed according to

$$\begin{aligned} A_{l,i}^1 &= \frac{1}{(\theta_l - i)!}\left[\frac{\partial^{\theta_l - i}}{\partial \zeta^{\theta_l - i}}\left(\Psi'(\zeta)(\zeta - G_l - H_l)^{\theta_l}\right)\right]_{\zeta = G_l + H_l} \\ &= \frac{1}{(\theta_l - i)!}\left[\frac{\partial^{\theta_l - i}}{\partial \zeta^{\theta_l - i}}\left(\prod_{\substack{j=1\\ j\neq l}}^{L'} \frac{1}{(\zeta - G_j - H_j)^{\theta_j}} \prod_{j=1}^{L'} \frac{1}{(\zeta - G_j + H_j)^{\theta_j}}\right)\right]_{\zeta = G_l + H_l}. \end{aligned} \tag{3.36}$$

$$\begin{aligned} A_{l,i}^2 &= \frac{1}{(\theta_l - i)!}\left[\frac{\partial^{\theta_l - i}}{\partial \zeta^{\theta_l - i}}\left(\Psi'(\zeta)(\zeta - G_l + H_l)^{\theta_l}\right)\right]_{\zeta = G_l - H_l} \\ &= \frac{1}{(\theta_l - i)!}\left[\frac{\partial^{\theta_l - i}}{\partial \zeta^{\theta_l - i}}\left(\prod_{j=1}^{L'} \frac{1}{(\zeta - G_j - H_j)^{\theta_j}} \prod_{\substack{j=1\\ j\neq l}}^{L'} \frac{1}{(\zeta - G_j + H_j)^{\theta_j}}\right)\right]_{\zeta = G_l - H_l}. \end{aligned} \tag{3.37}$$

Inserting Equation 3.35 into Equation 3.33 and then applying the inverse FT, we obtain the PDF of z_{re} conditioned on s as

$$\begin{aligned} p(z_{\text{re}}|s) &= F^{-1}\{\Psi(\upsilon i)\} \\ &= \prod_{k=1}^{L'}\left(\frac{-2H_k}{F_k}\right)^{\theta_k}\left[-\sum_{l=1}^{L'}\sum_{i=1}^{\theta_l}\frac{A_{l,i}^1}{(i-1)!} z_{\text{re}}'^{\,i-1} \mathrm{e}^{(G_l + H_l) z_{\text{re}}'} u(-z_{\text{re}}')\right. \\ &\quad \left. + \sum_{l=1}^{L'}\sum_{i=1}^{\theta_l}\frac{A_{l,i}^2}{(i-1)!} z_{\text{re}}'^{\,i-1} \mathrm{e}^{(G_l - H_l) z_{\text{re}}'} u(z_{\text{re}}')\right]. \end{aligned} \tag{3.38}$$

To compute the desired probability, Prob $(z_{\text{re}} < 0|s)$, it is necessary to integrate this PDF from $-\infty$ to 0, resulting in

$$\begin{aligned} \text{Prob}(z_{\text{re}} < 0|s) &= \int_{-\infty}^{0} p(z_{\text{re}}|s)\, \mathrm{d}z_{\text{re}} \\ &= \prod_{k=1}^{L'}\left(\frac{-2H_k}{F_k}\right)^{\theta_k} \sum_{l=1}^{L'}\sum_{i=1}^{\theta_l} (-1)^i A_{l,i}^1 (G_l + H_l)^{-i}. \end{aligned} \tag{3.39}$$

It should be noted that, according to this expression, it is not necessary to compute coefficients $A^2_{l,i}$ to obtain Prob $(z_{\text{re}} < 0|s)$.

The general probability expression derived can be written in a simpler closed form for two specific cases. If all the diversity branches have equal powers (EPs), $\theta_k = L$ for any k, $L' = 1$, and F_k, G_k, and H_k are all equal (index k can be dropped), leading to

$$\begin{aligned}
&\text{Prob}\,(z_{\text{re}} < 0|s) \\
&= \frac{1}{(2F \cdot H)^L} \sum_{k=1}^{L} \binom{2L-k-1}{L-k} \left(\frac{2H}{G+H}\right)^k \\
&= \left(\frac{1-|\mu'|^2}{4-4\,|\mu'|^2\,(\sin\varepsilon')^2}\right)^L \sum_{k=1}^{L} \binom{2L-k-1}{L-k} \left(\frac{2\sqrt{1-|\mu'|^2\,(\sin\varepsilon')^2}}{|\mu'|\cos\varepsilon' + \sqrt{1-|\mu'|^2\,(\sin\varepsilon')^2}}\right)^k.
\end{aligned} \tag{3.40}$$

If all the diversity branches are different, that is, $\theta_k = 1$ for any k and $L' = L$, then

$$A^1_{l,1} = \prod_{\substack{j=1 \\ j\neq l}}^{L} \frac{1}{\left(G_l + H_l - G_j - H_j\right)^{\theta_j}} \times \prod_{j=1}^{L} \frac{1}{\left(G_l + H_l - G_j + H_j\right)^{\theta_j}} \tag{3.41}$$

and

$$\text{Prob}\,(z_{\text{re}} < 0|s) = \prod_{k=1}^{L} \left(\frac{-2H_k}{F_k}\right) \times \sum_{l=1}^{L} -\frac{A^1_{l,1}}{(G_l + H_l)}. \tag{3.42}$$

Assuming the same model for the channel estimates of the different diversity branches as the one used previously for the no diversity case, that is, $\hat{\alpha}_i = a \cdot \alpha_i + \xi_i$, we present in Figure 3.3 performance curves for several equal receiving antennas when the channel estimation error variance is $E\left[|\xi_i|^2\right] = 0.1 \cdot N_0 + 0.001 \cdot \Omega_i^2$ (irreducible estimation error). Although all the curves show irreducible error floors, it is clear that even though the channel estimation error is the same for all cases, the use of multiple diversity branches can lower the error floor considerably.

3.2 Conventional Channel Estimation

The conventional approach for achieving coherent detection at the receiver comprises the transmission of known pilot symbols multiplexed with data symbols. Figure 3.4 shows a typical frame structure for this type of approach.

The frame is composed of N_{F} symbols with the first one being a pilot, s_{pilot}. The receiver computes the channel estimate for each pilot in frame j through

$$\hat{\alpha}(j \cdot N_{\text{F}} + 1) = \frac{s^*_{\text{pilot}}}{|s_{\text{pilot}}|^2} r(j \cdot N_{\text{F}} + 1), \quad j = 0, 1, \ldots, \tag{3.43}$$

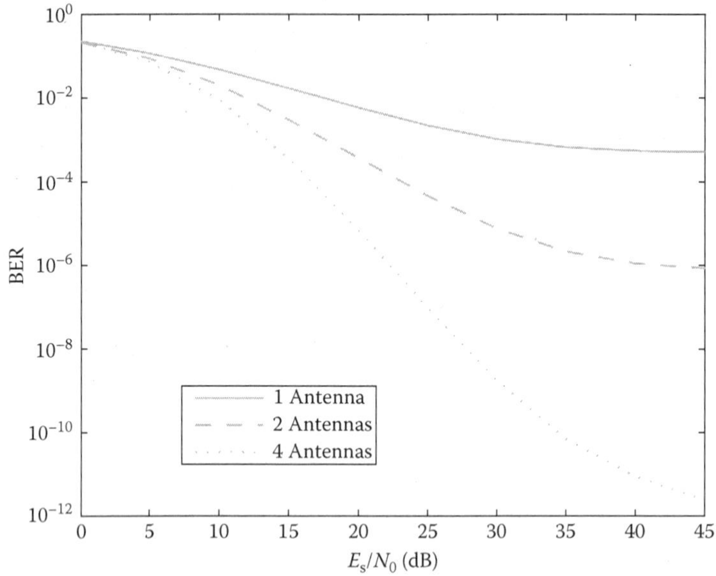

Figure 3.3 QPSK BER performance with several diversity branches for a channel estimation error variance given by $E\left[|\xi|^2\right] = 0.1 \cdot N_0 + 0.001 \cdot \Omega_i^2$.

where $r(j \cdot N_F + 1)$ is the received signal sample in the first position of frame j. The obtained sequence of pilot symbol channel estimates must then be interpolated to derive a channel estimate for each information symbol. This interpolation can be accomplished by several alternative methods. Considering that the channel coefficient for each information symbol t is estimated using a finite impulse response (FIR) filter that takes into account W neighbor pilot symbols, then each channel estimate can be computed using

$$\hat{\alpha}((u-1) \cdot N_F + t) = \sum_{j=-\lfloor (W-1)/2 \rfloor + u - 1}^{\lfloor W/2 \rfloor + u - 1} h_t^j \hat{\alpha}\left(j \cdot N_F + 1\right), \qquad (3.44)$$

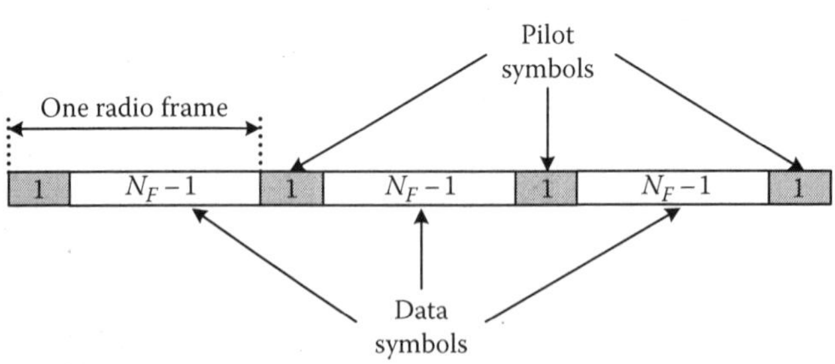

Figure 3.4 Frame structure for a PSAM system.

where h_k^j are the interpolation coefficients of the channel estimation FIR filter. If the autocorrelation of the fading process is known, the Wiener filter is the optimal filter, as was shown by Cavers in Ref. [Cavers 1991]. In this case, the objective of the optimum Wiener filter is to find the interpolation coefficients that minimize the channel estimation error variance, that is,

$$\mathbf{h}_t = \min_{h_t^j, j=-\lfloor (W-1)/2 \rfloor + u - 1, \ldots, \lfloor (W-1)/2 \rfloor + u - 1} \times \left\{ E\left[\left(\alpha((u-1)\cdot N_{\mathrm{F}} + t) - \hat{\alpha}((u-1)\cdot N_{\mathrm{F}} + t) \right)^2 \right] \right\}, \tag{3.45}$$

where $\mathbf{h}_t = \left[h_t^{-\lfloor (W-1)/2 \rfloor + u - 1}, \ldots, h_t^{\lfloor (W-1)/2 \rfloor + u - 1} \right]$. As shown in Ref. [Torrance and Hanzo 1995], these coefficients can be obtained as the solutions of

$$\sum_{j=-\lfloor (W-1)/2 \rfloor + u - 1}^{\lfloor W/2 \rfloor + u - 1} h_t^j R\left(|j - i| \right) + 2\sigma^2 h_t^i = R\left(i - u + 1 \right),$$
$$i = -\left\lfloor \frac{(W-1)}{2} \right\rfloor + u - 1, \ldots, \left\lfloor \frac{(W-1)}{2} \right\rfloor + u - 1, \tag{3.46}$$

which is a form of the well-known Wiener–Hopf equations, often used in estimation and prediction theory. In these equations, σ^2 is the noise variance and $R(\cdot)$ represents the autocorrelation function of the channel, which, considering stationary fading statistics, depends only on the difference between two symbol positions

$$R\left(|j - i| \right) = E\left\{ \alpha(j)\alpha^*(i) \right\} s. \tag{3.47}$$

In the case of Rayleigh fading, the channel coefficients are modeled as zero-mean complex Gaussian variables and, adopting Jake's isotropic scattering model [Jakes 1974], the corresponding autocorrelation function can be expressed as [Stuber 2001]

$$R(i) = \Omega^2 J_0(2\pi f_D T_s i), \tag{3.48}$$

where $J_0(\cdot)$ is the zeroth-order Bessel function of the first kind, T_s is the symbol period, f_D is the Doppler frequency, and Ω^2 is the average fading power. The autocorrelation of the fading process depends on the Doppler frequency, which is unknown unless the vehicle speed is estimated. To avoid the requirement for knowledge of the vehicle speed, adaptive algorithms can be used to perform the interpolation. A popular adaptive filter is the linear predictor, where the least-mean squares (LMS) or the recursive least-squares (RLS) algorithms can be used to minimize the prediction error. As an alternative, simpler interpolation methods are available such as a simple linear interpolation, a low-order Gaussian interpolation [Sampei and Sunaga 1993] or a low-pass sinc interpolator [Kim et al. 1997]. In the case of a low-order Gaussian interpolation, the channel estimate is computed using an FIR as defined in Equation 3.44 with a length $W = 3$ and the interpolation weights obtained as (in the case of second-order interpolation)

$$
\begin{aligned}
h_t^u &= \frac{1}{2}\left[\left(\frac{t-1}{N}\right)^2 + \frac{t-1}{N}\right], \\
h_t^{u-1} &= 1 - \left(\frac{t-1}{N}\right)^2, \\
h_t^{u-2} &= \frac{1}{2}\left[\left(\frac{t-1}{N}\right)^2 - \frac{t-1}{N}\right].
\end{aligned} \tag{3.49}
$$

If a sin c interpolator is employed, the channel estimates are also computed using an FIR defined as Equation 3.44 where the interpolation coefficients are obtained from the sin c function using

$$
h_t^j = \frac{\sin c\left(((t-1)/N) - j\right)}{\sum_{j'=-\lfloor (W-1)/2 \rfloor}^{\lfloor W/2 \rfloor} \sin c\left(((t-1)/N) - j'\right)}. \tag{3.50}
$$

In Ref. [Torrance and Hanzo 1995], a comparison between several interpolation schemes was performed. It was concluded that the higher complexity of the MMSE interpolator is not justifiable in terms of performance improvements when compared with a low-complexity first-order linear interpolator. A similar conclusion was achieved in Ref. [Kim et al. 1997] when comparing the low-complexity sin c and Gaussian interpolators with the MMSE interpolator. In that paper, it was shown that although the Gaussian interpolator had a low computational complexity it was very sensitive to the frequency of pilot bit insertion. In this perspective, the sin c interpolator was shown to be more robust and presented only a small performance degradation compared to the MMSE interpolator.

The conventional channel estimation method just described can be applied for SC as well as for OFDM systems. In the latter case, pilot symbols are multiplexed with data symbols using a frame size of N_F OFDM blocks in the time domain and ΔN_{Freq} subcarriers in the frequency domain. The interpolation required for estimating the channel at the data symbols will have to be performed both in frequency and in time. Nevertheless, in the following we will assume $\Delta N_{Freq} = 1$ (OFDM symbols composed of only pilots periodically inserted in the frame) to simplify the analysis, although in the next chapters we will also work with other more typical pilot spacings. In this case, the transmission is equivalent to N (number of subcarriers) parallel flat fading transmissions with frame size N_F and only interpolation in the time domain is required for each subcarrier. Owing to this equivalence, we can evaluate the performance of OFDM as an SC transmission in flat fading so that we can directly apply the BER expressions derived in Section 3.1 to evaluate the OFDM performance with conventional channel estimation. Before using the expressions, we need to know the second-order moments concerning the channel estimate. Note that looking at expression 3.44, we see that when using FIR interpolation the channel estimate, $\hat{\alpha}$, is a weighted sum of zero-mean complex Gaussian variables and thus it is also a zero-mean complex Gaussian variable. The variance of the channel estimate for symbol t in frame u can be written as

$$E\left[\left|\hat{\alpha}((u-1)\cdot N_F+t)\right|^2\right]$$
$$=\Omega^2\sum_{j=-\lfloor (W-1)/2\rfloor+u-1}^{\lfloor W/2\rfloor+u-1}\ \sum_{i=-\lfloor (W-1)/2\rfloor+u-1}^{\lfloor W/2\rfloor+u-1} h_t^{j-u+1}h_t^{i-u+1}J_0\left(2\pi f_D|i-j|N_F\cdot T_s\right)$$
$$+\frac{1}{|s_{pilot}|^2}N_0\sum_{j=-\lfloor (W-1)/2\rfloor+u-1}^{\lfloor W/2\rfloor+u-1}\left(h_t^{j-u+1}\right)^2, \tag{3.51}$$

where, in this case, due to the equivalence between OFDM and SC, T_S should be the OFDM symbol duration. The cross moment of r and $\hat{\alpha}$ is given by

$$E\left[r((u-1)\cdot N_F+t)\hat{\alpha}^*((u-1)\cdot N_F+t)|s\right]$$
$$=\Omega^2 s\sum_{j=-\lfloor (W-1)/2\rfloor+u-1}^{\lfloor W/2\rfloor+u-1} h_t^{j-u+1}J_0\left(2\pi f_D\left|(u-1-j)\cdot N_F+t-1\right|T_s\right). \tag{3.52}$$

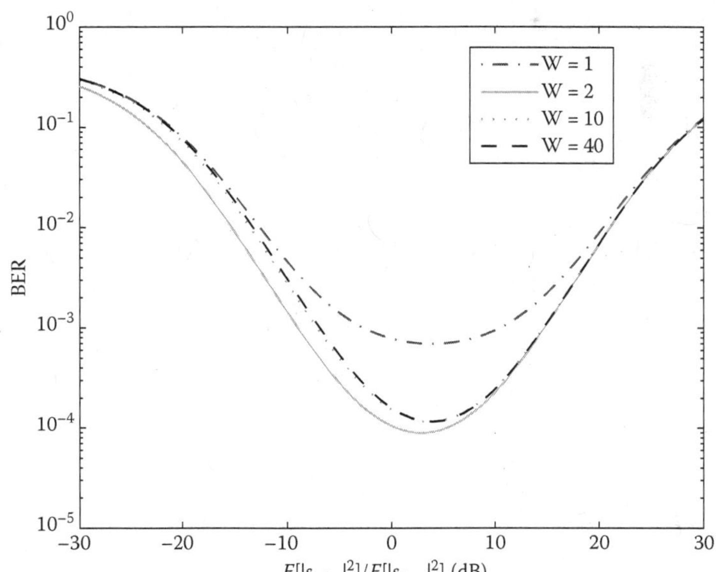

Figure 3.5 Impact of the fraction of the power assigned to pilots on the BER.

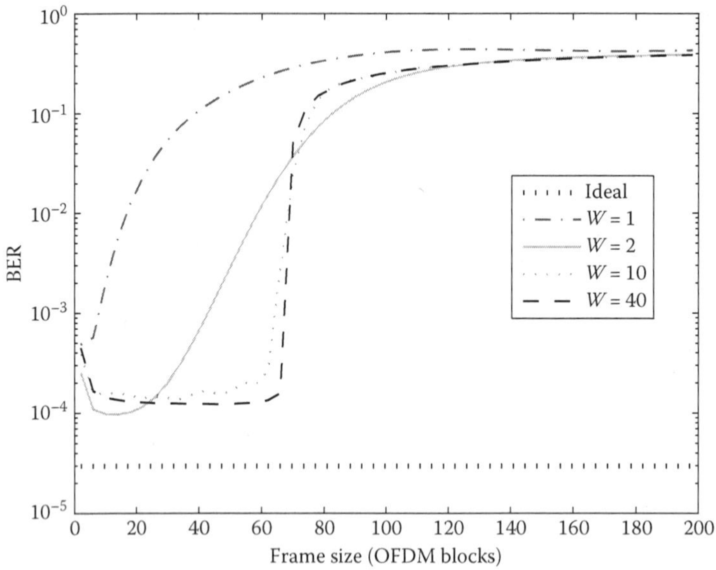

Figure 3.6 Impact of the frame length on the BER.

Figure 3.5 shows the impact of the fraction of power assigned to the pilots on the BER performance when we have $N = 512$ QPSK symbols per OFDM block, 7 OFDM blocks per frame, $f_D T_s = 1.4 \times 10^{-5}$, and a total E_s/N_0 of 25 dB. Clearly, the optimum pilot power should be around 3 dB above the data power, almost independent of the length of the interpolation filter W, whose optimum value is $W = 2$.

Figure 3.6 shows the impact of the frame length on the BER performance when we have $N = 512$ QPSK symbols per OFDM block, $f_D T_s = 1.4 \times 10^{-5}$, and a total E_s/N_0 of 25 dB. For $W = 1$ the BER increases as we increase the frame length, but for higher values of W there is a region of good values of the frame length that increases with W.

Figure 3.7 shows the impact of velocity on BER performance when we have $N = 512$ QPSK symbols per OFDM block and seven OFDM blocks per frame. As expected, increasing the Doppler frequency leads to performance degradation, especially for smaller values of W. For high values of W, the performance is almost independent of the Doppler frequency up to a certain value.

3.3 Channel Estimation with Implicit Pilots

3.3.1 Concept

As it was already mentioned, overhead due to training symbols for channel estimation can be high, leading to decrease of system capacity, especially in fast-varying scenarios

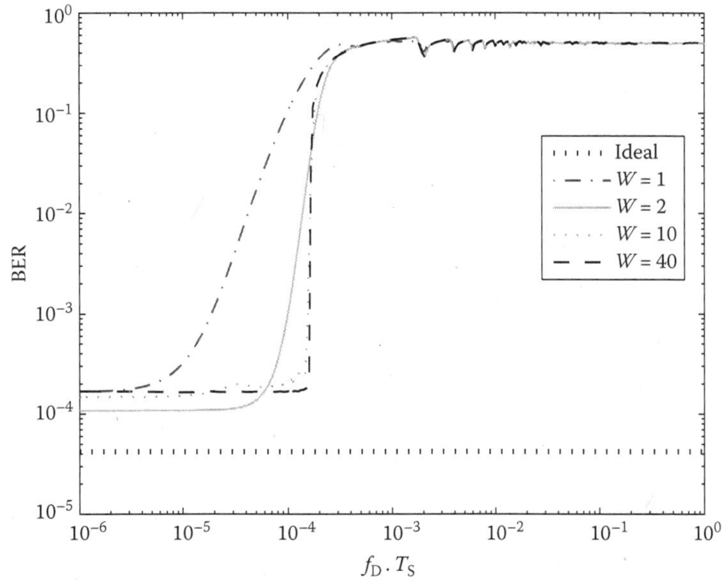

Figure 3.7 Impact of the Doppler effects on the BER.

and/or for bursty transmission. It would be desirable to reduce the overhead for channel estimation purposes. A promising technique to overcome this problem is to use implicit training or implicit pilots, also called superimposed pilots, where the training block is added to the data block instead of being multiplexed with it [Ho et al. 2001; Zhu et al. 2003; Orozco-Lugo et al. 2004; Lam et al. 2006; Meng et al. 2007; Lam et al. 2008]. This means that we can increase significantly the density of pilots (even to the extent of one pilot per data symbol), with zero pilot overhead.

In Ref. [Ohno and Giannakis 2002], a general framework for several approaches to low or zero pilot overhead in the context of OFDM is provided. In one approach, periodic pilot sequences are added to data symbols in the time domain for SC systems [Orozco-Lugo et al. 2004; Tugnait and Meng 2006; Josiam and Rajan 2007; Meng et al. 2007], or in the frequency domain for OFDM systems [Ho et al. 2001; Zhu et al. 2003]. The power level of the added pilots is chosen to minimize error rate degradation due to channel estimation errors and to loss of data power. The interference to pilots (and therefore to channel estimates) from data can be mitigated by time-averaging over many pilot sequence repetitions [Zhu et al. 2003; Orozco-Lugo et al. 2004; Tugnait and Meng 2006; Meng et al. 2007]. Once channel estimates are obtained in this way, pilots are subtracted from the received signal prior to equalization and data detection. Improved channel estimation and data detection performance can be obtained with iterative joint maximum likelihood or quasi-maximum likelihood data detection and channel estimation procedures [Josiam and Rajan 2007; Meng et al. 2007].

Another approach with zero pilot overhead is to add data-dependent pilot sequences to data such that interference to data is zero. Such data-dependent superimposed training (DDST) carried out in the time domain [Ghogho et al. 2005; Tugnait and He 2006] or frequency-domain superimposed pilot techniques (FDSPTs) carried out in the frequency domain [Lam et al. 2006; Lam et al. 2008] essentially replace a subset of data-carrying frequencies with pilots at those frequencies. Nulls are thus created in the apparent channel frequency response seen by the data leading to performance degradation, which must be dealt with by advanced equalization methods such as iterative equalization [Lam et al. 2006; Lam et al. 2008]. Moreover those schemes lead to increased envelope fluctuations on the transmitted signals.

Another problem associated with implicit training is that, by adding training to data signals, the envelope fluctuations of the transmitted signals are increased. This is especially important when a low-PMEPR, SC-based transmission is intended. Since the use of implicit pilots requires averaging over several blocks, these techniques are more sensitive to channel variations.

3.3.2 Transmitted Signals

We assume a frame structure similar to that of [Dahlman et al. 2005] and depicted in Figure 3.8. Each frame has N_T time-domain blocks, each comprising N_T size-N FFT blocks, each one preceded by a suitable CP. The pilot sequence is superimposed to the data sequence. In general, we have a sparse frequency-domain pilot grid, with subcarrier separation of ΔN_F and block separation of ΔN_T. For the sake of simplicity, in the following, we will assume that the channel is almost invariant within the frame.

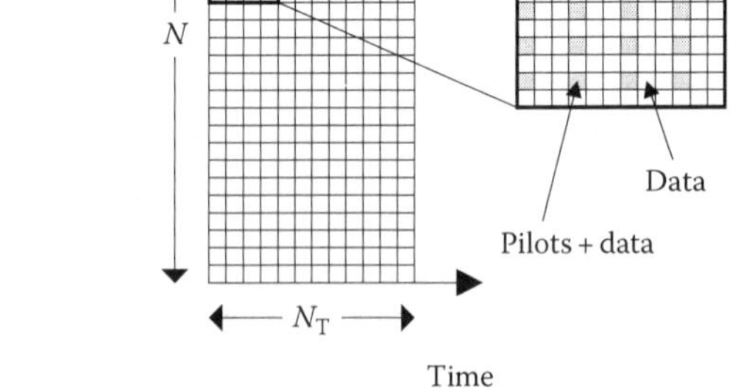

Figure 3.8 (a) Frame structure and (b) detail.

For both OFDM and SC-FDE schemes, the transmitted OFDM signal associated with the frame is

$$s^{Tx}(t) = \sum_{m=1}^{N_T} s_m^{Tx}(t - mT_B), \tag{3.53}$$

with T_B denoting the duration of each block. The mth block has the form

$$s_m^{Tx}(t) = \sum_{n=-N_G}^{N-1} s_{n,m}^{Tx} h_T(t - nT_S), \tag{3.54}$$

with T_S denoting the symbol duration, N_G denoting the number of samples at the CP, and $h_T(t)$ being the adopted reconstruction filter. As usual, $\{s_{n,m}^{Tx}; n = 0, 1, \ldots, N-1\} =$ IDFT $\left\{S_{k,m}^{Tx}; k = 0, 1, \ldots, N-1\right\}$, where $S_{k,m}^{Tx}$ is the symbol transmitted at the kth subcarrier, and $s_{-n,m}^{Tx} = s_{N-n,m}^{Tx}$ (i.e., the first N_G are the cyclic extension of $\{s_{n,m}^{Tx}; n = 0, 1, \ldots, N-1\}$). Clearly, $T_S = T/N$ and $T_B = T(N + N_G)/N$. The frequency-domain symbols to be transmitted are given by

$$S_{k,m}^{Tx} = S_{k,m} + S_{k,m}^{P}, \tag{3.55}$$

where $S_{k,m}$ is the data symbol transmitted by the kth subcarrier of the mth block, selected from a given constellation under an appropriate mapping rule, and $\left\{S_{k,m}^{P}; k = 0, 1, \ldots, N-1\right\}$ is the block of implicit pilots.

We will consider a receiver with L-order space diversity. The signal at the receiver input associated with the mth block and the lth receive antenna is sampled and the CP is removed, leading to the time-domain block $\{y_{n,m}^{Rx(l)}; n = 0, 1, \ldots, N-1\}$. If the CP is longer than the overall CIR, then the corresponding frequency-domain block, obtained after an appropriate size-N DFT operation, is $\left\{Y_{k,m}^{Rx(l)}; k = 0, 1, \ldots, N-1\right\}$, where

$$Y_{k,m}^{Rx(l)} = S_{k,m}^{Tx} H_{k,m}^{(l)} + N_{k,m}^{(l)} = \left(S_{k,m} + S_{k,m}^{P}\right) H_{k,m}^{(l)} + N_{k,m}^{(l)}, \tag{3.56}$$

with $H_{k,m}^{(l)}$ denoting the overall channel frequency response for the kth frequency of the mth time block, at the lth receive antenna, and $N_{k,m}^{(l)}$ denoting the corresponding channel noise.

Let us first assume that $S_{k,m} = 0$, that is, there is no data overlapping the training block, as in conventional schemes. In that case, we could estimate the channel frequency response as follows:

$$\tilde{H}_{k,m}^{(l)} = \frac{Y_{k,m}^{Rx(l)}}{S_{k,m}^{P}} = H_{k,m}^{(l)} + \frac{N_{k,m}^{(l)}}{S_{k,m}^{P}} = H_{k,m}^{(l)} + \varepsilon_{k,m}^{H(l)}. \tag{3.57}$$

If $S_{k,m} = 0$, that is, there is no data overlapping the training block, as in conventional pilot-multiplexing schemes, we could estimate the channel frequency response as follows:

$$\tilde{H}_{k,m}^{(l)} = \frac{Y_{k,m}^{Rx(l)}}{S_{k,m}^{P}} = H_{k,m}^{(l)} + \frac{N_{k,m}^{(l)}}{S_{k,m}^{P}} = H_{k,m}^{(l)} + \varepsilon_{k,m}^{H(l)}. \tag{3.58}$$

The channel estimation error $\varepsilon_{k,m}^{H(l)}$ is Gaussian distributed, with zero mean and

$$E\left[\left|\varepsilon_{k,m}^{H(l)}\right|^2\right] = E\left[\left|N_{k,m}^{(l)}\right|^2\right] E\left[\frac{1}{\left|S_{k,m}^{P}\right|^2}\right]. \tag{3.59}$$

Since the power assigned to the training block is proportional to $E\left[\left|S_{k,m}^{P}\right|^2\right] = 2\sigma_P^2$ (σ_P^2 is the variance of the real and imaginary parts of $S_{k,m}^{P}$) and $E\left[1/\left|S_{k,m}^{P}\right|^2\right] \geq 1/E\left[\left|S_{k,m}^{P}\right|^2\right]$, with equality for $\left|S_{k,m}^{P}\right|$ constant, the training blocks should have $\left|S_{k,m}^{P}\right|^2 = 2\sigma_P^2$ for all k. On the other hand, if we want to minimize the envelope fluctuations of the transmitted signal $\left|s_{n,m}^{P}\right|$ should also be constant. This can be achieved by employing Chu sequences, which have both $\left|s_{n,m}^{P}\right|$ and $\left|S_{k,m}^{P}\right|$ constant [Chu 1972]. The nth element of a length-N Chu sequence is

$$s_{n,m}^{P} = \begin{cases} \exp(j\pi rn^2/N), & N \text{ even} \\ \exp(j\pi rn(n-1)/N), & N \text{ odd}, \end{cases} \tag{3.60}$$

where r is relatively prime to N. This is particularly important for SC-FDE schemes, which are usually designed to be employed with amplifiers with lower backoff than the ones for OFDM schemes.

Since the CIR is shorter than the CP, which is just a fraction of the block duration, we could employ training blocks that are shorter than the standard data blocks. Alternatively, we could use the enhanced channel estimates $\left\{\tilde{H}_{k,m}^{(l)}; k = 0, 1, \ldots, N-1\right\} = \text{DFT}$ $\left\{\tilde{h}_{n,m}^{(l)} = \hat{h}_{n,m}^{(l)} w_n; n = 0, 1, \ldots, N-1\right\}$, where $w_n = 1$ for $0 \leq n \leq N_G - 1$ and 0 otherwise and $\left\{\hat{h}_{n,m}^{(l)}; n = 0, 1, \ldots, N-1\right\} = \text{IDFT}\left\{\hat{H}_{k,m}^{(l)}; k = 0, 1, \ldots, N-1\right\}$. In this case, the SNR at the channel estimates is improved by a factor N/N_G.

Let us consider now the use of implicit pilots, that is, $S_{k,m} \neq 0$ for the pilots. In the following, we will assume that

$$E\left[\left|S_{k,m}\right|^2\right] = 2\sigma_D^2, \tag{3.61}$$

and for the frequencies that have pilots,

$$E\left[\left|S_{k,m}^{P}\right|^2\right] = 2\sigma_T^2. \tag{3.62}$$

Clearly, we will have interference between data symbols and pilots. This leads to performance degradation for three reasons:

- The data symbols produce interference on pilots, which might lead to inaccurate channel estimates. This effect is negligible if $\sigma_D^2 \ll \sigma_T^2$.
- The pilots produce interference on data symbols, which might lead to performance degradation (even if the channel estimation was perfect). This effect is negligible if $\sigma_D^2 \gg \sigma_P^2$.
- The use of implicit pilots leads to increased envelope fluctuations on the transmitted signals. The complementary cumulative distribution function (CCDF) of the envelope of the transmitted signals is depicted in Figure 3.9, where we consider $N = 512$ QPSK constellations and square-root raised cosine filtering. Clearly, it is desirable to have $\eta_T = \sigma_T^2/\sigma_D^2 \ll 1$ (i.e., $\sigma_D^2 \gg \sigma_T^2$) when we want to have transmitted signals with reduced envelope fluctuations: the PMEPR of the transmitted signals increases as we increase the power associated with pilots, although it is significantly better than for OFDM schemes, even for high-power pilots. The PMEPR degradation due to the use of implicit pilots is lower for very small roll-off values.

To overcome these problems, we can employ pilots with relatively low power (i.e., $\sigma_T^2 \ll \sigma_D^2$) and average the pilots over a large number of blocks so as to obtain accurate channel estimates. This is very effective since the data symbols have usually zero mean and different data blocks are uncorrelated. Naturally, there are limitations on the length of

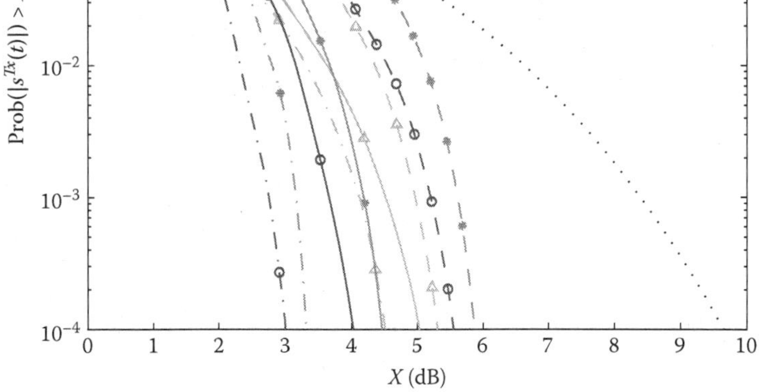

Figure 3.9 CCDF of $|s^{Tx}(t)|$ when $\Delta N_F = 1$.

this averaging window, since the channel should be constant within it (not to mention the associated delays). In the following, we average over N_T blocks, where N_T is the frame length. Once we have an accurate channel estimate, we can detect the data symbols, eventually removing first the signal associated with the pilots.

If the overall CIR (and the CP of each FFT block) has $N_G = NT_G/T$ samples, we will need N_G equally spaced frequency-domain pilots for the channel estimation. For pilot spacings in time and frequency ΔN_T and ΔN_F, respectively, the total number of pilots in the frame is

$$N_P^{\text{Frame}} = \frac{N}{\Delta N_F} \cdot \frac{N_T}{\Delta N_T}. \tag{3.63}$$

This means that we have a pilot multiplicity or redundancy of

$$N_R = \frac{N_P^{\text{Frame}}}{N_G} = \frac{N}{N_G \Delta N_F} \cdot \frac{N_T}{\Delta N_T}. \tag{3.64}$$

Therefore, the SNR associated with the channel estimation procedure is

$$\text{SNR}_{\text{est}} = \frac{N_R \sigma_T^2}{\sigma_N^2 + \sigma_D^2} = N_R \frac{\sigma_T^2}{\sigma_D^2} \text{SNR}_{\text{data}} \frac{1}{1 + \text{SNR}_{\text{data}}}, \tag{3.65}$$

where

$$\sigma_N^2 = \frac{1}{2} E\left[\left|N_{k,m}\right|^2\right], \tag{3.66}$$

and the SNR associated with data symbols is $\text{SNR}_{\text{data}} = \sigma_D^2/\sigma_N^2$. For moderate and high SNR values,

$$\text{SNR}_{\text{est}} \approx N_R \frac{\sigma_T^2}{\sigma_D^2}. \tag{3.67}$$

To avoid significant performance degradation due to channel estimation errors, SNR_{est} should be high. This could be achieved with $\sigma_T^2 \ll \sigma_D^2$, provided that $N_R \gg 1$.

3.3.3 Receiver Design

In this section we present an iterative receiver with decision-directed channel estimation for OFDM and SC-FDE schemes with implicit pilots.

3.3.3.1 OFDM Schemes

Let us first consider OFDM schemes. The extension to SC-FDE schemes will be made later. The receiver structure for OFDM schemes is the one depicted in Figure 3.10. Without loss of generality it is assumed that there is a pilot for each subcarrier of each block of the frame, that is, $\Delta N_F = \Delta N_T = 1$, leading to $N_P^{\text{Frame}} = NN_T$ and a pilot multiplicity or redundancy of $N_R = N_P^{\text{Frame}}/N_G = NN_T/N_G$.

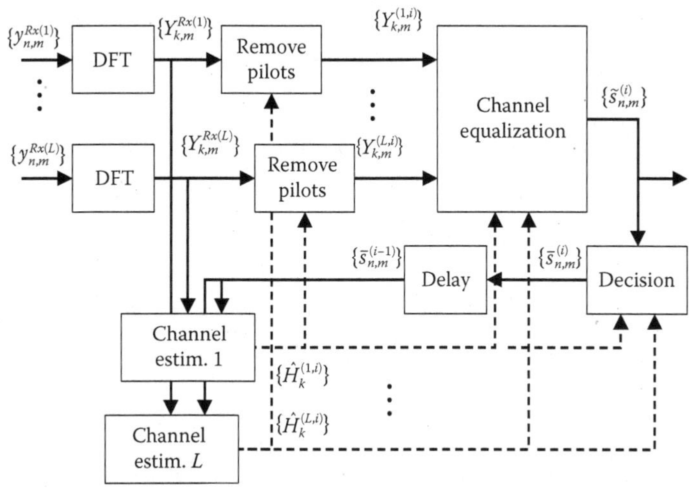

Figure 3.10 Receiver structure for OFDM schemes.

The estimation/detection procedure is as follows:

1. We first obtain the L channel frequency response estimates

$$\tilde{H}_k^{(l,1)} = \frac{1}{N_{\mathrm{T}}} \sum_{N_{\mathrm{T}}}^{m=1} \frac{Y_{k,m}^{Rx(l)}}{S_{k,m}^{P}}, \quad l = 1, 2, \ldots, L, \tag{3.68}$$

where $\left\{ Y_{k,m}^{Rx(l)}; k = 0, 1, \ldots, N-1 \right\}$ denotes the mth received frequency-domain block ($m = 1, 2, \ldots, N_{\mathrm{T}}$).

2. Each of the L channel estimates is enhanced by ensuring that the corresponding impulse response has duration N_{G}, that is, we use the channel estimation $\left\{ \hat{H}_k^{(l,1)}; k = 0, 1, \ldots, N-1 \right\} = \mathrm{DFT} \left\{ \hat{h}_n^{(l,1)} = \tilde{h}_n^{(l,1)} w_n; k = 0, 1, \ldots, N-1 \right\}$, where $\left\{ \tilde{h}_n^{(l,1)}; k = 0, 1, \ldots, N-1 \right\} = \mathrm{IDFT} \left\{ \tilde{H}_k^{(l,1)}; k = 0, 1, \ldots, N-1 \right\}$.

3. The pilots are removed from the received frequency-domain blocks, leading to the blocks $\left\{ Y_{k,m}^{(l,1)} = Y_{k,m}^{Rx(l)} - \hat{H}_k^{(l,1)} S_{k,m}^{P}; k = 0, 1, \ldots, N-1 \right\}$ and the N_{T} blocks of equalized samples (one for each block of the frame),

$$\tilde{S}_{k,m}^{(1)} = \frac{\sum_{l=1}^{L} Y_{k,m}^{(l,1)} \hat{H}_k^{(l,1)*}}{\sum_{l=1}^{L} \left| \hat{H}_k^{(l,1)} \right|^2}, \tag{3.69}$$

are generated.

4. The equalized blocks are submitted to a decision device so as to obtain the average values of the transmitted symbols $\left\{\overline{S}_{k,m}^{(l,2)}; k = 0, 1, \ldots, N-1\right\}$ that will be used in the next iteration.

5. For the second iteration, the pilots are removed from the received blocks and the average values of the data symbols will be used as training symbols for obtaining the channel frequency response estimate

$$\tilde{H}_k^{(l,2)} = \frac{\sum_{m=1}^{N_T} Y_{k,m}^{(l,1)} \overline{S}_{k,m}^{(1)*}}{\sum_{m=1}^{N_T} \left|\overline{S}_{k,m}^{(1)}\right|^2}. \tag{3.70}$$

6. As in step (2), the L enhanced channel estimates $\left\{\hat{H}_k^{(l,2)}; k = 0, 1, \ldots, N-1\right\} = \text{DFT}\left\{\hat{h}_n^{(l,2)} = \tilde{h}_n^{(l,1)} w_n; k = 0, 1, \ldots, N-1\right\}$, where $\left\{\tilde{h}_n^{(l,2)}; k = 0, 1, \ldots, N-1\right\} = \text{IDFT}\left\{\tilde{H}_k^{(l,2)}; k = 0, 1, \ldots, N-1\right\}$ s, $l = 1, 2, \ldots, L$, are computed.

7. Repeat steps (3)–(6) for each iteration of the receiver.

The average values associated with the data symbols used in Equation 3.70 are given by

$$\overline{S}_{k,m}^{(i)} = \tanh\left(\frac{L_{k,m}^{\mathrm{I}(i)}}{2}\right) + \mathrm{j}\tanh\left(\frac{L_{k,m}^{\mathrm{Q}(i)}}{2}\right), \tag{3.71}$$

where the LLRs of the "in-phase bit" and the "quadrature-phase bit," associated with $S_{n,m}^{\mathrm{I}(i)}$ and $S_{n,m}^{\mathrm{Q}(i)}$, respectively, are given by

$$L_{k,m}^{I(i)} = \frac{2\tilde{S}_{k,m}^{\mathrm{I}(i)}}{\sigma_{\mathrm{N}}^2} \sum_{l=1}^{L} \left|\hat{H}_k^{l,i}\right|^2, \tag{3.72}$$

and

$$L_{k,m}^{\mathrm{Q}(i)} = \frac{2\tilde{S}_{k,m}^{\mathrm{Q}(i)}}{\sigma_{\mathrm{N}}^2} \sum_{l=1}^{L} \left|\hat{H}_k^{l,i}\right|^2, \tag{3.73}$$

respectively.

If we do not perform the channel decoding in the feedback loop, the log-likelihood values can be computed on a symbol-by-symbol basis. However, as with turbo FDE defined in Chapter 2, we can improve significantly the performance if the channel decoding procedure is performed before each channel estimation iteration. In this case, a SISO channel decoder (Soft-In, Soft-Out) is employed in the feedback loop. The SISO block provides the LLRs of both the "information bits" and the "coded bits." The input of the SISO block is LLRs of the "coded bits" at the equalizer output, given by Equations 3.72 and 3.73.

3.3.3.2 SC-FDE Schemes

Let us consider now the receiver design for SC-FDE schemes with implicit pilots. The receiver structure is depicted in Figure 3.11. Clearly, this receiver is similar to the one considered for OFDM schemes, with the main difference being the fact that the FDE receives as input the average symbol values conditioned to the FDE output at the previous iteration, since our receiver is based on the IB-DFE described in Chapter 2.

The principles behind this receiver are similar to the ones behind the receiver for OFDM schemes with implicit pilots. The major difference is that in step (5), the conditional expectations of the data symbols will be used as training symbols for obtaining the channel frequency response estimate

$$\tilde{H}_k^{(l,2)} = \frac{\sum_{m=1}^{N_T} Y_{k,m}^{(l,1)} \overline{S}_{k,m}^{(1)*}}{\sum_{m=1}^{N_T} \left|\overline{S}_{k,m}^{(1)}\right|^2 + \alpha}, \tag{3.74}$$

where $\left\{\overline{S}_{k,m}^{(2)}; k = 0, 1, \ldots, N-1\right\} = \text{DFT}\left\{\overline{s}_{n,m}^{(2)}; n = 0, 1, \ldots, N-1\right\}$ $(m = 1, 2, \ldots, N_T)$. Since using $\alpha = 0$ (as in the OFDM case) might lead to noise enhancement effects in the channel estimates when $\sum_{m=1}^{N_T} \left|\overline{S}_{k,m}^{(1)}\right|^2$ is small, we will consider $\alpha = E\left[\left|N_{k,m}\right|^2\right] / E\left[\left|S_{k,m}\right|^2\right]$. If $N_T \gg 1$, then $\sum_{m=1}^{N_T} \left|\overline{S}_{k,m}^{(1)}\right|^2 \approx N_T 2\sigma_D^2$ and we can use $\alpha = 0$.

The structure of the FDE for the ith iteration and the mth block is depicted in Figure 3.12. As described in Chapter 2, the frequency-domain block at the output of the

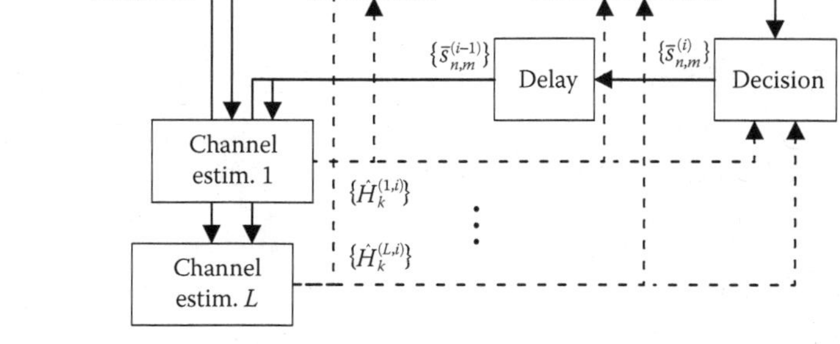

Figure 3.11 Receiver structure for SC-FDE schemes.

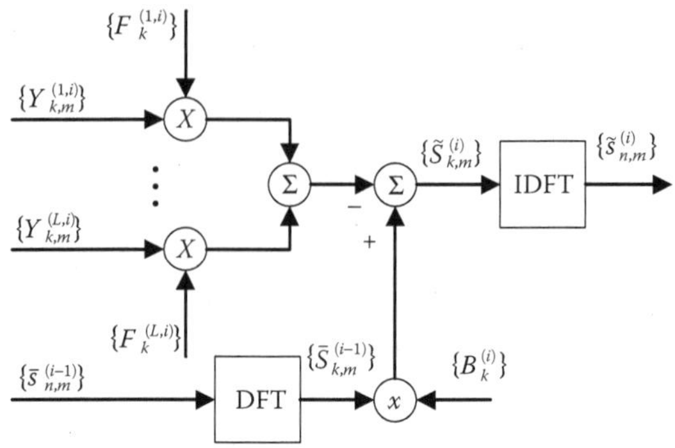

Figure 3.12 Detail of the Turbo FDE for the *i*th iteration.

equalizer is $\left\{\tilde{S}_{k,m}^{(i)}; k = 0, 1, \ldots, N-1\right\}$, with

$$\tilde{S}_{k,m}^{(i)} = \sum_{l=1}^{L} F_{k,m}^{(l,i)} Y_{k,m}^{(l,i)} - B_{k,m}^{(i)} \bar{S}_{k,m}^{(i-1)}, \tag{3.75}$$

where $\left\{F_{k,m}^{(l,i)}; k = 0, 1, \ldots, N-1\right\}$ are the feedforward coefficients associated wth the lth diversity antenna and $\left\{B_{k,m}^{(i)}; k = 0, 1, \ldots, N-1\right\}$ are the feedback coefficients, which can be obtained as in Chapter 2.

Once again, we can define an IFDE (iterative FDE) where, as with IB-DFE receivers, the log-likelihood values are computed on a symbol-by-symbol basis (i.e., we do not need to perform the channel decoding in the feedback loop) or a Turbo FDE that employs the channel decoder outputs instead of the uncoded "soft decisions" in the feedback loop. The main difference between IFDE and Turbo FDE is in the decision device: in the first case, the decision device is a symbol-by-symbol soft-decision (for QPSK constellation this corresponds to the hyperbolic tangent, as explained in Chapter 2); for the Turbo FDE, a SISO channel decoder (Soft-In, Soft-Out) was employed in the feedback loop.

3.3.4 Performance Results for OFDM Schemes

In the following we present a set of performance results concerning channel estimation using implicit pilots for OFDM modulations. The frame has $N_{\mathrm{T}} = 12$ FFT blocks, each with $N = 512$ subcarriers. The data symbols are selected from a QPSK constellation under a Gray mapping rule (similar results will be observed for other values of N, provided that $N \gg 1$). $\Delta N_{\mathrm{T}} = \Delta N_{\mathrm{F}} = 1$, that is, there is an implicit pilot for each symbol of each FFT block. The receiver has L diversity branches and the channel associated with each diversity antenna is characterized by the power delay profile (PDP) of Figure 3.13, with uncorrelated Rayleigh fading on the different paths and different diversity branches.

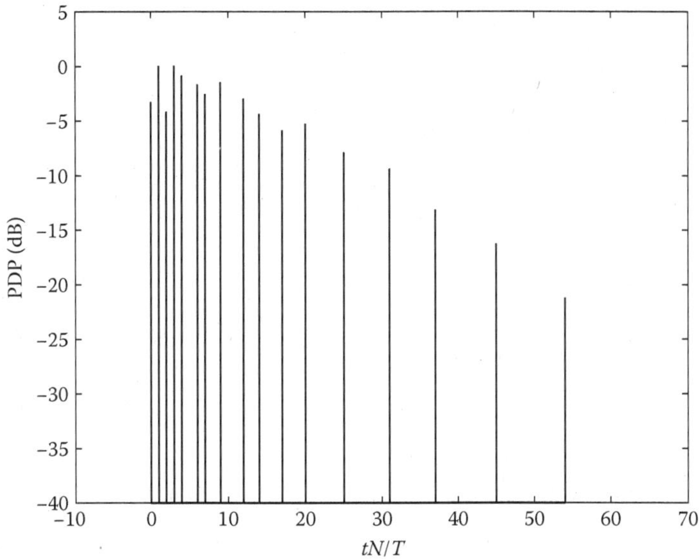

Figure 3.13 Adopted PDP.

The duration of the overall CIR is 12.5% of the duration of the useful part of the block (i.e., $T_G/T = N_G/N = 0.125$). This means that the pilot multiplicity or redundancy is $N_R = 96$. The channel is assumed to be invariant within the frame duration. Linear power amplification is considered at the transmitter and perfect synchronization is assumed at the receiver. The channel encoder is a rate-1/2 turbo code based on two identical recursive convolutional codes characterized by $G(D) = \left[1(1 + D^2)/(1 + D + D^2)\right]$. A random interleaver is employed within the turbo encoder and the coded bits are also interleaved before being mapped into a QPSK constellation and distributed by the symbols of the frame.

Our performance results are expressed as a function of E_b/N_0 or E_b^{Tot}/N_0, where N_0 is the one-sided power spectral density of the noise, E_b is the energy of the transmitted bits [i.e., the degradation due to the useless power spent on the CP (about 0.5 dB, in our case) is not included] and $E_b^{Tot} = E_b + 10\log_{10}((\sigma_T^2 + \sigma_D^2)/(\sigma_D^2))$ (dB) is the total bit energy, including the energy spent on the pilots. Since we are considering a rate-1/2 channel encoder, the energy of the corresponding information bits is 3 dB higher.

Let us consider first a conventional OFDM receiver where the channel estimation is made from the implicit pilots (i.e., just the first iteration). The turbo decoder has 12 iterations. Figure 3.14 shows the BER performance for $L = 1$ (no diversity), 2, and 4 and different values of $\eta_T = \sigma_T^2/\sigma_D^2 s$. We also included the performance with perfect channel estimation (and $\sigma_T^2 = 0$). Clearly, channel estimates based only on low-power pilots can be very poor, leading to significant performance degradation (the performance degradation is already high for $\eta_T = 1/4$). From this figure it might seem that we should spend significant power on the pilots. However, if we express the performance as a function of E_b^{Tot}/N_0 (i.e., including the power spent on the pilots) instead of just the power associated with data symbols [this corresponds to an additional degradation of $10\log_{10}\left(\left(\sigma_T^2+\sigma_D^2\right)/\sigma_D^2\right) = 10\log_{10}(1+\eta_T)$], it is clear that the power spent on the pilots should not be too high, as depicted in Figure 3.15. From Figures 3.14 and 3.15, it is also clear that the required pilot power is not too different for different values of L.

Let us consider now the iterative receivers with joint detection and channel estimation. The receiver has four iterations with channel estimation and detection procedures. For each detection/estimation iteration, we perform three iterations of the turbo decoder. To speed up the decoding procedure, the extrinsic values of the decoding procedure of the previous detection/estimation iteration are stored and used as *a priori* information for the next decoding procedure. The results with perfect channel estimation are obtained with 12 iterations of the turbo decoder. Figures 3.16 through 3.18 concern $L = 1$ (no diversity), 2, and 4, respectively. From these figures we can conclude that it is possible to have excellent performances, close to the ones with perfect channel estimation, even for pilots with relatively low power, even for $L = 4$.

It should be pointed out that, although the performance is better for large values of N_T (i.e., larger frames), our technique is still effective for values of N_T as small as 4 or even 2 (naturally, the power spent on the pilots should be higher for smaller values of N_T). This means that we can use implicit pilots for channel estimation even with time-varying channels.

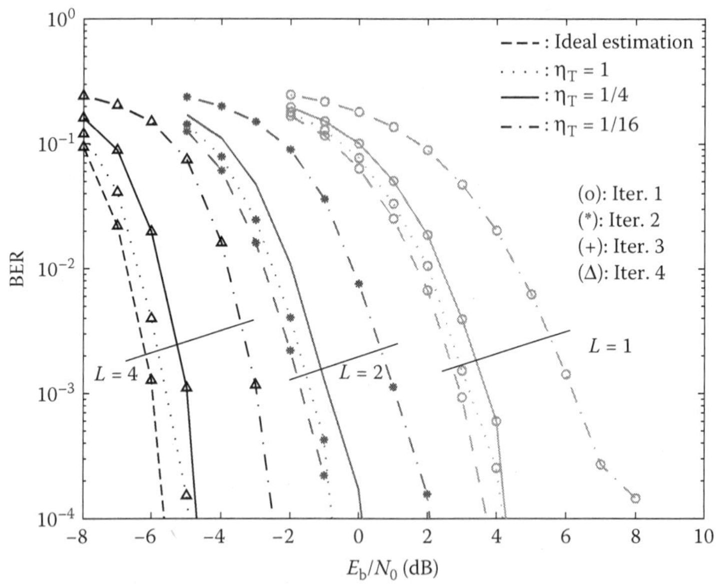

Figure 3.14 BER performance for conventional receivers.

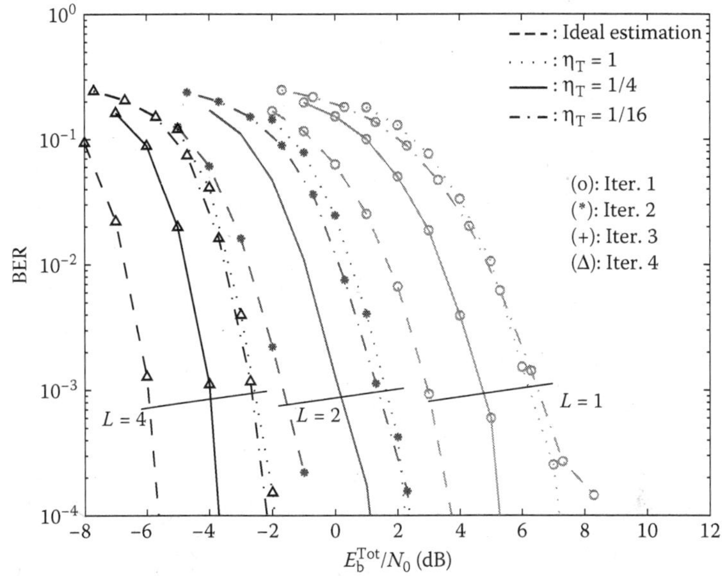

Figure 3.15 The same as Figure 3.14, but with BER as a function of E_b^{Tot}/N_0 (i.e., including the power spent on the pilots).

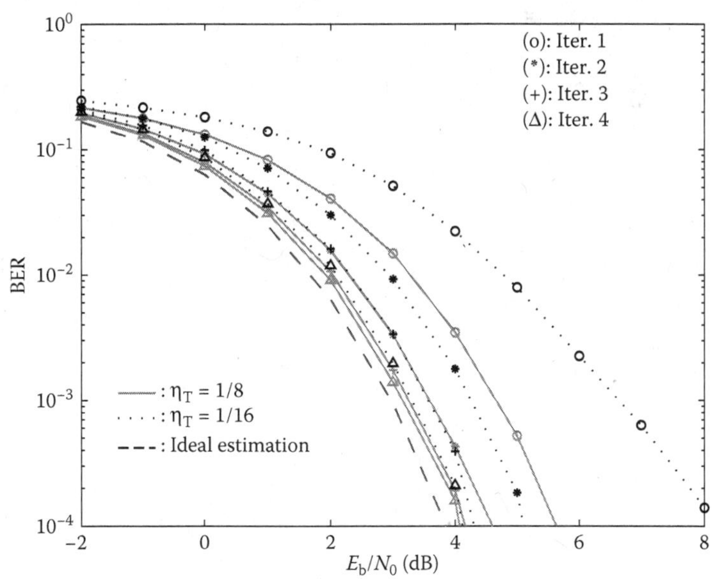

Figure 3.16 BER performance for the iterative receiver when $L = 1$.

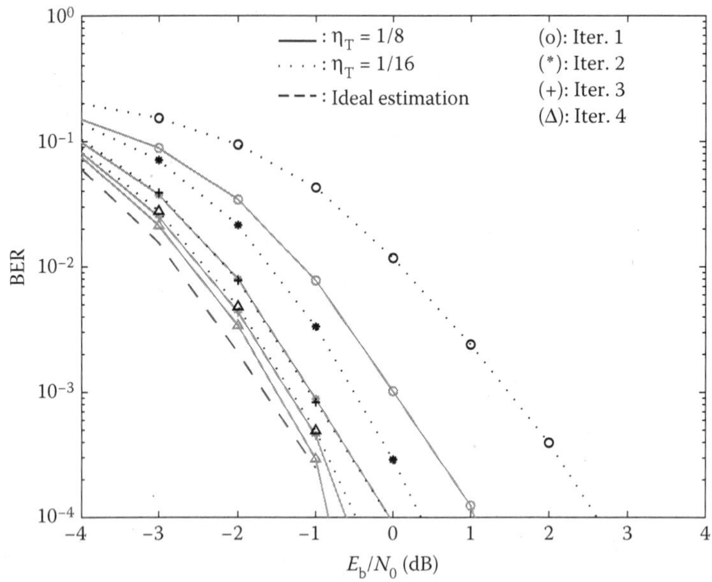

Figure 3.17 BER performance for the iterative receiver when $L = 2$.

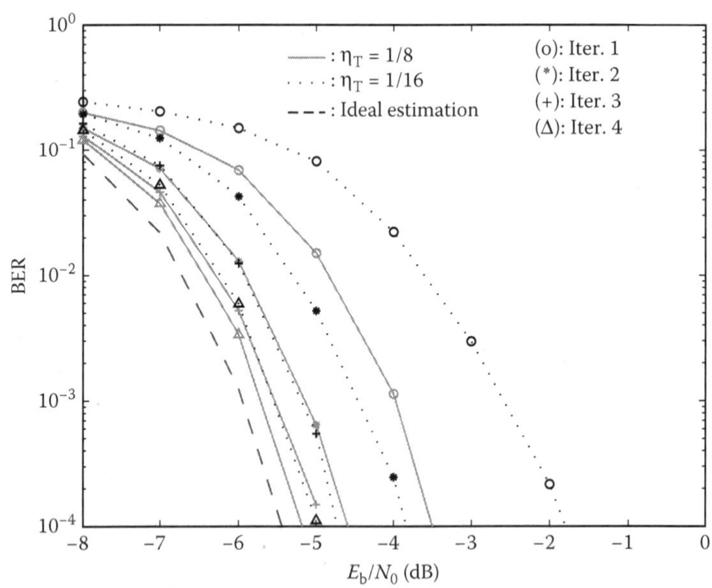

Figure 3.18 BER performance for the iterative receiver when $L = 4$.

3.3.5 Performance Results for SC-FDE Schemes

Let us consider now SC-FDE schemes employing implicit pilots for channel estimation purposes. The frame has $N_T = 12$ FFT blocks, each with $N = 512$ data symbols selected from a QPSK constellation under a Gray mapping rule (similar results were observed for other values of N, provided that $N \gg 1$). There is an implicit pilot for each symbol of each FFT block, that is, $\Delta N_T = \Delta N_F = 1$ (without loss of generality, we assume the same set of pilots for each block). The CIR is characterized by the PDP of Figure 3.13, with uncorrelated Rayleigh fading on the different paths (similar results were observed for other severely time-dispersive channels). Unless otherwise stated, the channel is assumed to be invariant within the frame duration. The duration of the useful part of the data blocks (N symbols) is 1 μs and the CP has 0.125 μs duration. This means that the pilot multiplicity or redundancy is $N_R = 96$. Linear power amplification is considered at the transmitter, and perfect synchronization is assumed at the receiver. The channel encoder is a convolutional code with generators $1 + D^2 + D^3 + D^5 + D^6$ and $1 + D + D^2 + D^3 + D^6$ and the coded bits are interleaved before being mapped into the constellation points and distributed by the symbols of the frame (the interleaving depth is $2NN_T$). Our performance results are expressed as a function of E_b/N_0, where N_0 is the one-sided power spectral density of the noise and E_b is the energy of the transmitted bits (i.e., the degradation due to the useless power spent on the CP is not included). Once again, if we consider the power spent on the pilots, there is an additional degradation of $10\log_{10}\left[\left(\sigma_T^2 + \sigma_D^2\right)/\sigma_D^2\right] = 10\ \log_{10}(1 + \eta_T)$, (4.77, 3, 1.76, 0.97, 0.67, 0.34, and 0.18 dB for $\eta_T = 2, 1, 1/2, 1/4, 1/6, 1/12$, and $1/24$, respectively).

Let us consider first a linear FDE (i.e., just the first iteration). Figure 3.19 shows the uncoded BER performance for different values of η_T. We also included the performance with perfect channel estimation (and $\sigma_T^2 = 0$). Clearly, channel estimates based only on low-power pilots can be very poor, leading to significant performance degradation (the performance degradation is already high for $\eta_T = 1/4$). As it was already pointed out, it is not desirable to have high-power pilots due to the large power spent on pilots and the envelope fluctuations of the transmitted signals.

Let us consider now the iterative receivers with joint detection and channel estimation described above.

Figures 3.20 and 3.21 concern an IFDE receiver (i.e., without channel decoding in the feedback loop); Figure 3.22 concerns Turbo FDE. We present the results for the first iteration (linear FDE) as well as for the fourth iteration (our simulations indicate that the gains associated with the iterative procedures are usually very small beyond four iterations). For the sake of comparisons, we also included the MFB defined in Chapter 2.

From these figures, we can conclude that it is possible to have good performance even when $\eta_T = 1/12$ or $1/24$, in spite of the poor performance with pilot-based only channel estimation (iteration 1). Best performance is achieved with a pilot power of about 8 dB below data power ($\eta_T = 1/6$). At this level, pilot power is still well below the noise power in the region of interest; hence, the interference effect on the data is minor. At the same time, the pilot power and processing gain (N_R) are enough to provide good

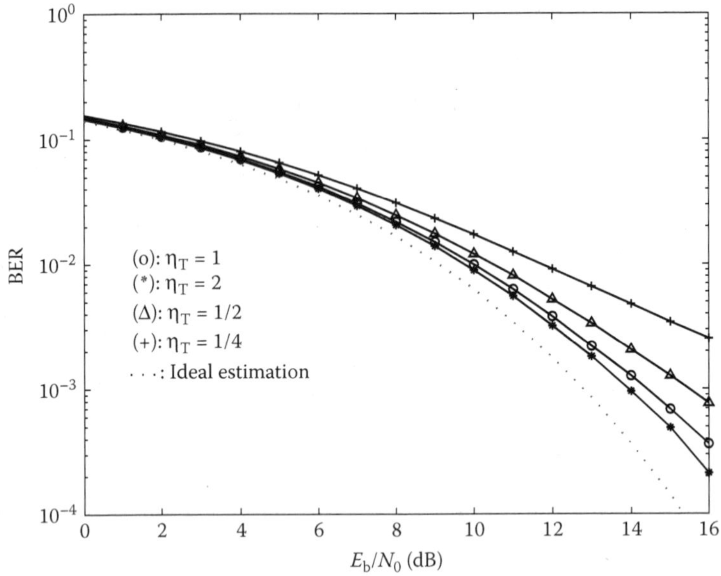

Figure 3.19 Uncoded BER performance for a linear FDE when $L = 1$.

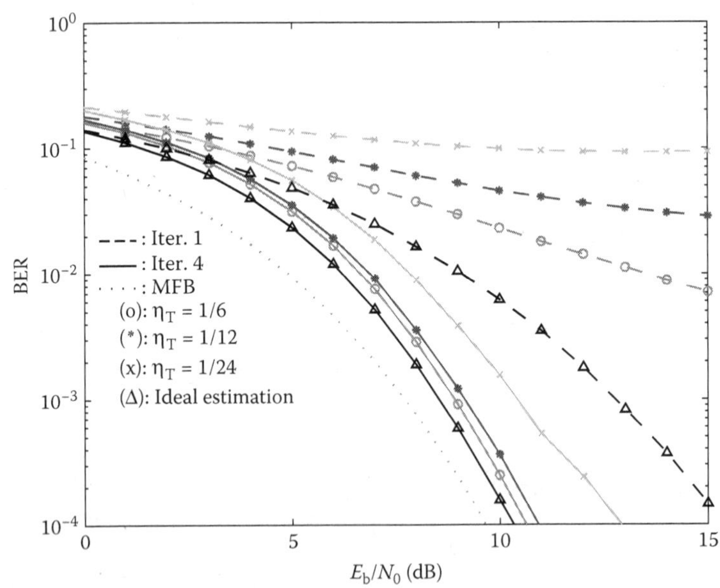

Figure 3.20 Uncoded BER performance for an IFDE when $L = 1$.

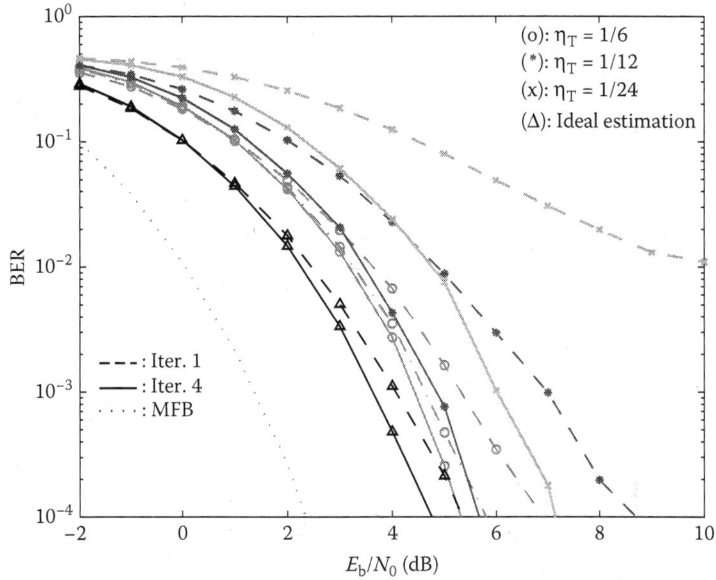

Figure 3.21 Coded BER performance for an IFDE when $L = 1$.

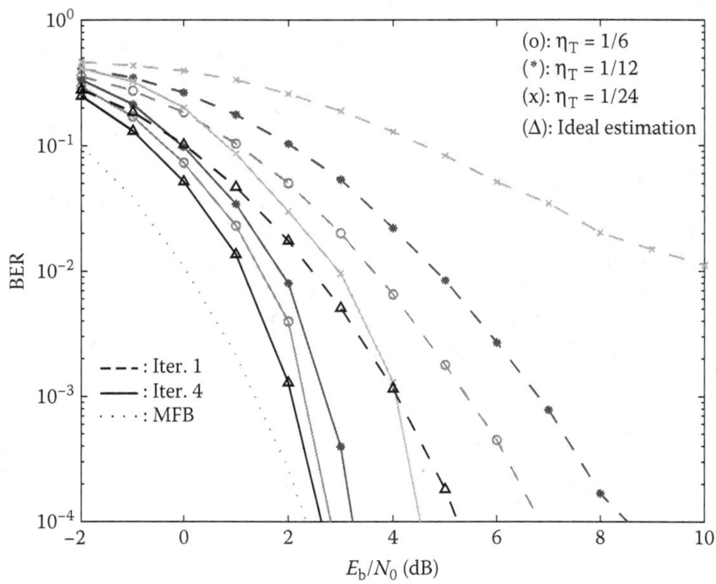

Figure 3.22 Coded BER performance for a Turbo FDE when $L = 1$.

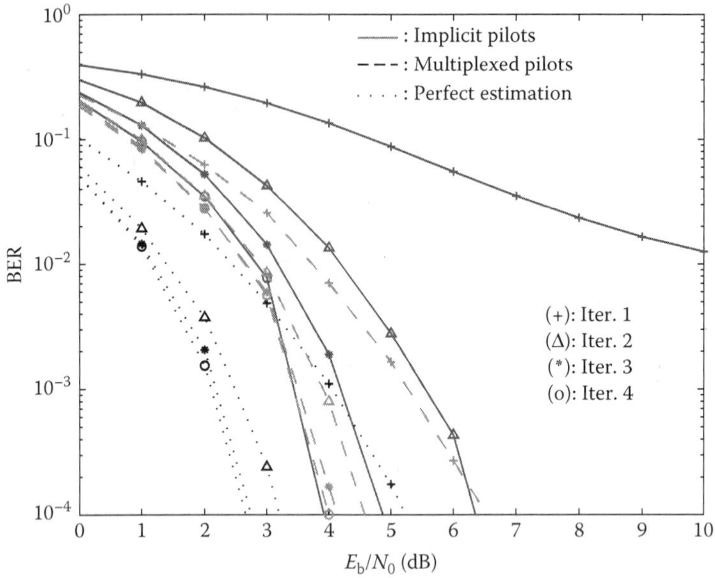

Figure 3.23 Coded BER performance of a Turbo FDE when $M_T = 12$, $L = 1$ and $\eta_T = 1/24$ for multiplexed pilots and implicit pilots.

channel estimation, with performance close to the one with perfect channel estimation.* For smaller pilot powers the performance can be rather poor, since decision errors lead to unreliable decision-directed channel estimates. The performance for the Turbo FDE is much better, with gains between 1 and 2 dB when compared with the IFDE.

In Figure 3.23, we consider our Turbo FDE when $L = 1, M_T = 12$, and $\eta_T = 1/24$. In this figure, we compare the performance when implicit pilots are employed with the performance when multiplexed pilots are employed. For the case of multiplexed pilots, we assume that we add to each frame an FFT block with only pilots and whose power is the same as the total power spent in implicit pilots (similar results could be obtained for a "genie-aided receiver" where we remove completely the interference between data and implicit pilots). From this figure, it can be seen that, although we are considering low-power pilots, our performance can be close to the performance with multiplexed pilots.

Figure 3.24 shows the required E_b/N_0 for BER $= 10^{-4}$ for IFDE and Turbo FDE (both with four iterations). In this figure, we have both the case where we do not consider the power spent on the pilots (taken directly from the previous figures) and the case where the degradation due to this power is included. As expected, the required E_b/N_0 decreases

* As with the remaining results, we did not consider the power spent on pilots for the case with perfect channel estimation as well as for computing the MFB.

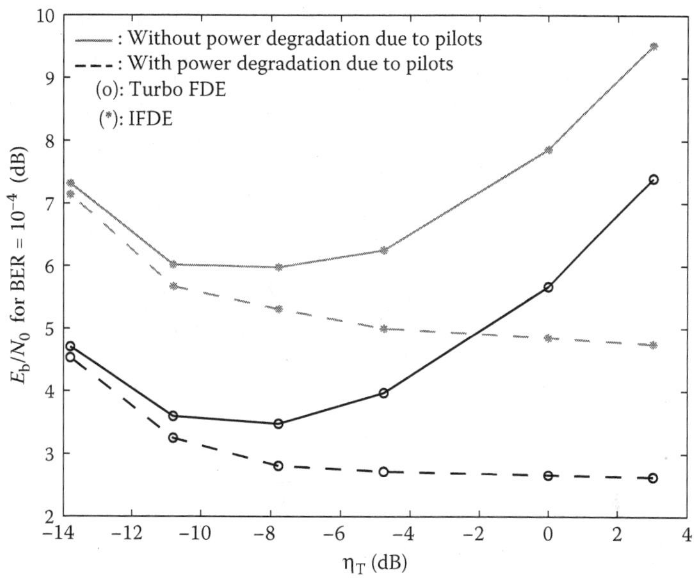

Figure 3.24 Required E_b/N_0 for BER = 10^{-4} when we consider or not the power spent on the pilots ($L = 1$).

as we increase the power spent on the pilots. However, when we include the degradation due to the power spent on the pilots, there is clearly an optimum value of η_T in the vicinity of -10 dB (i.e., 1/10). However, it should be pointed out that the most adequate value of η_T should take into account aspects such as the PMEPR of the transmitted signals (in this case, probably it would be better to adopt a smaller value of η_T).

Let us consider now the case where we have space diversity at the receiver (i.e., $L > 1$). Figure 3.25 shows the BER performance for our Turbo FDE with four detection/channel estimation iterations and different values of L. From this figure it is clear that, in spite of working with lower SNR, the degradation associated with employing low-power pilots decreases as we increase the diversity order: for $L = 4$ the degradation for $\eta_T = 1/24$ is about 0.5 dB, while this degradation is above 2 dB for $L = 1$.

It should be pointed out that high N_T values are required (i.e., large frame lengths) when implicit pilots are employed. Since the channel should be almost constant over the frame (N_T FFT blocks), this might lead to difficulties for fast varying channels. One could ask if these techniques are still valid for smaller frames. To answer this, we consider smaller values of N_T in Figure 3.26. Clearly, we can still employ implicit pilots for $N_T = 6$ or even $N_T = 3$, provided that our Turbo FDE is employed and we increase the power associated with the pilots (η_T should be 1/6 or 1/3). In Figure 3.27, we show the performance of our Turbo FDE receiver in the presence of Doppler effects (we considered Jakes' Doppler spectrum for each path), $L = 1$ and different values of N_T. As expected, the use of large values of N_T is only recommendable for almost static channels; for larger

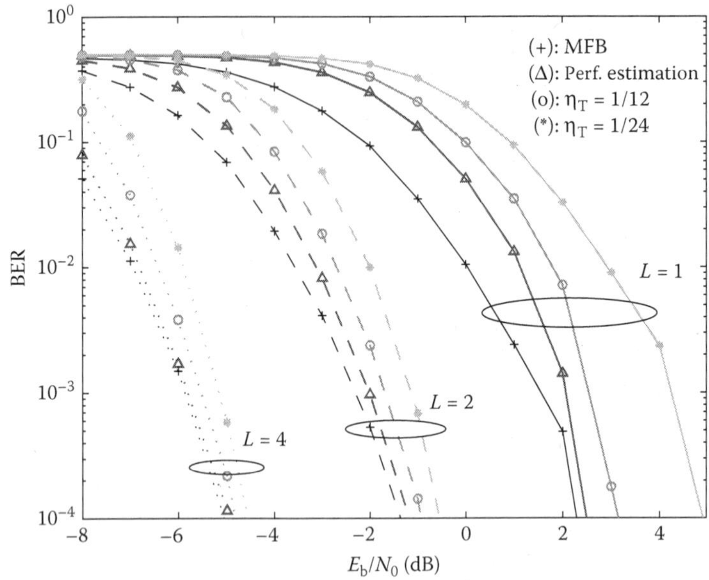

Figure 3.25 Coded BER performance for a Turbo FDE with receiver diversity.

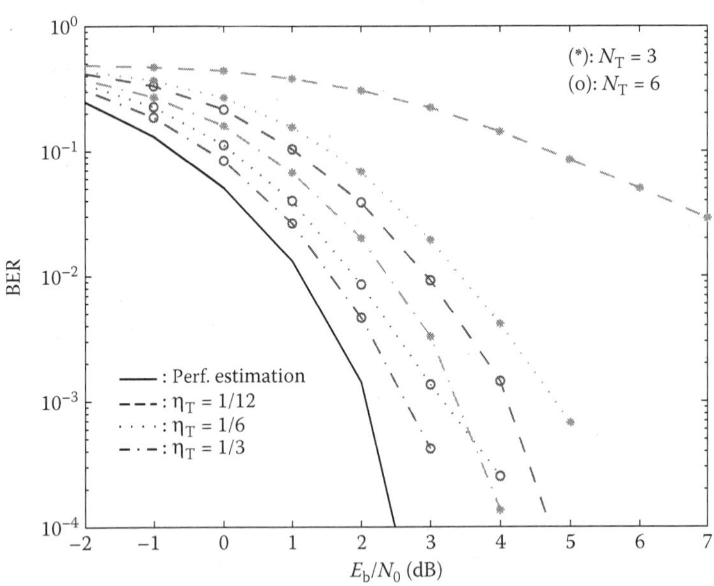

Figure 3.26 Coded BER performance for a Turbo FDE with $L = 1$ and different values of N_T.

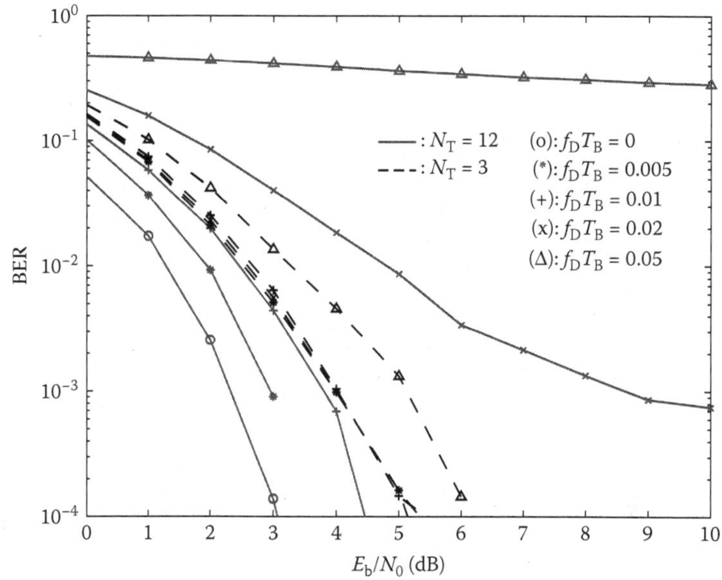

Figure 3.27 Coded BER performance for a Turbo FDE with $L = 1$ in the presence of Doppler effects [$T_B = (N + N_G)T_S$ is the overall block duration].

speeds it is better to use smaller values of N_T (although the required value of η_T increases, we assumed that $\eta_T = 1/12$ for $N_T = 12$ and 1/3 for $N_T = 3$).

3.4 Channel Estimation for CDMA Transmissions

In this section we will briefly study channel estimation in CDMA systems. Figure 3.28 shows a typical CDMA transmitter chain. An information sequence is first encoded using a powerful channel code such as a turbo code [Berrou et al. 1993]. Rate matching (usually puncturing) is applied to the encoded sequence for achieving the desired bit rate. After rate matching, the sequence is segmented into P physical channels [each physical channel will be spreaded by a different orthogonal variable spreading code (OVSF)], which are individually interleaved. The sequence of symbols goes into the modulation mapper that maps each consecutive pair of bits into a QPSK symbol. The modulated symbols are then spreaded and scrambled and the resulting physical channels are summed. Before transmission, pilot symbols are added to the sequence.

The two pilot transmission methods previously described, namely data multiplexed and superimposed pilots, can be both employed in CDMA systems as considered by third-generation WCDMA networks. For example, in the downlink connection of UMTS, a pilot channel, named common pilot channel (CPICH) [3GPP 2004a], is transmitted in parallel with the data channels for channel estimation purposes (as shown in Figure 3.29), and time-multiplexed pilot symbols are also transmitted in dedicated traffic channels

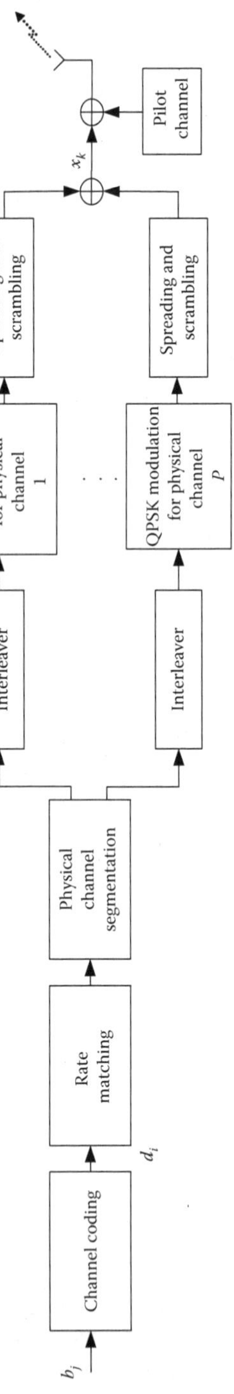

Figure 3.28 CDMA transmitter chain.

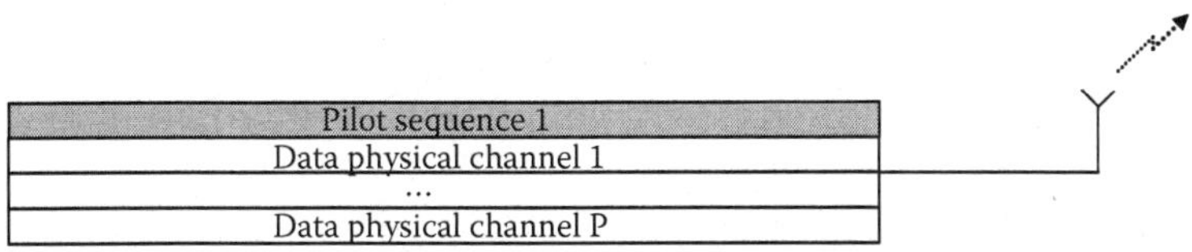

Figure 3.29 Transmission of a pilot channel orthogonal to the data channel in the CDMA systems.

[3GPP 2004a]. These time-multiplexed pilot bits can become useful when the mobile is at a cell edge. In this situation, the use of the CPICH for channel estimation can become insufficient as it is not power controlled, whereas the dedicated channels are. The transmission of a pilot channel parallel to the data channels without interference between them is possible in CDMA systems by reserving an orthogonal spreading code specifically for it. For example, in the downlink connection of UMTS, the CPICH [3GPP 2004a] uses a reserved channelization code with a spreading factor of 256. However, the orthogonality between data and pilot channel is destroyed in a multipath environment, although this interference can be substantially reduced by using adequate signal processing schemes at the receiver, as we will show later on in Chapter 4.

Figure 3.30 shows a basic receiver for a CDMA system whose transmitter follows a structure similar to the one presented in Figure 3.28. It is composed of a RAKE (described in Chapter 2) followed by all the blocks that accomplish the reverse functions of the ones in the transmitter.

Assuming the transmission of a pilot channel, channel estimation can be performed using an approach similar to the one described in Section 3.3 for superimposed pilots, taking into account in this case the specificities of CDMA. Considering that a spreading factor of $\mathrm{SF}_{\mathrm{pilot}}$ is employed for the pilot channel, using spreading sequence $c_{\mathrm{pilot},t'}$ ($t' = 1, \ldots, \mathrm{SF}_{\mathrm{pilot}}$) then, the channel coefficients can be estimated by first despreading the pilot channel, multiplying the despreaded pilot symbols, $s_{\mathrm{pilot},t'}$, by its conjugates and then filtering the noisy channel estimates with a moving average filter with size W. Following this procedure, the channel estimate in each pilot symbol position t can be expressed as

$$\hat{\alpha}(t) = \frac{1}{W} \sum_{i=k-\lfloor W/2 \rfloor}^{k+\lceil W/2 \rceil - 1} \hat{\alpha}^{\mathrm{noisy}}(i), \tag{3.76}$$

where

$$\hat{\alpha}^{\mathrm{noisy}}(i) = \frac{1}{\mathrm{SF}_{\mathrm{pilot}}} \cdot \frac{s^*_{\mathrm{pilot},t}}{\left|s_{\mathrm{pilot},t}\right|^2} \cdot \sum_{t'=1}^{\mathrm{SF}_{\mathrm{pilot}}} r_{t'} \cdot c^*_{\mathrm{pilot},t'}, \tag{3.77}$$

and $r_{t'}$ is the signal sample received in chip position t'. It is important to note that the moving average filter is approximately equivalent to the MMSE filter when $f_{\mathrm{D}} T_{\mathrm{s}} \ll 1$ and $W \ll 1/f_{\mathrm{D}} T_{\mathrm{s}}$ since the solution to Equation 3.46 becomes $h_t^i \approx 1/W$, $\forall i$. Given that the data channels can have different data rates compared to the pilot symbol rate

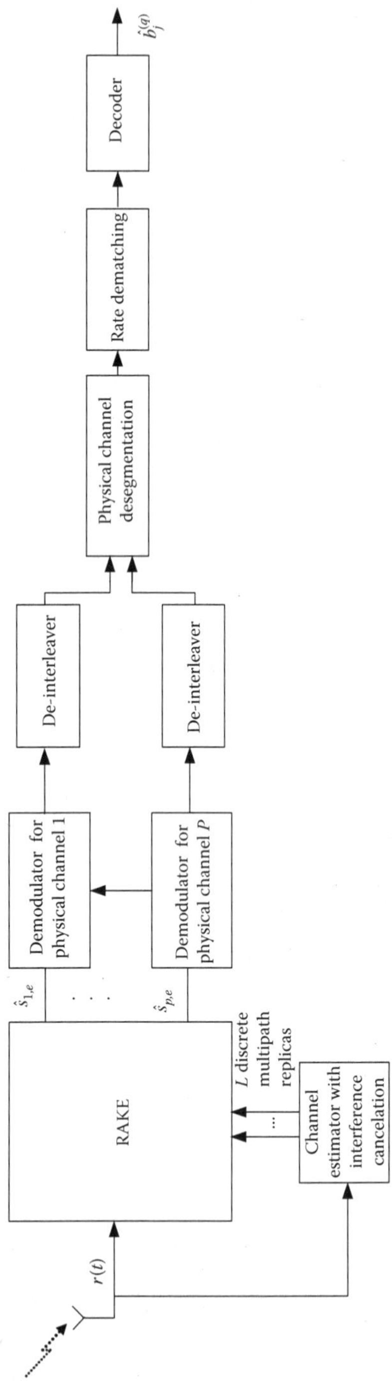

Figure 3.30 Basic CDMA receiver structure.

(the spreading factor of the pilot channel is usually higher than that of the data channels), an interpolation can be performed over the channel estimates for matching the rates. If the environment is frequency selective, that is, if it has several resolvable paths, then the estimation process has to be employed for each of the paths.

To evaluate the impact of the pilot channel-based channel estimation method on a CDMA system, we present in Figures 3.31 and 3.32 several performance curves for the Vehicular A environment (all taps with different powers) obtained using expressions 3.6 and 3.42. Although the expressions do not take into account possible intersymbolic interference due to multipath, they provide very close bounds for high spreading factors; these expressions can also be accurate when interference cancelation scheme is employed. To simplify the analysis, we assume a filter window size of $W = 1$, that is, no filtering is applied except for the averaging operation inherent in the despreading of the pilot symbols, as can be seen in Equation 3.77. Moreover, no interpolation is performed for matching the pilot symbols rate and the data symbols rate, which means that the channel estimate obtained for a pilot symbol (using Equation 3.77) is directly used for all the data symbols inside that pilot symbol duration. The moments involving the channel estimate can be expressed as

$$E\left[\hat{\alpha}(t)\hat{\alpha}^*(t)\right] = \Omega^2 \frac{\left|c_{\text{pilot}}\right|^2}{\text{SF}_{\text{pilot}}} \sum_{j=1}^{\text{SF}_{\text{pilot}}} J_0\left(2\pi f_D \left|(t-1)\cdot\text{SF} + \frac{\text{SF}+1}{2} - j\right| T_{\text{chip}}\right) \tag{3.78}$$

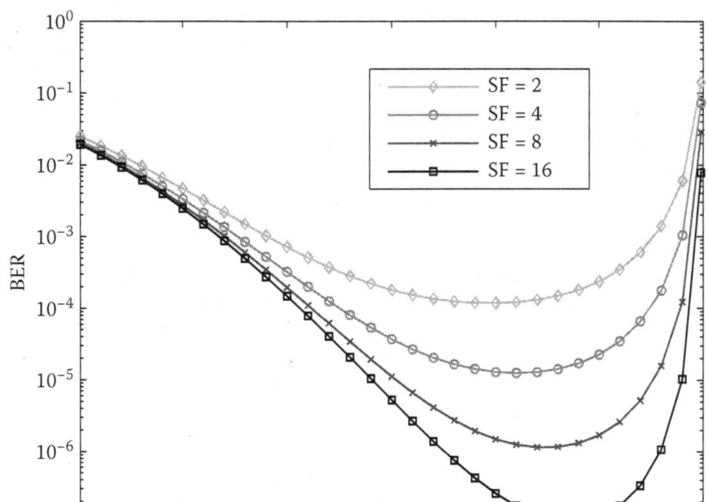

Figure 3.31 BER versus fraction of pilot channel power for uncoded CDMA transmission in Vehicular A environment, with 1 physical channel, $W = 1$, total $E_c/N_0 = 15$ dB, and $f_D \cdot T_{\text{Chip}} = 6 \times 10^{-5}$.

and

$$E\left[r(t)\hat{\alpha}^*(t)|s\right] = \Omega^2 \frac{\left|c_{\text{pilot}}\right|^4}{\text{SF}^2_{\text{pilot}}} \sum_{j=1}^{\text{SF}_{\text{pilot}}} \sum_{j'=1}^{\text{SF}_{\text{pilot}}} J_0\left(2\pi f_{\text{D}}|j-j'|T_{\text{chip}}\right) + \frac{\left|c_{\text{pilot}}\right|^2}{\text{SF}_{\text{pilot}}\left|s_{\text{pilot}}\right|^2} N_0, \tag{3.79}$$

where we have assumed that both the spreading sequence and the pilot symbols have constant energy (position index can be dropped in $\left|c_{\text{pilot},t'}\right|^2$ and $\left|s_{\text{pilot},t}\right|^2$).

Figure 3.31 shows the BER as a function of the fraction of total transmitted power allocated to the pilot channel, for different spreading factors. The average energy per chip, E_c/N_0, considered was 15 dB. This value incorporates data channel chip energy added to pilot channel chip energy. We can observe that the lower the spreading factor, the higher the BER since E_c/N_0 and the BW are always the same for all the curves, which means that E_s/N_0 is in fact lower for lower spreading factors. It is also visible that if either an excessively low or an excessively high fraction of the total power is allocated to the pilot, severe performance degradation results. The first situation corresponds to higher E_s/N_0 values for the data symbols but lower E_s/N_0 values for the pilot symbols, which reduces the reliability of the channel estimates. In the second situation, although the energy applied to the pilots is high and allows a more accurate channel estimation,

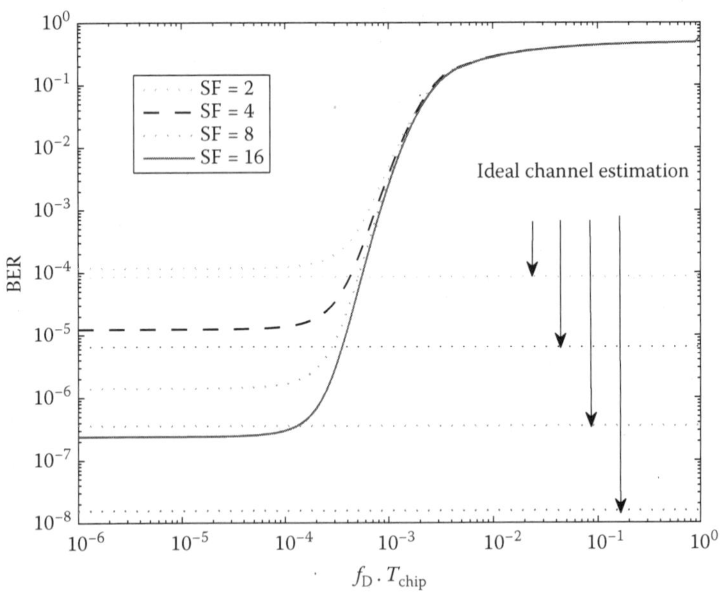

Figure 3.32 BER versus normalized Doppler frequency ($f_D \cdot T_{Chip}$) for uncoded CDMA transmission in Vehicular A environment, with –10 dB pilot channel power and total $E_c/N_0 = 15$ dB.

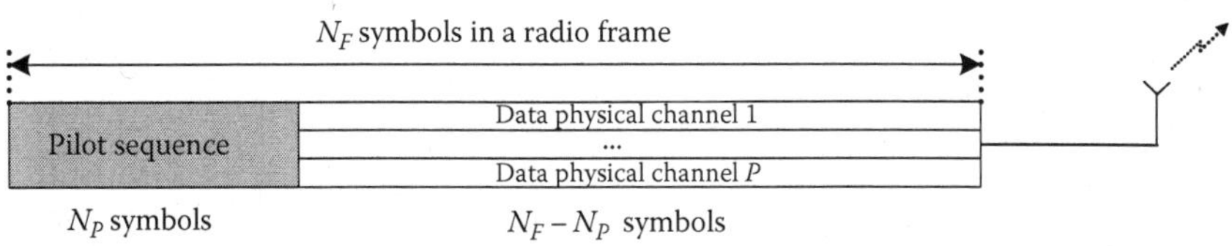

Figure 3.33 Possible frame structure for time-multiplexed pilot transmission in CDMA systems.

the E_s/N_0 for the data symbols has to be reduced to maintain the same total transmitted power, which results in a decreased performance.

Regarding Figure 3.32, it evaluates the impact of mobility on channel estimation based on pilot symbols for CDMA systems. With this aim, the BER is shown as a function of the normalized Doppler frequency, $f_D \cdot T_{\text{Chip}}$, for different spreading factors and considering an average total E_c/N_0 of 15 dB. For normalized Doppler rates below 10^{-4}, the performance is almost insensitive to changes in the velocity of the user since the channel is almost constant during a window of SF_{pilot} chips. Above this value, the channel stops behaving as constant inside the time interval of SF_{pilot} chips and the BER starts to deteriorate very quickly, making the transmissions unreliable. Also, higher differences in the perfect channel estimation curves are visible in the graph when the spreading factor increases. This is due to the fact that for a specific E_c/N_0 value, the channel estimates will be the same for different spreading factors, with the same level of noise. Given that higher spreading factors correspond to higher E_s/N_0 values, a larger performance degradation is visible for these cases since the channel estimation error will be proportionally higher relative to simple thermal noise.

As previously stated, alternatively or additionally to the use of pilot channels it is also possible to transmit data multiplexed pilots in CDMA systems. The procedure employed can be the one shown in Figure 3.33, where a frame of N_F symbols is used, with the first N_P symbols being pilots and the following $N_F - N_P$ symbols being data symbols (these can correspond to several orthogonal data channels/physical channels). All the symbols in the frame are spreaded as usual. At the receiver, channel estimates can be computed using these pilot symbols. The channel estimates for the data symbols located between two consecutive sets of pilot symbols can be computed through interpolation using an FIR filter, as described in Section 3.2.

Chapter 4

Hierarchical Constellations

Hierarchical constellations constitute a simple and flexible technique for achieving multiresolution in a wireless system. Although several modulations can be used for obtaining multiresolution, such as M-ary phase shift keying (M-PSK) and multilevel differential amplitude and phase shift keying (M-DAPSK), in this chapter we focus particularly on M-QAM, which we will refer to as multilevel hierarchical quadrature amplitude modulation (M-HQAM). The chapter starts with a description of the construction of these constellations. The performance of M-HQAM in different propagation environments, namely in additive white Gaussian noise (AWGN), flat Rayleigh, and multipath, is analyzed in Section 4.2. The impact of imperfect channel estimation is evaluated in Section 4.3 where some of the different channel estimation strategies discussed in Chapter 3 are applied.

4.1 Constellation Design

In hierarchical constellations there are two or more classes of bits with different error protection levels onto which different streams of information can be mapped. By using nonuniformly spaced signal points (where the distances along the I- or Q-axis between adjacent symbols are different) it is possible to modify the different error protection levels. Figure 4.1 shows the example of a 16-HQAM constellation.

As in any 16-QAM constellation, four bits are required for selecting a symbol. Two bits select one of the four inner QPSK constellations, while the other two bits select the symbol inside that inner QPSK constellation. Since we are assuming a square constellation, there are two classes of bits with different error protection. The basic idea is that the constellation can be viewed as a 16-QAM constellation if the channel conditions are good enough or as a QPSK constellation otherwise. In the latter case, the received bit rate is reduced to half. Uniformly or nonuniformly spaced signal points can be used depending on the desired differentiation between the error protection levels. Working with inner distances, D_1 and D_2, defined in Figure 4.1, each symbol s of the constellation can be written as

$$s = \left(\pm\frac{D_2}{2} \pm \frac{D_1}{2}\right) + \left(\pm\frac{D_2}{2} \pm \frac{D_1}{2}\right) i. \tag{4.1}$$

Instead of defining a hierarchical constellation through the inner constellation distances, it is usually more convenient to characterize it using ratios between these distances. For example, the 16-HQAM constellation of Figure 4.1 can be defined using $k_1 = D_1/D_2 (0 < k_1)$. Changing this parameter is equivalent to changing distances D_1 and D_2 by modifying the degree of protection of the different bits. If $k_1 = 0.5$, the resulting constellation corresponds to a uniform 16-QAM constellation. This approach can be easily extended to any square M-QAM constellation resulting in the following general expression for a symbol:

$$s = \sum_{l=1}^{1/2\cdot\log_2 M} \left(\pm\frac{D_l}{2}\right) + \sum_{l=1}^{1/2\cdot\log_2 M} \left(\pm\frac{D_l}{2}\right) i, \tag{4.2}$$

where the number of possible classes of bits with different error protection is $1/2 \cdot \log_2 M$. These constellations can be characterized by the ratios between the distances of the inner

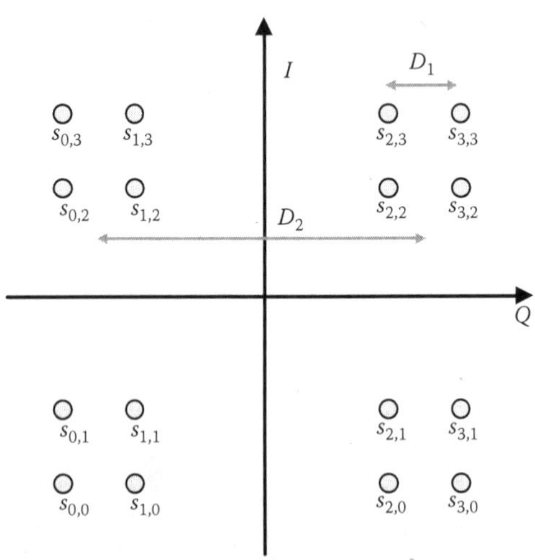

Figure 4.1 Signal constellation for 16-QAM hierarchical modulation.

nested constellations as

$$k_i = \frac{D_i}{D_{i+1}}, \quad i = 1, \ldots, 1/2 \cdot \log_2 M - 1. \tag{4.3}$$

The previous description is adequate for square constellations. However, we can also employ rectangular constellations. A rectangular constellation is obtained when the number of bits mapped to the I and Q branches is not the same or when the inner distances along these axes are different. In any case, the number of different error protection levels becomes $\log_2 M$. As an example, Figure 4.2 shows a rectangular 32-HQAM constellation. It can be constructed as a hierarchical 8-PAM constellation with three different protection levels along the I-axis combined with a hierarchical 4-PAM constellation with two different protection levels along the Q-axis. The constellation can be characterized using different nonuniformity ratios for the I and Q branches, namely $k_1^{\mathrm{I}} = D_1^{\mathrm{I}}/D_2^{\mathrm{I}}$, $k_2^{\mathrm{I}} = D_2^{\mathrm{I}}/D_3^{\mathrm{I}}$, and $k_1^{\mathrm{Q}} = D_1^{\mathrm{Q}}/D_2^{\mathrm{Q}} (0 < k_1^{\mathrm{I}}, k_2^{\mathrm{I}}, k_1^{\mathrm{Q}})$ with the case $k_1^{\mathrm{I}} = k_2^{\mathrm{I}} = k_1^{\mathrm{Q}} = 0.5$ corresponding to a uniform constellation. Each symbol can be

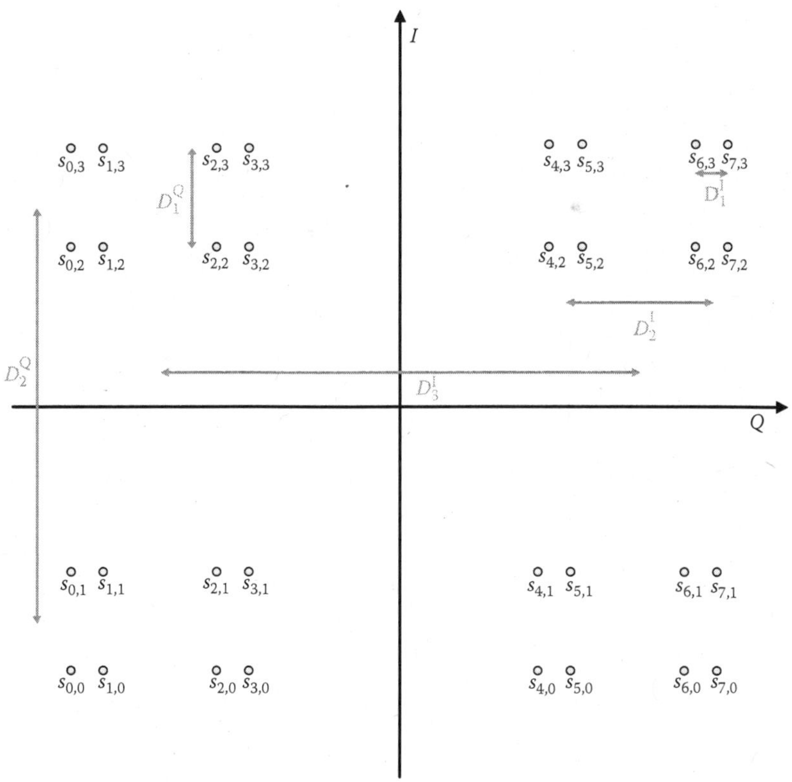

Figure 4.2 Signal constellation for a 32-QAM hierarchical constellation.

written as

$$s = \left(\pm\frac{D_3^{\mathrm{I}}}{2} \pm \frac{D_2^{\mathrm{I}}}{2} \pm \frac{D_1^{\mathrm{I}}}{2}\right) + \left(\pm\frac{D_2^{\mathrm{Q}}}{2} \pm \frac{D_1^{\mathrm{Q}}}{2}\right) i. \tag{4.4}$$

Extending this construction approach to any M-QAM rectangular constellation, the required nonuniformity ratios become

$$\begin{cases} k_i^{\mathrm{I}} = \dfrac{D_i^{\mathrm{I}}}{D_{i+1}^{\mathrm{I}}}, & i = 1, \ldots, \log_2 M_{\mathrm{PAM}}^{\mathrm{I}} - 1 \\ k_j^{\mathrm{Q}} = \dfrac{D_j^{\mathrm{Q}}}{D_{j+1}^{\mathrm{Q}}}, & j = 1, \ldots, \log_2 M_{\mathrm{PAM}}^{\mathrm{Q}} - 1 \end{cases}, \tag{4.5}$$

with the symbols expressed as

$$s = \sum_{l=1}^{\log_2 M_{\mathrm{PAM}}^{\mathrm{I}}} \left(\pm\frac{D_l^{\mathrm{I}}}{2}\right) + \sum_{l=1}^{\log_2 M_{\mathrm{PAM}}^{\mathrm{Q}}} \left(\pm\frac{D_l^{\mathrm{Q}}}{2}\right) i, \tag{4.6}$$

where $M_{\mathrm{PAM}}^{\mathrm{I}}$ and $M_{\mathrm{PAM}}^{\mathrm{Q}}$ are the number of projected symbols onto the I- and Q-axis of the constellation, that is, the sizes of the equivalent PAM constellations. For a square constellation, we have $M_{\mathrm{PAM}}^{\mathrm{I}} = M_{\mathrm{PAM}}^{\mathrm{Q}} = \sqrt{M}$. After defining a square/rectangular constellation through these ratios, we can build the constellation with a specified average symbol energy E_s by computing the corresponding inner distances D_i^{I} and D_i^{Q}. To show how these are obtained, we must start by writing the expression for the average symbol energy, which, due to symmetry, only requires taking into account the symbols inside one quadrant, resulting in

$$\begin{aligned}
E_s = E\left[|s|^2\right] &= \frac{4}{M}\sum_{l=1}^{M/4} |s_l|^2, \\
&= \frac{2}{M_{\mathrm{PAM}}^{\mathrm{I}}} \sum_{l=1}^{M_{\mathrm{PAM}}^{\mathrm{I}}/2} \left(s_l^{\mathrm{I}}\right)^2 + \frac{2}{M_{\mathrm{PAM}}^{\mathrm{Q}}} \sum_{l=1}^{M_{\mathrm{PAM}}^{\mathrm{Q}}/2} \left(s_l^{\mathrm{Q}}\right)^2, \\
&= \frac{2}{M_{\mathrm{PAM}}^{\mathrm{I}}} \sum_{l=1}^{M_{\mathrm{PAM}}^{\mathrm{I}}/2} \left(\sum_{i=1}^{\log_2 M_{\mathrm{PAM}}^{\mathrm{I}}} \left(\pm\frac{D_i^{\mathrm{I}}}{2}\right)\right)^2 + \frac{2}{M_{\mathrm{PAM}}^{\mathrm{Q}}} \sum_{l=1}^{M_{\mathrm{PAM}}^{\mathrm{Q}}/2} \left(\sum_{i=1}^{\log_2 M_{\mathrm{PAM}}^{\mathrm{Q}}} \left(\pm\frac{D_i^{\mathrm{Q}}}{2}\right)\right)^2, \\
&= \frac{1}{2M_{\mathrm{PAM}}^{\mathrm{I}}} \left[\sum_{l=1}^{M_{\mathrm{PAM}}^{\mathrm{I}}/2} \sum_{i=1}^{\log_2 M_{\mathrm{PAM}}^{\mathrm{I}}} \left(D_i^{\mathrm{I}}\right)^2 + \sum_{l=1}^{M_{\mathrm{PAM}}^{\mathrm{I}}/2} \sum_{k=1}^{\log_2 M_{\mathrm{PAM}}^{\mathrm{I}}} \sum_{i=1}^{\log_2 M_{\mathrm{PAM}}^{\mathrm{I}}} \left(\pm D_k^{\mathrm{I}}\right)\left(\pm D_i^{\mathrm{I}}\right) \right], \\
&\quad + \frac{1}{2M_{\mathrm{PAM}}^{\mathrm{Q}}} \left[\sum_{l=1}^{M_{\mathrm{PAM}}^{\mathrm{Q}}/2} \sum_{i=1}^{\log_2 M_{\mathrm{PAM}}^{\mathrm{Q}}} \left(D_i^{\mathrm{Q}}\right)^2 + \sum_{l=1}^{M_{\mathrm{PAM}}^{\mathrm{Q}}/2} \sum_{k=1}^{\log_2 M_{\mathrm{PAM}}^{\mathrm{Q}}} \sum_{i=1}^{\log_2 M_{\mathrm{PAM}}^{\mathrm{Q}}} \left(\pm D_k^{\mathrm{Q}}\right)\left(\pm D_i^{\mathrm{Q}}\right) \right],
\end{aligned}$$

$$= \frac{1}{4}\left[\sum_{i=1}^{\log_2 M_{\text{PAM}}^{\text{I}}} \left(D_i^{\text{I}}\right)^2 + \sum_{i=1}^{\log_2 M_{\text{PAM}}^{\text{Q}}} \left(D_i^{\text{Q}}\right)^2\right]. \tag{4.7}$$

We can combine Equation 4.7 with the definition of the nonuniformity ratios Equation 4.5 and obtain an expression that only depends on one of the inner distances on each axis:

$$E_s = \frac{1}{4}\left[\left(D^{\text{I}}_{\log_2 M_{\text{PAM}}^{\text{I}}}\right)^2 + \sum_{i=1}^{\log_2 M_{\text{PAM}}^{\text{I}}-1}\left(D^{\text{I}}_{\log_2 M_{\text{PAM}}^{I}} \prod_{j=i}^{\log_2 M_{\text{PAM}}^{\text{I}}-1} k_j^{\text{I}}\right)^2 \right.$$
$$\left. + \left(D^{\text{Q}}_{1/2\cdot\log_2 M_{\text{PAM}}^{\text{Q}}}\right)^2 + \sum_{i=1}^{\log_2 M_{\text{PAM}}^{\text{Q}}-1}\left(D^{\text{Q}}_{\log_2 M_{\text{PAM}}^{\text{Q}}} \prod_{j=i}^{\log_2 M_{\text{PAM}}^{\text{Q}}-1} k_j^{\text{Q}}\right)^2\right]. \tag{4.8}$$

The first two terms in Equation 4.8 correspond to the average symbol energy in the I branch, while the other two refer to the average symbol energy in the Q branch. Therefore, we can also write

$$E_s = E_s^{\text{I}} + E_s^{\text{Q}}, \tag{4.9}$$

with

$$E_{\text{s}}^{\text{I/Q}} = \frac{\left(D^{\text{I/Q}}_{\log_2 M_{\text{PAM}}^{\text{I/Q}}}\right)^2}{4}\left[1 + \sum_{i=1}^{\log_2 M_{\text{PAM}}^{\text{I/Q}}-1} \prod_{j=i}^{\log_2 M_{\text{PAM}}^{\text{I/Q}}-1} \left(k_j^{\text{I/Q}}\right)^2\right]. \tag{4.10}$$

The desired inner distances can then be easily computed using

$$D^{\text{I/Q}}_{\log_2 M_{\text{PAM}}^{\text{I/Q}}} = \sqrt{\frac{4E_{\text{s}}^{\text{I/Q}}}{1 + \sum_{l=1}^{\log_2 M_{\text{PAM}}^{\text{I/Q}}-1} \prod_{j=l}^{\log_2 M_{\text{PAM}}^{\text{I/Q}}-1} \left(k_j^{\text{I/Q}}\right)^2}},$$
$$D_i^{\text{I/Q}} = D^{\text{I/Q}}_{\log_2 M_{\text{PAM}}^{\text{I/Q}}} \prod_{j=i}^{\log_2 M_{\text{PAM}}^{\text{I/Q}}-1} k_j^{\text{I/Q}}, \quad i < \log_2 M_{\text{PAM}}^{\text{I/Q}}. \tag{4.11}$$

Although in this chapter we focus on the QAM constellation, we will briefly show how we can apply a similar construction approach and obtain other hierarchical constellations, such as hierarchical PSK (M-HPSK). PSK symbols lie on the same circumference on the complex plan; only the phase differs between them. Following the procedure described in Ref. [Pursley and Shea 1999], we can start with a BPSK constellation with two symbols spaced by π radians, and split each symbol into two new ones spaced by some angle θ_2 as shown in Figure 4.3. This results in a hierarchical QPSK constellation where the closest symbols are spaced by an angle θ_2 and the others are spaced by at least $\pi - \theta_2$. We can

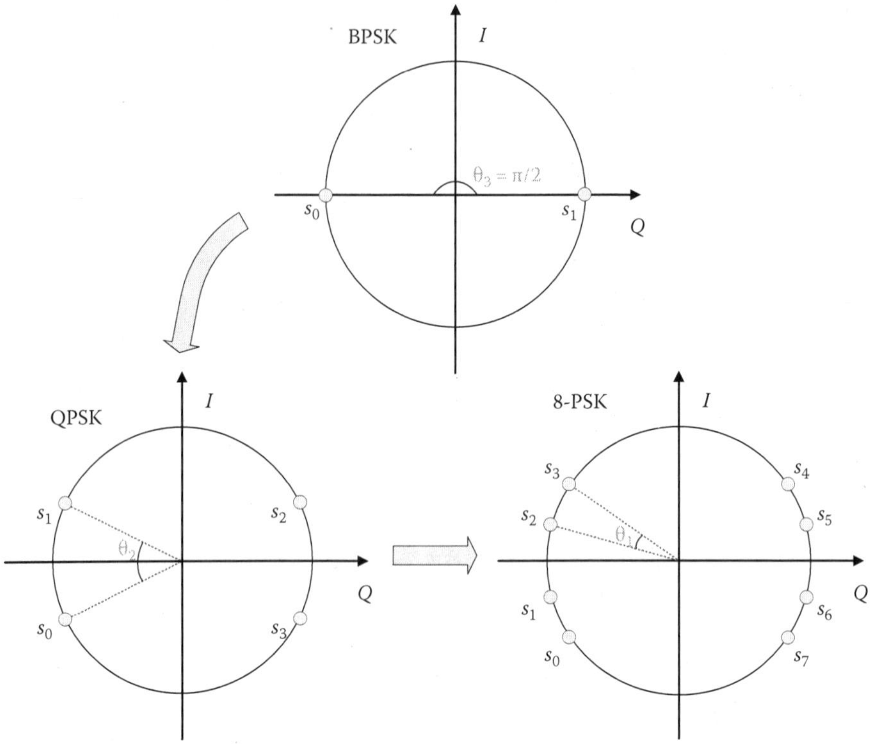

Figure 4.3 Signal constellation construction for an 8-PSK hierarchical modulation.

then split each symbol again into two new ones and space them by an angle θ_1, in order to obtain an 8-HPSK constellation with three differently protected bits.

A complex 8-HPSK constellation symbol can be written as

$$s = \sqrt{E_s} \exp\left[i\left(\frac{\pi}{2} \pm \frac{\pi}{2} \pm \frac{\theta_2}{2} \pm \frac{\theta_1}{2}\right)\right], \tag{4.12}$$

with the nonuniformity ratios defined by following a similar approach to HQAM as $k_1 = \theta_1/\theta_2$ and $k_2 = \theta_2/\theta_3$. Typically, a constellation is defined with $\theta_3 = \pi/2$ to minimize the error probability for the most protected bit (MPB). Generalizing to higher-order constellations, we can write an M-HPSK symbol as

$$\begin{aligned} s &= \sqrt{E_s} \exp\left[i\left(\frac{\pi}{2} + \sum_{l=1}^{\log_2 M}\left(\pm\frac{\theta_l}{2}\right)\right)\right] \\ &= \sqrt{E_s} \exp\left[i\left(\frac{\pi}{2} \pm \frac{\pi}{2} + \sum_{l=1}^{\log_2 M-1}\left(\pm\frac{\theta_l}{2}\right)\right)\right], \end{aligned} \tag{4.13}$$

with

$$k_i = \frac{\theta_i}{\theta_{i+1}}, \quad i = 1, \ldots, 1/2 \cdot \log_2 M - 1. \tag{4.14}$$

4.2 BER Analysis of Hierarchical *M*-QAM Constellations

In this section, we will derive expressions for computing the bit error probability of M-HQAM constellations in several channels. We will assume independent mapping of the bits to the I and Q branches of the constellation so that half of each stream goes for the in-phase branch and the other for the quadrature branch of the modulator. This means that we can analyze an M-QAM constellation as two independent $\sqrt{M}$-PAM constellations for obtaining the bit error probability. Owing to the separation between the I and Q branches, the following expressions are also directly applicable to rectangular constellations, only requiring the substitution of $\sqrt{M}$ by either $M_{\text{PAM}}^{\text{I}}$ or $M_{\text{PAM}}^{\text{Q}}$ as well as using the corresponding inner constellation distances D_i^{I} and D_i^{Q}. We will also admit that the resulting bit sequence in each branch is Gray encoded and mapped to the respective $\sqrt{M}$-PAM constellation symbols. This encoding is performed according to the procedure described in Ref. [Vitthaladevuni and Alouini 2003]. First, the constellation symbols are represented in the horizontal axis and are labeled from left to right with integers starting from 0 to $\sqrt{M} - 1$. These labels are subsequently converted into their binary representation so that each symbol s_j can be represented by a $1/2 \cdot \log_2 M$-digit binary sequence: $b_j^1, b_j^2, \ldots, b_j^{1/2 \cdot \log_2 M}$. The corresponding Gray code is then computed using

$$\begin{aligned} g_j^1 &= b_j^1 \\ g_j^i &= b_j^i \oplus b_j^{i-1}, \quad i = 2, 3, \ldots, 1/2 \cdot \log_2 M, \end{aligned} \tag{4.15}$$

where $\oplus$ represents modulo-2 addition.

4.2.1 AWGN Channel

We will start by analyzing the performance of 16-HQAM in AWGN channels and later we provide the general expression for any M-HQAM constellation.

4.2.1.1 16-HQAM

Figure 4.4 shows a 4-PAM constellation representing the I or Q branch of a 16-HQAM constellation with the respective decision regions for each of the two bit types.

Considering the transmission of a PAM symbol, s, the received signal in the presence of AWGN can be modeled as

$$r = s + n, \tag{4.16}$$

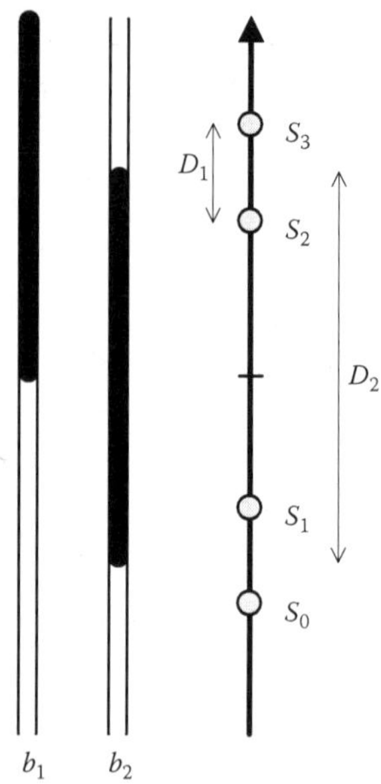

Figure 4.4 Decision regions for a 16-QAM constellation projected into the *I* branch. The dark areas correspond to "1" and the white areas to "0".

where n is a Gaussian variable with zero mean and variance $\sigma^2 = N_0/2$ ($N_0/2$ is the two-sided noise power spectrum density). Admitting equiprobable symbols, $\text{Prob}(s_l) = 1/\sqrt{M}$ for any l, the general bit error probability for bit class m can be expressed as

$$P_b(b_m) = \sum_{l=1}^{\sqrt{M}} \text{Prob}(b_m, s_l) = \sum_{l=1}^{\sqrt{M}} \text{Prob}(b_m \mid s_l)\text{Prob}(s_l) = \frac{1}{\sqrt{M}} \sum_{l=1}^{\sqrt{M}} \text{Prob}(b_m \mid s_l). \tag{4.17}$$

According to the decision borders shown in Figure 4.4, the conditional error probabilities of the different bit classes can be computed as

$$\begin{aligned} P_e\{b_1 \mid s_1\} = P_e\{b_1 \mid s_4\} &= \text{Prob}\left\{n < -\left(\frac{D_2}{2} + \frac{D_1}{2}\right)\right\} \\ &= Q\left(\sqrt{\frac{2}{N_0}}\left(\frac{D_2}{2} + \frac{D_1}{2}\right)\right), \end{aligned} \tag{4.18}$$

$$\begin{aligned} P_e\{b_1 \mid s_2\} = P_e\{b_1 \mid s_3\} &= \text{Prob}\left\{n < -\left(\frac{D_2}{2} - \frac{D_1}{2}\right)\right\} \\ &= Q\left(\sqrt{\frac{2}{N_0}}\left(\frac{D_2}{2} - \frac{D_1}{2}\right)\right), \end{aligned} \tag{4.19}$$

$$P_e\{b_2 \mid s_1\} = P_e\{b_2 \mid s_4\} = \text{Prob}\left\{-\left(D_2 + \frac{D_1}{2}\right) < n < -\frac{D_1}{2}\right\}$$
$$= Q\left(\sqrt{\frac{2}{N_0}}\frac{D_1}{2}\right) - Q\left(\sqrt{\frac{2}{N_0}}\left(D_2 + \frac{D_1}{2}\right)\right), \quad (4.20)$$
$$P_e\{b_2 \mid s_2\} = P_e\{b_2 \mid s_3\} = \text{Prob}\left\{n > \frac{D_1}{2}\right\} + \text{Prob}\left\{n < -\left(D_2 - \frac{D_1}{2}\right)\right\}$$
$$= Q\left(\sqrt{\frac{2}{N_0}}\frac{D_1}{2}\right) + Q\left(\sqrt{\frac{2}{N_0}}\left(D_2 - \frac{D_1}{2}\right)\right). \quad (4.21)$$

The overall probability of error for bit 1 is then given as

$$P_b(b_1) = \frac{1}{2}\cdot\left[Q\left(\sqrt{\frac{2}{N_0}}\left(\frac{D_2}{2} + \frac{D_1}{2}\right)\right) + Q\left(\sqrt{\frac{2}{N_0}}\left(\frac{D_2}{2} - \frac{D_1}{2}\right)\right)\right], \quad (4.22)$$

while probability of error for bit 2 is

$$P_b(b_2) = \frac{1}{2}\cdot\left[2\cdot Q\left(\sqrt{\frac{2}{N_0}}\frac{D_1}{2}\right) - Q\left(\sqrt{\frac{2}{N_0}}\left(D_2 + \frac{D_1}{2}\right)\right)\right.$$
$$\left. + Q\left(\sqrt{\frac{2}{N_0}}\left(D_2 - \frac{D_1}{2}\right)\right)\right]. \quad (4.23)$$

As an example, Figure 4.5 shows the BER performance of the different bits for $k_1 = 0.3$ (MPB denotes most protected bit, while LPB denotes the least protected bits) and compares it against the performance of QPSK and of a conventional nonhierarchical 16-QAM constellation. It is visible that the performance of the LPBs has worsened compared with the average BER of a typical 16-QAM modulation. However, the performance of the MPBs improved, becoming halfway between the average BER of 16-QAM and QPSK.

4.2.1.2 M-HQAM

The derivation presented for 16-HQAM can be easily extended to higher-level constellations as was done in Refs. [Vitthaladevuni and Alouini 2001; Vitthaladevuni and Alouini 2003; Vitthaladevuni and Alouini 2004]. It is simple to verify that for any constellation size, the BER is always computed as a weighted sum of $Q(x)$ functions. From the constellation construction pattern, an explicit closed-form expression was obtained for the BER of the different bits in an M-HQAM constellation in Ref. [Vitthaladevuni and Alouini 2004]. Using the characterization of M-HQAM constellations given previously,

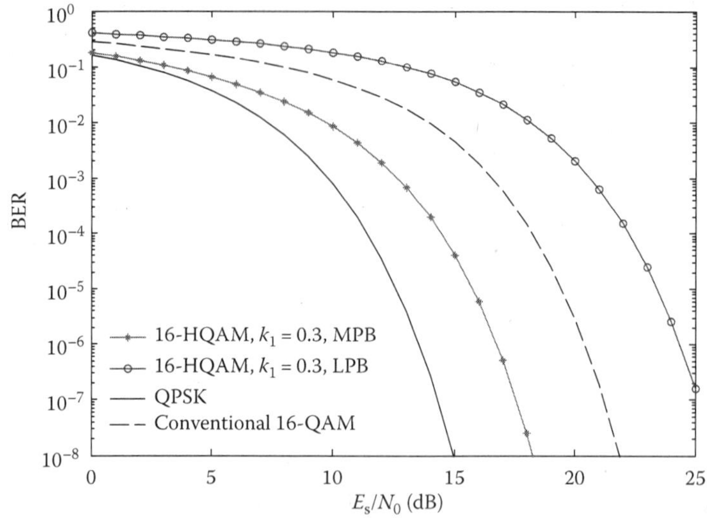

Figure 4.5 BER performance of 16-HQAM, $k_1 = 0.3$, in AWGN.

we can rewrite the general expression from Ref. [Vitthaladevuni and Alouini 2004] as

$$P_b(b_m) = \frac{2}{\sqrt{M}} \sum_{j=0}^{\sqrt{M}/2-1} \left[g_j^m + (-1)^{g_j^m} \sum_{l=1}^{2^{(m-1)}} \left[(-1)^{l+1} \times Q\left(\sqrt{\frac{2}{N_0}} \left[\mathbf{B}_m(l) - \mathbf{D}_s(j) \right] \right) \right] \right], \quad m = 1, \ldots, \log_2 M/2, \quad (4.24)$$

where

$$\mathbf{B}_m(l) = \frac{\mathbf{D}_s\left((2l-1)2^{1/2\cdot\log_2 M - m} - 1\right) + \mathbf{D}_s\left((2l-1)2^{1/2\cdot\log_2 M - m}\right)}{2}, \quad (4.25)$$

represents the decision borders and

$$\mathbf{D}_s(j) = \sum_{i=1}^{1/2\cdot\log_2 M} (2b_j^i - 1)\frac{D_{1/2\cdot\log_2 M - i + 1}}{2}, \quad (4.26)$$

are coordinates of the symbols.

On the basis of these expressions, Figure 4.6 shows the BER performance of the different bit types for a 64-HQAM constellation with $k1 = 0.3$ and $k2 = 0.4$. As a reference performance curves of conventional QPSK, 16-QAM, and 64-QAM are also shown. In the legends of graphs, IPB denotes intermediate protected bits. The use of a nonuniform constellation with a hierarchical configuration allows us to obtain two-thirds of the bits with better performances than the average 64-QAM BER at the cost of a degradation of the remainder one-third (the least protected ones).

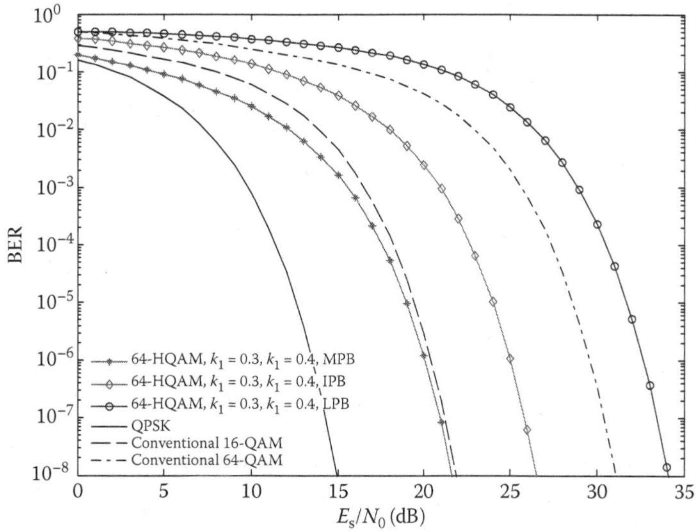

Figure 4.6 BER performance of 64-HQAM, $k_1 = 0.3$ and $k_2 = 0.4$, in AWGN.

4.2.2 Flat Fading Channel with Perfect Channel Estimation

The presence of a fading channel will affect the BER performance of the constellations. We will start by obtaining BER expressions for a flat Rayleigh fading channel and then extend it to other fading statistics. Let us consider that the received signal is distorted in amplitude and phase by the channel according to

$$r = \alpha \cdot s + n, \tag{4.27}$$

where s is the M-HPAM transmitted symbol and n is a Gaussian variable with mean 0 and variance $\sigma^2 = N_0/2$. Regarding the channel coefficient, α, we will start by modeling it as a zero mean complex Gaussian variable, which means that $|\alpha|$ has a Rayleigh distribution, its phase follows a uniform distribution and $|\alpha|^2$ has an exponential distribution. Since the BER expressions obtained previously for AWGN channel are written as a weighted sum of $Q(\cdot)$ functions, it is simple to extend them for flat Rayleigh fading channels. We only need to note that, apart from a phase rotation that can be compensated assuming perfect channel knowledge and coherent detection at the receiver, the channel will result in a multiplication of the constellation symbols by $|\alpha|$, which is equivalent to a multiplication of the instantaneous symbol energy by $|\alpha|^2$. Therefore, the $Q(.)$ function in Equation 4.24 can be rewritten as $Q\left(\sqrt{(2/N_0)|\alpha|^2[\mathbf{B}_m(l) - \mathbf{D}_s(j)]}\right)$ and the resulting overall expression will correspond to the conditional probability $P_b(b_m \mid \gamma)$ with $\gamma = |\alpha|^2$. To obtain the average BER for each bit, it is only necessary to average that expression over the PDF of $|\alpha|^2$, that is,

$$P_b(b_m) = \int_0^{+\infty} P_b(b_m \mid \gamma) p(\gamma)\, d\gamma. \tag{4.28}$$

Since $\gamma = |\alpha|^2$ follows an exponential distribution, we can write

$$p(\gamma) = \frac{1}{\gamma} e^{\gamma/\bar{\gamma}}, \quad \gamma \geq 0, \tag{4.29}$$

where $\bar{\gamma} = E[\gamma]$. It can be shown that

$$\int_0^{+\infty} Q\left(z \cdot \sqrt{\gamma}\right) p(\gamma)\, d\gamma = \begin{cases} \frac{1}{2} - \frac{1}{2}\sqrt{\frac{z^2\bar{\gamma}/2}{1+z^2\bar{\gamma}/2}}, & z > 0 \\ \frac{1}{2} + \frac{1}{2}\sqrt{\frac{z^2\bar{\gamma}/2}{1+z^2\bar{\gamma}/2}}, & z < 0 \end{cases}, \tag{4.30}$$

which, combined with Equations 4.28 and 4.24, results in the following BER expression for the individual bits:

$$P_b(b_m \mid \kappa) = \frac{2}{\sqrt{M}} \sum_{j=0}^{\sqrt{M}/2-1} \left[g_j^m + (-1)^{g_j^m} \sum_{l=1}^{2^{(m-1)}} \left[(-1)^{l+1} \right.\right.$$
$$\times \left(\frac{1}{2} + (-1)^{1/2+1/2\,\mathrm{sign}(\mathbf{B}_m(l)-\mathbf{D}_s(j))} \right.$$
$$\left.\left.\left. \times \frac{1}{2}\sqrt{\frac{1/N_0\left[\mathbf{B}_m(l) - \mathbf{D}_s(j)\right]^2 \bar{\gamma}}{1+(1/N_0)\left[\mathbf{B}_m(l) - \mathbf{D}_s(j)\right]^2 \bar{\gamma}}} \right) \right] \right], \tag{4.31}$$

with sign(x) being the sign function defined as

$$\mathrm{sign}(x) = \begin{cases} 1, & x > 0 \\ 0, & x = 0 \\ -1, & x < 0 \end{cases}. \tag{4.32}$$

As examples, Figures 4.7 and 4.8 show the BER performances of 16-HQAM and 64-HQAM constellations in flat Rayleigh fading channels as a function of the nonuniformity parameters k_1 (16-HQAM) and k_1 and k_2 (64-HQAM). It is clear that by changing the values of the nonuniformity parameters it is possible to improve the error protection of some of the bit streams at the cost of some performance degradation of the others. When $k_1 = 0$ (16-QAM) or $k_1 = k_2 = 0$ (64-QAM), the constellations reduce to QPSK and only the MPB stream can be reliably extracted (the other bits have BERs of 0.5). It is important to note also that $k_1 = 0.5$ (16-QAM) or $k_1 = k_2 = 0.5$ (64-QAM) does not

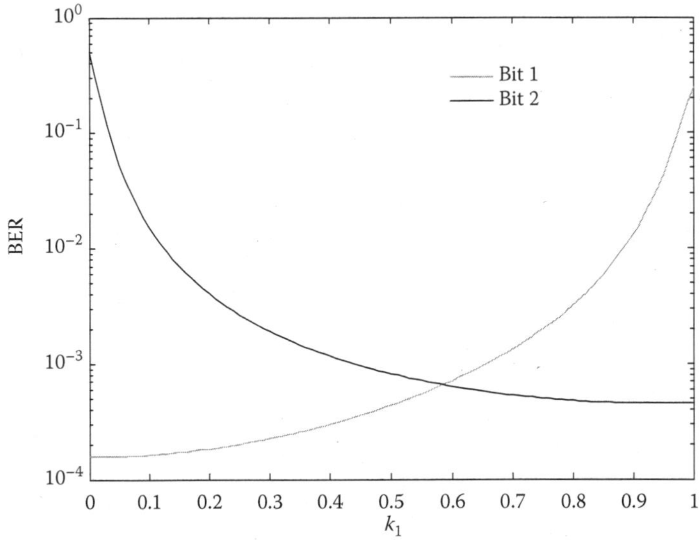

Figure 4.7 BER performances of hierarchical 16-QAM constellations as a function of the nonuniformity ratios k_1 for $E_s/N_0 = 35$ dB.

correspond to equal protection of all the bits; it only means that the constellations are uniform. For example, for the conditions of Figure 4.7, equal protection for the different bit streams is obtained when k_1 is close to 0.6.

We can also extend the BER expressions to other fading environments. In a Nakagami fading environment, the PDF of $a = |\alpha|$ can be written as

$$p(a) = \frac{2}{\Gamma(m)}\left(\frac{m}{\overline{\gamma}}\right)a^{2m-1}e^{-m\cdot a^2/\overline{\gamma}}, \quad a \geq 0, \tag{4.33}$$

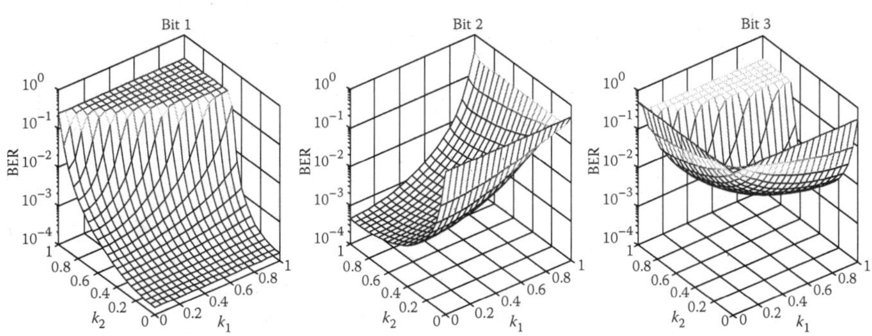

Figure 4.8 BER performances of hierarchical 64-QAM constellations as a function of the nonuniformity ratios k_1 and k_2 for $E_s/N_0 = 35$ dB.

where $\bar{\gamma} = E\left[|\alpha|^2\right]$, m is the Nakagami severity of the fading parameter

$$m = \frac{\bar{\gamma}^2}{E\left[\left(|\alpha|^2 - \bar{\gamma}\right)^2\right]}, \quad m \geq \frac{1}{2}, \tag{4.34}$$

and $\Gamma(\cdot)$ is the gamma function defined as

$$\Gamma(x) = \int_0^{+\infty} t^{x-1} \mathrm{e}^{-t}\, \mathrm{d}t, \quad x > 0. \tag{4.35}$$

It is easy to verify that in this case, $\gamma = |\alpha|^2$ will follow a gamma distribution described as

$$p(\gamma) = \frac{m^m}{\Gamma(m)\,\bar{\gamma}^m} \gamma^{m-1} \mathrm{e}^{-(m\cdot\gamma/\bar{\gamma})}, \quad \gamma \geq 0. \tag{4.36}$$

Using this distribution it is possible to show that

$$\int_0^{+\infty} Q\left(z \cdot \sqrt{\gamma}\right) p(\gamma)\, \mathrm{d}\gamma = \begin{cases} \frac{1}{2}\sqrt{\frac{z^2\bar{\gamma}/2}{\pi}} \frac{m^m}{\left(m + z^2\bar{\gamma}/2\right)^{m+1/2}} \frac{\Gamma(m+1/2)}{\Gamma(m+1)} \\ \quad \times {}_2F_1\left(1, m+1/2; m+1; \frac{m}{m + z^2\bar{\gamma}/2}\right), & z > 0 \\ 1 - \frac{1}{2}\sqrt{\frac{z^2\bar{\gamma}/2}{\pi}} \frac{m^m}{\left(m + z^2\bar{\gamma}/2\right)^{m+1/2}} \frac{\Gamma(m+1/2)}{\Gamma(m+1)} \\ \quad \times {}_2F_1\left(1, m+1/2; m+1; \frac{m}{m + z^2\bar{\gamma}/2}\right), & z < 0, \end{cases} \tag{4.37}$$

where ${}_2F_1\,(\cdot;\cdot;\cdot;\cdot)$ denotes Gauss' hypergeometric function. As a result, the BER expression for the individual bits becomes

$$\begin{aligned} &P_b(b_m \mid \kappa) \\ &= \frac{2}{\sqrt{M}} \sum_{j=0}^{\sqrt{M}/2-1} \left[g_j^m + (-1)^{g_j^m} \sum_{l=1}^{2^{(m-1)}} \left[\frac{1}{2}(-1)^{l+1} \left(1 - \mathrm{sign}\left(\mathbf{B}_m(l) - \mathbf{D}_s(j)\right) \right.\right.\right. \\ &\quad + (-1)^{1/2-1/2\mathrm{sign}(\mathbf{B}_m(l)-\mathbf{D}_s(j))} \sqrt{\frac{1/N_0 \left[\mathbf{B}_m(l) - \mathbf{D}_s(j)\right]^2 \bar{\gamma}}{\pi}} \\ &\quad \times \frac{m^m}{\left(m + (1/N_0)\left[\mathbf{B}_m(l) - \mathbf{D}_s(j)\right]^2 \bar{\gamma}\right)^{m+1/2}} \frac{\Gamma(m+1/2)}{\Gamma(m+1)} \\ &\quad \left.\left.\left. \times {}_2F_1\left(1, m+1/2; m+1; \frac{m}{m + (1/N_0)\left[\mathbf{B}_m(l) - \mathbf{D}_s(j)\right]^2 \bar{\gamma}}\right)\right)\right]\right]. \end{aligned} \tag{4.38}$$

Another important distribution frequently used for characterizing fading channels is the Rice distribution. In this case, $a = |\alpha|$ is composed of two components: one fixed (representing a LOS component), a_0, and one with Rayleigh distribution, $a - a_0$. The PDF of a Rice random variable is expressed as

$$p(a) = \frac{2a}{\sigma_a^2} \mathrm{e}^{-((a^2+a_0^2)/\sigma_a^2)} I_0\left(\frac{a \cdot a_0}{\sigma_a^2/2}\right), \quad a \geq 0, \tag{4.39}$$

with $I_0(\cdot)$ being the zeroth-order modified Bessel function of the first kind and $\sigma_a^2 = E[(a-a_0)^2]$. As a consequence, $\gamma = |\alpha|^2$ follows a noncentral chi-square distribution given by

$$p(\gamma) = \frac{1}{\sigma_a^2} \mathrm{e}^{-((\gamma+a_0^2)/\sigma_a^2)} I_0\left(\frac{\sqrt{\gamma} \cdot a_0}{\sigma_a^2/2}\right), \quad \gamma \geq 0. \tag{4.40}$$

Averaging function $Q(\cdot)$ over this distribution yields

$$\int_0^{+\infty} Q\left(z \cdot \sqrt{\gamma}\right) p(\gamma)\, \mathrm{d}\gamma = \begin{cases} \frac{1}{2} Q(u,v) - \frac{1}{4}\left[1 + \sqrt{\frac{w}{1+w}}\right] \mathrm{e}^{-((u^2+v^2)/2)} I_0(u \cdot v), & z > 0 \\ 1 - \frac{1}{2} Q(u,v) + \frac{1}{4}\left[1 + \sqrt{\frac{w}{1+w}}\right] \mathrm{e}^{-((u^2+v^2)/2)} I_0(u \cdot v), & z < 0, \end{cases} \tag{4.41}$$

with

$$w = \frac{z^2\bar{\gamma}/2}{K+1}$$
$$u = \sqrt{K\left(\frac{1+2w}{2+2w} - \left(\frac{w}{1+w}\right)^{1/2}\right)} \tag{4.42}$$
$$v = \sqrt{K\left(\frac{1+2w}{2+2w} + \left(\frac{w}{1+w}\right)^{1/2}\right)},$$

where $Q(u, v)$ is the Marcum Q-function and K is the Ricean factor defined as $K = a_0^2/\sigma_a^2$. Therefore, for a Rice fading channel, the BER expression for the individual bits becomes

$$P_b(b_m \mid \kappa) = \frac{2}{\sqrt{M}} \sum_{j=0}^{\sqrt{M}/2-1} \left[g_j^m + (-1)^{g_j^m} \sum_{l=1}^{2^{(m-1)}} \left[\frac{1}{2}(-1)^{l+1} \right.\right.$$
$$\times \left(1 - \mathrm{sign}\left(\mathbf{B}_m(l) - \mathbf{D}_s(j)\right) + (-1)^{1/2-1/2\mathrm{sign}(\mathbf{B}_m(l)-\mathbf{D}_s(j))} \right. \tag{4.43}$$
$$\left.\left.\left.\times \left(Q(u,v) - \frac{1}{2}\left[1 + \sqrt{\frac{w}{1+w}}\right] \mathrm{e}^{-((u^2+v^2)/2)} I_0(u \cdot v)\right)\right)\right]\right],$$

with

$$w = \frac{1/N_0 \left[\mathbf{B}_m(l) - \mathbf{D}_s(j)\right]^2 \bar{\gamma}}{K+1}. \tag{4.44}$$

4.2.3 MFB in Multipath Rayleigh Fading Channels

Although it is possible to obtain exact expressions for the performance of hierarchical constellations in flat fading environments, as we have shown in the previous section, extending them to MCs becomes more difficult due to the problem of dealing with ISI. Nevertheless, instead of trying to obtain exact expressions for this type of environment, it can be satisfactory just to have some analytical expressions representing an ideal performance that may not be possible to achieve in practice but can work as a very important benchmark for the evaluation of a specific receiver. The MFB can be regarded as a lower limit on the BER for a particular communication channel and is derived by assuming perfect channel knowledge and the transmission of only one pulse, that is, no ISI occurs. In this subsection, we will derive the MFB for the BERs of the different types of bits (in terms of error protection level) of M-HQAM constellations for diversity reception in time-discrete multipath Rayleigh fading channels with correlated paths.

Let us consider the case of a transmission over an N_{rx}th order diversity branch multipath Rayleigh fading channel where all branches can have different fading powers. The corresponding transmit/receive block diagram is shown in Figure 4.9.

Assuming a discrete MC for each diversity branch n composed of L_n discrete taps, the respective response at time t to an impulse applied at $t - \tau$ can be modeled as

$$c_n(\tau, t) = \sum_{i=1}^{L_n} \alpha_{i,n}(t)\delta(\tau - \tau_{i,n}), \quad n = 1 \ldots N_{rx}, \tag{4.45}$$

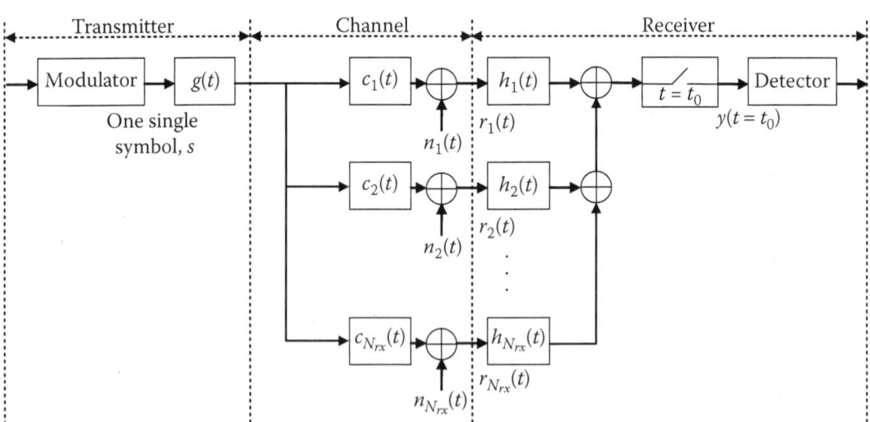

Figure 4.9 Baseband transmit/receive scheme.

where $\alpha_{i,n}(t)$ is a zero-mean complex Gaussian random process, $\delta(t)$ is the Dirac function, and $\tau_{i,n}$ is the delay (assumed constant) of the ith tap. Each channel tap autocovariance function can be expressed as [Stuber 2001]

$$R_{\alpha_{i,n}}(\tau) = E\{\alpha_{i,n}(t)\alpha_{i,n}(t+\tau)\} = \Omega_{i,n}^2 J_0(2\pi f_D \tau), \tag{4.46}$$

where $J_0(\cdot)$ is the zeroth-order Bessel function of the first kind, f_D is the Doppler frequency, and $\Omega_{i,n}$ is the root-mean-square value of the magnitude of tap i in diversity branch n. Regarding the cross-correlation between different taps belonging to the same or to different diversity branches, no restriction is imposed, that is, all the taps can be correlated.

For the derivation of the MFB, we assume the transmission of only one pulse $s.g(t)$, where s is an M-HQAM symbol and $g(t)$ is the impulse response of the transmit filter. Therefore, the signal received at each diversity branch can be written as

$$r_n(t) = s\sum_{i=1}^{L_n} \alpha_{i,n}(t)g(t-\tau_{i,n}) + n_n(t), \tag{4.47}$$

where $n_n(t)$ is a zero-mean complex Gaussian random process with power spectral density N_0, representing AWGN. As previously stated, the MFB is obtained considering perfect channel knowledge. The receiver filters employed in each diversity branch are expressed as

$$h_n(t) = \sum_{i=1}^{L_n} \alpha_{i,n}^*(t_0 - t)g^*(t_0 - t - \tau_{i,n}), \tag{4.48}$$

with t_0 representing the sampling instant. After the filter, the sampled signals of the different diversity branches are added, leading to

$$y(t=t_0) = s\cdot\sum_{n=1}^{N_{rx}}\sum_{i=1}^{L_n}\sum_{i'=1}^{L_n}\int_{-\infty}^{+\infty} \alpha_{i,n}(\tau)\alpha_{i',n}^*(\tau)g(\tau-\tau_{i,n})g^*(\tau-\tau_{i',n})d\tau + \sum_{n=1}^{N_{rx}} w_n, \tag{4.49}$$

where w_n represents noise samples, that is,

$$w_n = \sum_{i=1}^{L_n}\int_{-\infty}^{+\infty} n_n(\tau)\alpha_{i,n}^*(\tau)g^*(\tau-\tau_{i,n})\,d\tau. \tag{4.50}$$

4.2.3.1 Slow Fading Channel

Although the transmitter pulse $g(t)$ can have unlimited duration, typically it becomes almost zero outside a narrow interval, usually related to the symbol period. If the channel is slowly time-varying, it can be considered approximately constant inside this interval and Equation 4.49 becomes

$$y(t = t_0) = s \cdot \sum_{n=1}^{N_{rx}} \sum_{i=1}^{L_n} \sum_{i'=1}^{L_n} \alpha_{i,n}\alpha^*_{i',n} R\left(\tau_{i,n} - \tau_{i',n}\right) + \sum_{n=1}^{N_{rx}} w_n$$
$$= y_{\text{signal}} + y_{\text{noise}}, \tag{4.51}$$

with $R(\tau)$ representing the autocorrelation function of the transmit filter,

$$R(\tau) = \int_{-\infty}^{+\infty} g(\tau')g^*(\tau' + \tau)\, d\tau'. \tag{4.52}$$

The instantaneous received signal-to-noise power ratio is given by the ratio between the expected value of the signal component over the expected value of the noise component of Equation 4.51 conditioned to the channel and can be written as

$$\text{SNR} = \frac{E\left[\left|y_{\text{signal}}\right|^2 \middle| \alpha_{\cdot,\cdot}\right]}{E\left[\left|y_{\text{noise}}\right|^2 \middle| \alpha_{\cdot,\cdot}\right]} = \frac{E_s}{N_0}\kappa, \tag{4.53}$$

with

$$\kappa = \sum_{n=1}^{N_{rx}} \sum_{i=1}^{L_n} \sum_{i'=1}^{L_n} \alpha_{i,n}\alpha^*_{i',n} R(\tau_{i,n} - \tau_{i',n})$$
$$= \mathbf{z}^{\mathrm{H}} \boldsymbol{\Sigma} \mathbf{z}. \tag{4.54}$$

and

$$E_s = E\left[|s|^2\right]. \tag{4.55}$$

The last equality in Equation 4.54 corresponds to the matrix representation of κ. In this matrix format, $\mathbf{z}$ is an $L_{\text{total}} \times 1$ vector (with $L_{\text{total}} = \sum_{n=1}^{N_{rx}} L_n$) containing random variables $\alpha_{i,n}$, H denotes conjugate transpose, and Σ is an $L_{\text{total}} \times L_{\text{total}}$ matrix constructed according to

$$\boldsymbol{\Sigma} = \begin{bmatrix} \begin{matrix} R(0) & \cdots & R(\tau_{L_1,1} - \tau_{1,1}) \\ \vdots & \ddots & \vdots \\ R(\tau_{1,1} - \tau_{L_1,1}) & \cdots & R(0) \end{matrix} & & \mathbf{0} \\ & \ddots & \\ \mathbf{0} & & \begin{matrix} R(0) & \cdots & R(\tau_{L_{N_{rx}},N_{rx}} - \tau_{1,N_{rx}}) \\ \vdots & \ddots & \vdots \\ R(\tau_{1,N_{rx}} - \tau_{L_{N_{rx}},N_{rx}}) & \cdots & R(0) \end{matrix} \end{bmatrix}. \tag{4.56}$$

Since $R(\tau_{i,n} - \tau_{i',n}) = R^*(\tau_{i',n} - \tau_{i,n})$ we can easily recognize that $\mathbf{S}$ is Hermitian. For obtaining the MFB it is necessary to find the PDF of κ. To accomplish this, we will apply a characteristic function-based method that requires writing κ in the form of a sum of uncorrelated random variables with known PDFs. Since different taps can be correlated we need to transform $\mathbf{z}$ into a vector of uncorrelated components. Defining $\mathbf{\Psi}$ as the covariance matrix of $\mathbf{z}(\mathbf{\Psi} = \mathrm{Cov}[\mathbf{z}])$, and knowing that this matrix is Hermitian and positive semidefinite, it is possible to find at least one matrix $\mathbf{Q}$ so that $\mathbf{\Psi}$ can be decomposed into $\mathbf{\Psi} = \mathbf{Q}\mathbf{Q}^{\mathrm{H}}$. In particular, if we apply the Cholesky decomposition, $\mathbf{Q}$ will be a lower triangular matrix. After finding $\mathbf{Q}$ we can define a new vector $\mathbf{z}' = \mathbf{Q}^{-1}\mathbf{z}$ whose covariance matrix corresponds to the identity matrix, as it is easy to verify. As a consequence, all components of $\mathbf{z}'$ will be uncorrelated unit-variance complex Gaussian variables. Introducing this transformation into Equation 4.54 yields

$$\begin{aligned} \kappa &= \mathbf{z}'^{\mathrm{H}}\mathbf{Q}^{\mathrm{H}}\mathbf{\Sigma}\mathbf{Q}\mathbf{z}' \\ &= \mathbf{z}'^{\mathrm{H}}\mathbf{\Sigma}'\mathbf{z}', \end{aligned} \tag{4.57}$$

with

$$\mathbf{\Sigma}' = \mathbf{Q}^{\mathrm{H}}\mathbf{\Sigma}\mathbf{Q}. \tag{4.58}$$

Matrix $\mathbf{\Sigma}'$ is still Hermitian, which means that it can be decomposed as

$$\mathbf{\Sigma}' = \mathbf{\Phi}\mathbf{\Lambda}\mathbf{\Phi}^{\mathrm{H}}, \tag{4.59}$$

where $\mathbf{\Lambda}$ is a diagonal matrix whose elements are the eigenvalues λ_i $(i = 1, \ldots, L_{\mathrm{total}})$ of $\mathbf{\Sigma}'$, and $\mathbf{\Phi}$ is a matrix whose columns are the eigenvectors of $\mathbf{\Sigma}'$. These eigenvectors are orthogonal ($\mathbf{\Phi}\mathbf{\Phi}^{\mathrm{H}} = \mathbf{I}$). We can then rewrite Equation 4.57 as

$$\begin{aligned} \kappa &= \mathbf{z}'^{\mathrm{H}}\mathbf{\Phi}\mathbf{\Lambda}\mathbf{\Phi}^{\mathrm{H}}\mathbf{z}' \\ &= \mathbf{z}''^{\mathrm{H}}\mathbf{\Lambda}\mathbf{z}'' \\ &= \sum_{i=1}^{L_{\mathrm{total}}} \lambda_i \left|z_i''\right|^2, \end{aligned} \tag{4.60}$$

where we have defined another vector, $\mathbf{z}'' = \mathbf{\Phi}^{\mathrm{H}}\mathbf{z}'$, whose components are still uncorrelated unit-variance complex Gaussian variables (as is simple to verify). According to Equation 4.60, κ can be expressed as the sum of independent random variables with exponential distributions (z_i'' is Gaussian and, therefore, $\lambda_i \left|z_i''\right|^2$ follows an exponential distribution). The characteristic function of κ is then simply the product of the respective individual characteristic functions

$$E\{\mathrm{e}^{-\mathrm{j}u\kappa}\} = \prod_{i=1}^{L_{\mathrm{total}}} \frac{1}{1 + \mathrm{j}\lambda_i \upsilon}. \tag{4.61}$$

If there are L' distinct eigenvalues, each with a multiplicity of θ_i, $i = 1, \ldots, L'$, then by partial fractions decomposition and using the change of variables $s = j\upsilon$ we can rewrite Equation 4.61 as

$$E\{e^{-s\kappa}\} = \prod_{i=1}^{L'} \frac{1}{(1+s\lambda_i)^{\theta_i}} = \sum_{i=1}^{L'}\sum_{k=1}^{\theta_i} \frac{A_{i,k}}{\lambda_i^{\theta_i-k}(\theta_i-k)!(1+s\lambda_i)^k}, \quad (4.62)$$

with

$$A_{i,k} = \left[\frac{\partial^{\theta_i-k}}{\partial s^{\theta_i-k}}\left(\prod_{\substack{j=1\\ j\neq i}}^{L'} \frac{1}{(1+s\lambda_j)^{\theta_j}}\right)\right]_{s=-(1/\lambda_i)}. \quad (4.63)$$

Reversing the change of variables and computing the inverse FT, we obtain the PDF of κ as

$$p(\kappa) = \sum_{i=1}^{L'}\sum_{k=1}^{\theta_i} \frac{A_{i,k}}{\lambda_i^{\theta_i}(\theta_i-k)!(k-1)!}\kappa^{k-1}e^{-(\kappa/\lambda_i)}. \quad (4.64)$$

On the basis of the BER expressions for AWGN presented in Section 4.2.1 and observing that κ acts as a simple scaling factor on the constellation symbols, we can write the individual BERs conditioned to κ as

$$P_b(b_m \mid \kappa) = \frac{2}{\sqrt{M}} \sum_{j=0}^{\sqrt{M}/2-1}\left[g_j^m + (-1)^{g_j^m}\sum_{l=1}^{2^{(m-1)}}\left[(-1)^{l+1}\times Q\left(\sqrt{\frac{2}{N_0}}\kappa[\mathbf{B}_m(l)-\mathbf{D}_s(j)]\right)\right]\right],$$

$$m = 1, \ldots, \log_2 M/2, \quad (4.65)$$

The average BER can be computed by averaging Equation 4.65 over the PDF of κ

$$P_b(b_m) = \int_{-\infty}^{+\infty} P_b(b_m \mid \kappa)\, p(\kappa)\, d\kappa. \quad (4.66)$$

To obtain an explicit expression for Equation 4.66, we need to make use of the following integral expression:

$$\int_0^{+\infty} Q(\sqrt{x})\frac{x^{L-1}}{(L-1)!\bar{x}_i^L}e^{-(x/\bar{x}_i)}\,dx = \left[\frac{1-\mu}{2}\right]^L \sum_{r=0}^{L-1}\binom{L-1+r}{r}\left[\frac{1+\mu}{2}\right]^r, \quad (4.67)$$

with

$$\mu = \sqrt{\frac{\bar{x}_i/2}{1+\bar{x}_i/2}}. \tag{4.68}$$

The average BER can subsequently be written as

$$\begin{aligned} P_b(b_m) &= \frac{2}{\sqrt{M}} \sum_{j=0}^{\sqrt{M}/2-1} \left[g_j^m + (-1)^{g_j^m} \sum_{l=1}^{2^{(m-1)}} \left[\frac{1}{2}(-1)^{l+1} \left\{ 1 - \mathrm{sign}\left(\mathbf{B}_m(l) - \mathbf{D}_s(j)\right) \right.\right.\right. \\ &\quad + 2 \cdot (-1)^{1/2-\mathrm{sign}(\mathbf{B}_m(l)-\mathbf{D}_s(j))/2} \sum_{i=1}^{L'} \sum_{k=1}^{\theta_i} \frac{A_{i,k}}{\lambda_i^{\theta_i-k}(\theta_i-k)!} \left[\frac{1-\mu_i}{2}\right]^k \\ &\quad \left.\left.\left. \times \sum_{r=0}^{k-1} \binom{k-1+r}{r} \left[\frac{1+\mu_i}{2}\right]^r \right\} \right] \right], \end{aligned} \tag{4.69}$$

where

$$\mu_i = \sqrt{\frac{E_s/N_0 \left[\mathbf{B}_m(l) - \mathbf{D}_s(j)\right]^2 \lambda_i}{1 + (E_s/N_0)\left[\mathbf{B}_m(l) - \mathbf{D}_s(j)\right]^2 \lambda_i}}. \tag{4.70}$$

Using the MFB expressions, we can evaluate the effect of the correlation coefficient between taps, $\rho = E[\alpha_i \alpha_{i'}^*]/\sqrt{E[|\alpha_i|^2]E[|\alpha_{i'}|^2]}$ (assumed the same for all the taps), on the performance of M-HQAM constellations. Figure 4.10 presents MFB curves for a 16-HQAM constellation with $k_1 = 0.4$, one receiving antenna and assuming a root-square-raised cosine (RRC) filter with a roll-off BW factor $\beta = 0.22$ and $T_s = 260$ ns. The environment considered is Vehicular A from Ref. [ETSI 1998b], whose tap delays and respective powers are shown in Appendix A. As a reference, results for flat Rayleigh fading are also shown in the same graph. Curves for the MPBs and for the LPBs are presented. As expected, an increase in correlation causes a degradation of the performance due to a loss of the diversity effect from the usage of different taps. Nevertheless, this degradation is small until $\rho = 0.5$ and even for higher correlation values, such as 0.9, the performance is still relatively far from flat Rayleigh (which does not have any diversity gain).

Figure 4.11 compares the MFB with the Monte Carlo simulated performance of an IB-DFE receiver (described in Chapter 2). We consider a 16-HQAM constellation with $k_1 = 0.5$ and a channel with 64 EP taps and two receive antennas. The results show that the IB-DFE performance can become asymptotically close to the MFB for severely time-dispersive channels with rich multipath propagation combined with diversity techniques.

4.2.3.2 Fast Fading Channel

If the channel is too fast ($f_N > 0.01$, with f_N being the normalized Doppler rate, which is defined as $f_N = f_D T_S$, where f_D is the Doppler frequency) to be considered constant

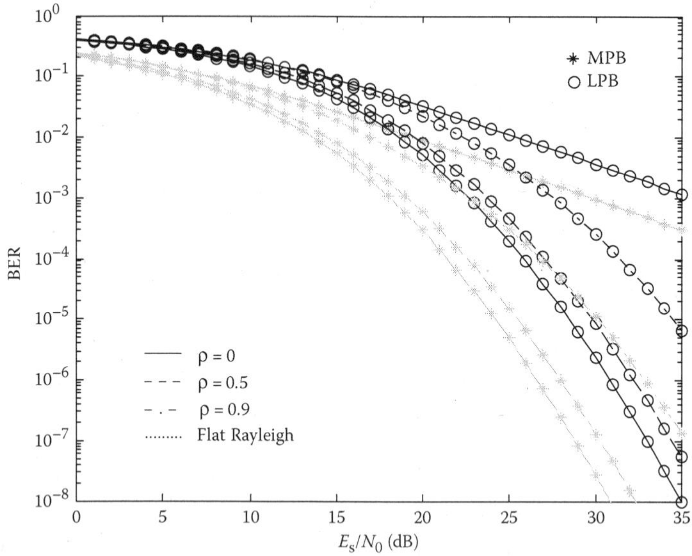

Figure 4.10 Impact of correlated taps on the BER of a 16-HQAM constellation with $k_1 = 0.4$ in Vehicular A environment.

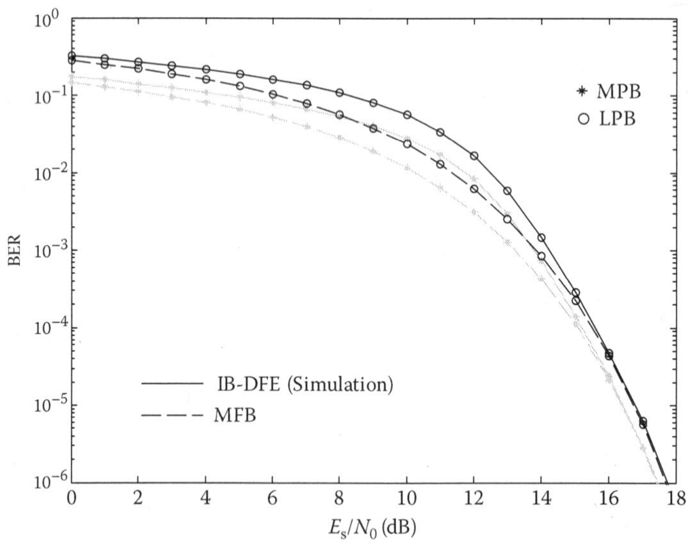

Figure 4.11 IB-DFE performance versus MFB for a 16-HQAM constellation with $k_1 = 0.5$ in a 64-tap environment with two receiver antennas.

during the approximate pulse duration, then a different approach has to be employed for deriving the MFB. The integral in Equation 4.49 becomes much more difficult to compute and an alternative approach needs to be used instead. A simple way to deal with it is to divide it into the sum of smaller ones so that the channel can be seen as almost constant inside the smaller time intervals. To simplify the derivation, we will consider the existence of only one diversity branch ($N_{rx} = 1$). Nevertheless, the extension to more diversity branches is straightforward. Applying the approximation just described, Equation 4.49 becomes

$$\begin{aligned} y(t = t_0) &\approx s \cdot \sum_{i=1}^{L} \sum_{i'=1}^{L} \sum_{l=-N/2}^{N/2-1} \alpha_i\left(\frac{\Delta t}{2} + l\Delta t\right) \alpha_{i'}^*\left(\frac{\Delta t}{2} + l\Delta t\right) \\ &\quad \times \int_{l\Delta t}^{(l+1)\Delta t} g(\tau - \tau_i) g^*(\tau - \tau_{i'}) \mathrm{d}\tau + w \\ &= s \cdot \sum_{i=1}^{L} \sum_{i'=1}^{L} \sum_{l=-N/2}^{N/2-1} \alpha_i\left(\frac{\Delta t}{2} + l\Delta t\right) \alpha_{i'}^*\left(\frac{\Delta t}{2} + l\Delta t\right) \\ &\quad \times R(\tau_i - \tau_{i'}, l\Delta t - \tau_i, (l+1)\Delta t - \tau_i) + w, \end{aligned} \tag{4.71}$$

where Δt is a sufficiently small interval chosen so that $\alpha_i(t)$ is approximately constant and the total number of intervals N is chosen so that $g(t) \approx 0$ for $|t| > N/2 \cdot \Delta t$. As for $R(\tau, t_{\text{lower}}, t_{\text{upper}})$ it corresponds to the partial autocorrelation function of the transmit filter, that is,

$$R(\tau, t_{\text{lower}}, t_{\text{upper}}) = \int_{t_{\text{lower}}}^{t_{\text{upper}}} g(\tau') g^*(\tau' + \tau) \, \mathrm{d}\tau. \tag{4.72}$$

In this case, the instantaneous received signal-to-noise power ratio and variable κ are still given by Equations 4.53 and 4.54, respectively. However, the elements in $\mathbf{z}$ and $\mathbf{\Sigma}$ are different. The former is defined as

$$\mathbf{z} = \begin{bmatrix} \alpha_1(\Delta t/2 - N/2\Delta t) \\ \vdots \\ \alpha_L(\Delta t/2 - N/2\Delta t) \\ \vdots \\ \alpha_1(\Delta t/2 + (N/2 - 1)\Delta t) \\ \vdots \\ \alpha_L(\Delta t/2 + (N/2 - 1)\Delta t) \end{bmatrix}. \tag{4.73}$$

Note that even if the different paths are uncorrelated, some of the elements of $\mathbf{z}$ are always correlated, namely the ones belonging to the same taps, due to the inherent time correlation of the channel. For example, if we assume a Jakes spectrum for the taps, each channel tap autocovariance function can be expressed as

$$R_{\alpha_i}(\tau) = E\left\{\alpha_i(t)\alpha_i^*(t+\tau)\right\} = \Omega_i^2 J_0(2\pi f_D \tau), \tag{4.74}$$

where $J_0(\cdot)$ is the zeroth-order Bessel function of the first kind, f_D is the Doppler frequency, and Ω_i^2 is the average fading power of tap i. In this case, the components of the covariance matrix of $\mathbf{z}$ can be computed as

$$E\left[z_u z_{u'}^*\right] = \begin{cases} \Omega^2_{u-\lfloor \frac{u-1}{L} \rfloor L}, & u = u', \\ \rho_{u-\lfloor \frac{u-1}{L} \rfloor L, u'-\lfloor \frac{u'-1}{L} \rfloor L} \Omega_{u-\lfloor \frac{u-1}{L} \rfloor L} \Omega_{u'-\lfloor \frac{u'-1}{L} \rfloor L}, & \\ \quad u \neq u' \wedge u - \left\lfloor \frac{u-1}{L} \right\rfloor L \neq u' - \left\lfloor \frac{u'-1}{L} \right\rfloor L, & \\ \Omega^2_{u-\lfloor \frac{u-1}{L} \rfloor L} J_0\left[2\pi f_D \left(\left\lceil \frac{u}{L} \right\rceil - \left\lceil \frac{u'}{L} \right\rceil\right) \Delta t\right], & \\ \quad u - \left\lfloor \frac{u-1}{L} \right\rfloor L = u' - \left\lfloor \frac{u'-1}{L} \right\rfloor L. & \end{cases} \tag{4.75}$$

where $\rho_{ii'}$ is the correlation between paths i and i'.

It is possible to transform $\mathbf{z}$ into a vector composed of independent random variables using the same method employed previously for the slow fading channel. As for matrix $\mathbf{\Sigma}$, it is built as a diagonal matrix whose diagonal elements are smaller matrices $\mathbf{\Sigma}_l$ $(l = -N/2, \ldots, N/2-1)$, that is,

$$\mathbf{\Sigma} = \begin{bmatrix} \mathbf{\Sigma}_{-N/2} & & 0 \\ & \ddots & \\ 0 & & \mathbf{\Sigma}_{N/2-1} \end{bmatrix},$$

with

$$\mathbf{\Sigma}_l = \begin{bmatrix} R(\tau_1 - \tau_1, l\Delta t - \tau_1, & & R(\tau_L - \tau_1, l\Delta t - \tau_L, \\ (l+1)\Delta t - \tau_1) & \cdots & (l+1)\Delta t - \tau_L) \\ \vdots & \ddots & \vdots \\ R(\tau_1 - \tau_L, l\Delta t - \tau_1, & & R(\tau_L - \tau_L, l\Delta t - \tau_L, \\ (l+1)\Delta t - \tau_1) & \cdots & (l+1)\Delta t - \tau_L) \end{bmatrix}. \tag{4.76}$$

Note that since $R(\tau_i - \tau_{i'}, l\Delta t - \tau_i, (l+1)\Delta t - \tau_i) = R^*(\tau_{i'} - \tau_i, l\Delta t - \tau_{i'}, (l+1)\Delta t - \tau_{i'})$, matrix $\mathbf{\Sigma}$ is still Hermitian. As a result, the subsequent derivation and final BER expression (Equation 4.69) is the same as the one presented for the slow fading channel.

As an example, Figure 4.12 presents MFB results for several normalized Doppler rates, f_N. The constellation is a 16-HQAM with $k_1 = 0.4$ and the environment is Vehicular A. The size of each partition interval, Δt, is chosen so as to ensure that $f_D \Delta t < 0.01$, with a minimum of four partitions. Curves for $f_N = 0$ and $f_N = 1 \times 10^{-3}$ are almost

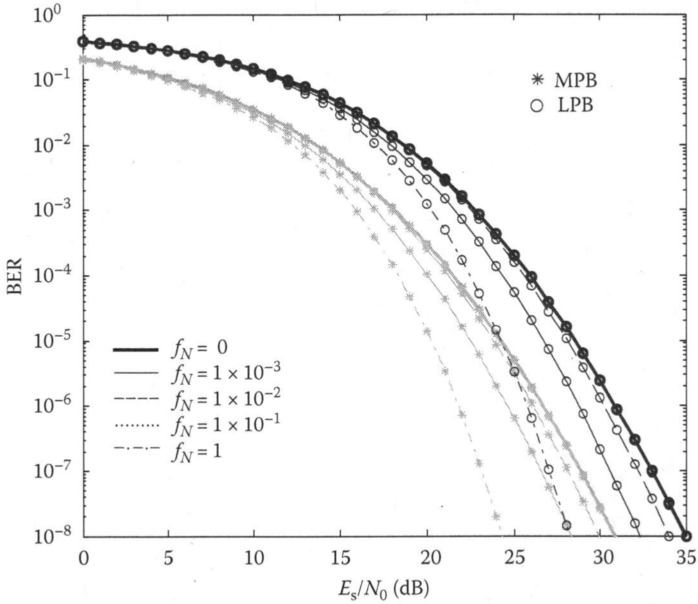

Figure 4.12 Impact of the normalized Doppler rate on the BER of a 16-HQAM constellation with $k_1 = 0.4$ in Vehicular A environment.

overlapping with $f_N = 1 \times 10^{-2}$ being also very close, which means that, for normalized Doppler rates below this value, the channel can be assumed constant. As f_N increases, the channel becomes faster and the MFB performance starts to improve substantially. This is a consequence of the fact that, theoretically, in fast fading channels, each individual bit can obtain some diversity gain even without multipath. Although this might suggest that increasing the mobility of the user (or alternatively increasing the symbol period) automatically improves the BER, in fact, for very high speeds the channel estimation is extremely difficult and these bounds become unattainable.

4.3 Channel Estimation Issues

In the previous section, the performances were all obtained assuming perfect channel state information (CSI) at the receiver. However, as addressed in Chapter 3, channel estimation is always imperfect in real systems and can have a significant impact on their performances. Even for flat fading channels, CSI is required to rescale the received symbols so that they can correspond to the symbols of the original constellation. Although in QPSK it is only important to compensate for phase fluctuations caused by the channel, in QAM modulations it is also required to know the amplitude fluctuations. Thus, M-QAM is more sensitive to the accuracy of channel estimation than QPSK. In this section, we will derive performance expressions for M-HQAM constellations when channel estimation is imperfect. We will start by obtaining a BER expression for single-antenna transmissions and then extend it to cases of diversity reception with nonidentical channels. We will also study efficient channel estimation strategies for this type of modulations.

4.3.1 Flat Rayleigh Fading with Imperfect Channel Estimation

4.3.1.1 No Diversity

Similarly to the model employed for QPSK transmissions in Section 3.1.1, in a flat Rayleigh channel each received signal sample can be expressed as

$$r = \alpha \cdot s + n, \tag{4.77}$$

where α is the channel coefficient, s is the transmitted M-HQAM symbol, and n represents additive white thermal noise. Both α and n are modeled as complex Gaussian random variables with $E[\alpha] = 0$, $E\left[|\alpha|^2\right] = 2\sigma_\alpha^2$ (average fading power), $E[n] = 0$, and $E\left[|n|^2\right] = N_0$ ($N_0/2$ is the two-sided noise power spectrum density). For a fixed transmitted symbol, r is also a complex Gaussian random variable with $E[r] = 0$ and

$$E\left[|r|^2|s\right] = 2\sigma_r^2 = 2|s|^2\sigma_\alpha^2 + N_0. \tag{4.78}$$

Since we are considering coherent detection and independent mapping of the bits to the I and Q branches, the receiver multiplies the received samples by the conjugate of the channel coefficients and obtains decision variables for the I and Q branches according to

$$\begin{cases} z_{\text{re}} = \text{Re}\left\{r \cdot \hat{\alpha}^*\right\} \\ z_{\text{im}} = \text{Im}\left\{r \cdot \hat{\alpha}^*\right\} \end{cases}. \tag{4.79}$$

Owing to the symmetry of both branches, we can develop our study using the decision variable for only one of them, as we did for QPSK in Section 3.1.1. Therefore, in the following derivations we will work with z_{re} only.

Assuming an unbiased linear channel estimator, channel estimate $\hat{\alpha}$ will be a complex Gaussian random variable with $E[\hat{\alpha}] = 0$ and $E\left[\left|\hat{\alpha}\right|^2\right] = 2\sigma_{\hat{\alpha}}^2$ ($\sigma_{\hat{\alpha}}^2$ depends on the channel estimation algorithm employed). Both r and $\hat{\alpha}$ will be correlated with the joint moment expressed as

$$E\left[r \cdot \hat{\alpha}^* \mid s\right] = sE\left[\alpha \cdot \hat{\alpha}^* \mid s\right] + E\left[n \cdot \hat{\alpha}^* \mid s\right]. \tag{4.80}$$

As we explained in Section 3.1.1, the reliability of the channel estimates can depend on the symbol position in the transmitted frame. As a result, the error probability of each different bit type b_m ($m = 1, \ldots, 1/2 \log_2 M$), which is our objective, will also depend on the position in the frame. Therefore, the BER must be averaged over all possible locations t in a frame of size N_F as

$$\begin{aligned} P_b(b_m) &= \sum_{t=1}^{N_\mathrm{F}} P_e(b_m, t) \\ &= \frac{1}{N_\mathrm{F}} \sum_{t=1}^{N_\mathrm{F}} P_b(b_m \mid t), \end{aligned} \tag{4.81}$$

where $P_e(b_m, t)$ represents the probability of bit b_m being incorrect and its position in the frame being t and $P_b(b_m \mid t)$ is the BER for that position.

As explained in Section 4.2, the BER performance of M-HQAM constellations with independent mapping of the bits to the in-phase and quadrature axis can be obtained from a simpler $\sqrt{M}$-HPAM constellation (two for rectangular QAM) when perfect channel estimation is assumed. However, when considering imperfect channel estimation, we cannot reduce the evaluation of an M-HQAM constellation to a simpler $\sqrt{M}$-PAM constellation. This happens due to the existence of phase error even after channel compensation, which adds interference from the quadrature components to the in-phase components and vice versa. Figure 4.13 shows how this cross-quadrature interference is produced. It presents the transmitted constellation and the received constellation after compensation with the conjugate of the channel estimate. To simplify the diagram, the fading amplitude is assumed to be perfectly estimated and thus only phase error exists.

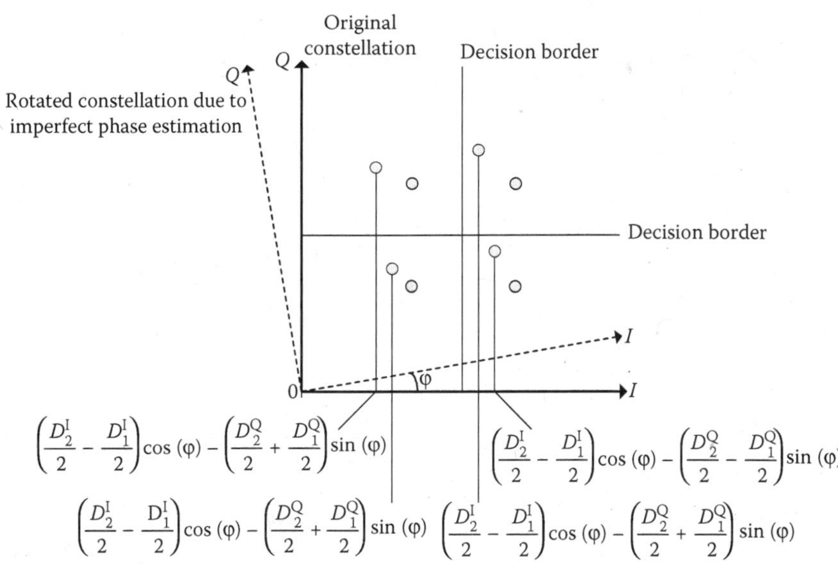

Figure 4.13 Impact of cross-quadrature interference originated by imperfect channel phase estimation in a 16-QAM constellation (the upper right quadrant is shown).

This error is equal to

$$\varphi = \theta - \hat{\theta}, \tag{4.82}$$

where θ is the phase of the channel coefficient and $\hat{\theta}$ is the phase of the channel estimate.

As shown in Figure 4.13, the presence of a phase error causes a rotation of the constellation. This rotation degrades the contribution of the in-phase component of the transmitted symbol from $(D_2^{\rm I}/2 \pm D_1^{\rm I}/2)$ to $(D_2^{\rm I}/2 \pm D_1^{\rm I}/2)\cos(\varphi)$ and results in the presence of interference from its quadrature component of $-(D_2^{\rm Q}/2 \pm D_1^{\rm Q}/2)\sin(\varphi)$ on the decision variable $z_{\rm re}$.

In Section 4.2.1, we presented an explicit closed-form expression for the bit error probability of M-HQAM constellations in AWGN channels, which was obtained from the pattern of construction of M-HQAM constellations in Ref. [Vitthaladevuni and Alouini 2004]. That expression assumes that the decision variable, $z_{\rm re}$, is compared against the decision borders of the constellation. In the problem we are dealing with in this section, the decision variable, $z_{\rm re}$, is similar (apart from imperfect channel compensation) and we also need to compare it against the decision borders of the constellation before making a decision about the information bits. The only difference is that in this situation, the decision borders will be weighted by $|\hat{\alpha}|^2$ and the probability of $z_{\rm re}$ being above or below the borders will be different, that is, will not be computed with the $Q(x)$ function. Therefore, we can directly extend expression 4.24 for this case, combining it with Equation 4.81, and obtain the BER expression for imperfect channel estimation as

$$\begin{aligned} P_b(b_m) &= \frac{1}{N_{\rm F}}\frac{2}{\sqrt{M}} \\ &\quad\times \sum_{t=1}^{N_{\rm F}} \sum_{j=0}^{\sqrt{M}/2-1} \left[1 - g_j^m - (-1)^{g_j^m} \sum_{l=1}^{2^{(m-1)}} \left[(-1)^{l+1} {\rm Prob}\left\{ z_{\rm re} < |\hat{\alpha}|^2 \mathbf{B}_m(l) \mid s_j, t \right\} \right] \right], \\ &\quad m = 1, \ldots, 1/2 \cdot \log_2 M, \end{aligned} \tag{4.83}$$

where $\mathbf{B}_m(l)$ is given by Equation 4.25. In this expression, we are taking advantage of the symmetries in the constellations that allow us to compute the error probability for each bit b_m by averaging the conditional BERs over all existing constellation symbols in only one of the quadrants. In addition, we are also assuming equiprobable transmitted symbols, that is, $P(s_{j,f}) = 1/M$. In Equation 4.83, ${\rm Prob}\{z_{\rm re} < |\hat{\alpha}|^2 \mathbf{B}_m(l) \mid s_j, t\}$ is the probability of the decision variable being lower than $|\hat{\alpha}|^2 \mathbf{B}_m(l)$ conditioned on the fact that the transmitted symbol has a projection onto the real axis equal to s_j and its position in the frame is t. We can rewrite this probability in the following form:

$$\begin{aligned} &{\rm Prob}\left\{ z_{\rm re} < |\hat{\alpha}|^2 \mathbf{B}_m(l) \mid s_j, t \right\} \\ &\quad = \sum_{f=0}^{\sqrt{M}/2-1} {\rm Prob}\left\{ z_{\rm re} < |\hat{\alpha}|^2 \mathbf{B}_m(l), s_{j,f} \mid {\rm Re}(s_{j,f}) = s_j, t \right\} \end{aligned}$$

$$= \sum_{f=0}^{\sqrt{M}/2-1} \text{Prob}\{z_{\text{re}} < |\hat{\alpha}|^2 \mathbf{B}_m(l) \mid s_{j,f}, t\} \, \text{Prob}\{s_{j,f} \mid \text{Re}(s_{j,f}) = s_j, t\}$$

$$= \frac{2}{\sqrt{M}} \sum_{f=0}^{\sqrt{M}/2-1} \text{Prob}\{z_{\text{re}} < |\hat{\alpha}|^2 \mathbf{B}_m(l) \mid s_{j,f}, t\}. \tag{4.84}$$

Inserting Equation 4.84 into Equation 4.83 leads to

$$P_b(b_m) = \frac{1}{N_{\text{F}}} \frac{2}{\sqrt{M}} \sum_{t=1}^{N_{\text{F}}} \sum_{j=0}^{\sqrt{M}/2-1} \left[1 - g_j^m - (-1)^{g_j^m} \sum_{l=1}^{2^{(m-1)}} \left[(-1)^{l+1} \frac{2}{\sqrt{M}} \right.\right.$$
$$\left.\left. \times \sum_{f=0}^{\sqrt{M}/2-1} \text{Prob}\left\{ z_{\text{re}} < |\hat{\alpha}|^2 \, \mathbf{B}_m(l) \mid s_{j,f}, t \right\} \right]\right]. \tag{4.85}$$

This means that, to compute the analytical BER it is necessary to obtain the expression for $\text{Prob}\{z_{\text{re}} < |\hat{\alpha}|^2 \mathbf{B}_m(l) \mid s_{j,f}, t\}$. In the following derivations, we will drop indexes t (symbol position in the frame), j (in-phase branch label of the constellation symbol), and f (quadrature branch label of the constellation symbol) for simplicity of notation. While for QPSK, there is only one decision border located over the origin, in the case of M-HQAM constellations, there are several decision borders. One is in the origin of the I- and Q-axis $[\mathbf{B}_m(l) = 0]$, which is similar to the case of QPSK and does not depend on channel estimate $\hat{\alpha}$. The other decision borders depend on channel estimate $[|\hat{\alpha}|^2 \cdot \mathbf{B}_m(l)]$ and, consequently, are random variables. To avoid decision borders depending explicitly on the channel estimate, probability expression 4.84 can be rewritten as

$$\begin{aligned} \text{Prob}\left(z_{\text{re}} < |\hat{\alpha}|^2 \, \mathbf{B}_m(l) \mid s \right) &= \text{Prob}\left(\text{Re}\left\{ \left(r - \hat{\alpha} \cdot \mathbf{B}_m(l) \right) \hat{\alpha}^* \right\} < 0 \mid s \right) \\ &= \text{Prob}\left(\text{Re}\left\{ z' \right\} < 0 \mid s \right). \end{aligned} \tag{4.86}$$

where we have defined a modified complex decision variable, z', which is related to z through

$$\begin{aligned} z' &= \left(r - \hat{\alpha} \cdot \mathbf{B}_m(l) \right) \hat{\alpha}^* \\ &= r' \hat{\alpha}^*. \end{aligned} \tag{4.87}$$

Since r and $\hat{\alpha}$ are complex random Gaussian variables and $\mathbf{B}_m(l)$ is a constant, then r' also has a Gaussian distribution, and probability 4.86 can be computed using the same expression derived for QPSK in Section 3.1.1, that is,

$$\text{Prob}\left(Re\left\{ z' \right\} < 0 \mid s \right) = \frac{1 - |\mu'|^2}{2\left(1 - |\mu'|^2 (\sin \varepsilon')^2 + |\mu'| \cos \varepsilon' \sqrt{1 - |\mu'|^2 (\sin \varepsilon')^2} \right)}, \tag{4.88}$$

where μ' is now defined based on the modified variables

$$\mu' = \frac{E\left[r'\hat{\alpha}^*|s\right]}{\sqrt{E\left[|r'|^2|s\right]E\left[|\hat{\alpha}|^2|s\right]}} = |\mu'|e^{-\varepsilon' j}. \tag{4.89}$$

In Equation 4.89, the second moment of r' is given by

$$E\left[|r'|^2 \mid s\right] = 2|s|^2\sigma_\alpha^2 + N_0 - 2\mathbf{B}_m(l)Re\left\{E\left[r\cdot\hat{\alpha}^* \mid s\right]\right\} + \mathbf{B}_m(l)^2 E\left[|\hat{\alpha}|^2\right], \tag{4.90}$$

and the cross moment of r' and $\hat{\alpha}$ is

$$E\left[r'\hat{\alpha}^* \mid s\right] = E\left[r\cdot\hat{\alpha}^* \mid s\right] - \mathbf{B}_m(l)E\left[|\hat{\alpha}|^2\right]. \tag{4.91}$$

Using the derived expressions, several performance curves for a 16-HQAM constellation with $k_1 = 0.4$ and imperfect channel estimation are shown in Figure 4.14. Similarly to Chapter 3, channel estimates are modeled as $\hat{\alpha} = a\cdot\alpha + \xi$, where a is a complex value denoting the bias of the estimate (which will be assumed as $a = 1$) and ξ is a zero-mean complex Gaussian variable representing the channel estimation error. As observed for QPSK in Chapter 3, channel estimates with irreducible errors lead to performances with irreducible BER floors, which, for higher-order constellations such as 16-HQAM, is more critical due to the higher sensitivity of the LPBs. The BER floors cease to exist if the channel estimates are only corrupted by a noise with variance proportional to N_0.

Figure 4.15 shows the effect of the nonuniformity parameter k_1 of a 16-HQAM on the E_S/N_0 degradation due to imperfect channel estimation, that is, the extra E_S/N_0

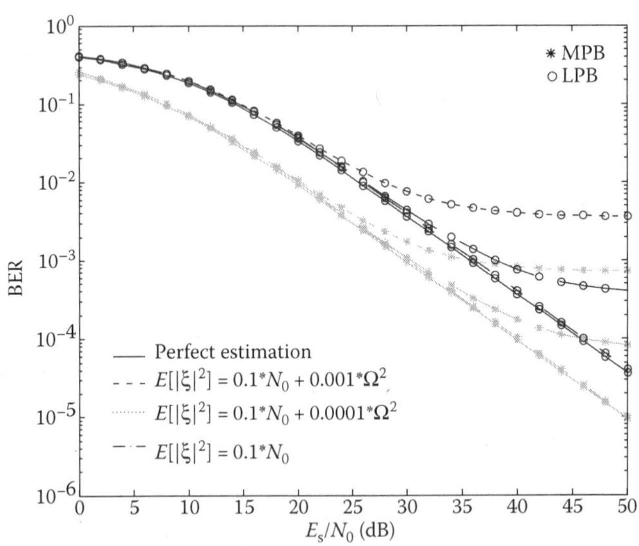

Figure 4.14 BER performance of a 16-HQAM constellation, $k_1 = 0.4$, in a flat Rayleigh channel for several channel estimation error variances.

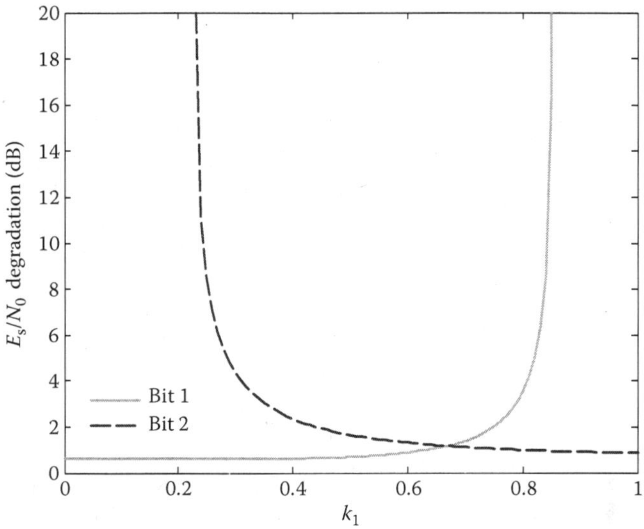

Figure 4.15 **E_S/N_0 degradation due to imperfect channel estimation versus the nonuniformity parameter k_1 for a 16-HQAM. (Target BER = 10^{-3}, $E[|\xi|^2] = 0.1 \cdot N_0 + 0.0001 \cdot \Omega^2$.)**

required for achieving a BER target of 10^{-3}. The channel estimation error variance considered is $E\left[|\xi|^2\right] = 0.1 \cdot N_0 + 0.0001 \cdot \Omega^2$. Note that in this legend we do not employ the terminology MPB and LPB for bit 1 and bit 2 since it might be misleading as for $k_1 > 0.65$ the two bit types swap roles in terms of error protection. Nevertheless, independently of which particular bit is the most protected one, it is visible that it will have a lower sensitivity to channel estimation errors, with degradations often below 1 dB. As for the LPB, a very nonuniform constellation (with $k_1 < 0.3$ or $k_1 > 0.8$) results in a very high sensitivity to channel estimation errors that can make a BER of 10^{-3} unattainable.

4.3.1.2 With Diversity

Let us consider the transmission of an M-HQAM symbol over L diversity Rayleigh channels, where different diversity branches can have different fading powers. In this case we have L different received versions of the transmitted signal:

$$r_k = \alpha_k \cdot s + n_k, \quad k = 1 \ldots L. \tag{4.92}$$

The receiver multiplies each of the received signals by the complex conjugate of the respective channel coefficient estimate, $\hat{\alpha}_k$, and adds the results of the L products. This corresponds to performing MRC of the received replicas. The decision variable can be expressed as

$$z_{\text{re}} = \text{Re}\left\{\sum_{k=1}^{L} r_k \cdot \hat{\alpha}_k^*\right\}. \tag{4.93}$$

Assuming an unbiased linear channel estimator, channel estimates, $\hat{\alpha}_k$, are still complex Gaussian random variables with $E[\hat{\alpha}_k] = 0$ and $E[|\hat{\alpha}_k|^2] = 2\sigma^2_{\hat{\alpha}_k}$. The general BER expression will be almost the same as Equation 4.85; only the borders will have slightly different weights,

$$P_b(b_m) = \frac{1}{N_{\text{F}}} \frac{2}{\sqrt{M}} \sum_{t=1}^{N_{\text{F}}} \sum_{j=0}^{\sqrt{M}/2-1} \left[1 - g_j^m - (-1)^{g_j^m} \sum_{l=1}^{2^{(m-1)}} \left[(-1)^{l+1} \frac{2}{\sqrt{M}} \right.\right.$$
$$\left.\left. \times \sum_{f=0}^{\sqrt{M}/2-1} \text{Prob}\left\{ z_{\text{re}} < \mathbf{B}_m(l) \sum_{k=1}^{L} \left|\hat{\alpha}_k\right|^2 \mid s_{j,f}, t \right\} \right] \right], \tag{4.94}$$

that is, they will be located at $\mathbf{B}_m(l)\sum_{k=1}^{L}|\hat{\alpha}_k|^2$. According to Equation 4.94, the probability that we need is $\text{Prob}\{z_{\text{re}} < \mathbf{B}_m(l)\sum_{k=1}^{L}|\hat{\alpha}_k|^2 \mid s\}$ (once again we dropped indexes t, j, and f for simplicity of notation), which can be rewritten as

$$\text{Prob}\left\{ z_{\text{re}} < \mathbf{B}_m(l) \sum_{k=1}^{L} \left|\hat{\alpha}_k\right|^2 \mid s \right\} = \text{Prob}\left(\text{Re}\left\{ \sum_{k=1}^{L} \left(r_k - \mathbf{B}_m(l) \cdot \hat{\alpha}_k\right) \hat{\alpha}_k^* \right\} < 0 \mid s \right)$$
$$= \text{Prob}\left(\sum_{k=1}^{L} z'_{\text{re}_k} < 0 \mid s \right), \tag{4.95}$$

where

$$z'_{\text{re}_k} = Re\{z'_k\} = \text{Re}\left\{r'_k \cdot \hat{\alpha}_k^*\right\} \tag{4.96}$$

and

$$r'_k = r_k - \mathbf{B}_m(l) \cdot \hat{\alpha}_k. \tag{4.97}$$

With the probability expression written using modified complex decision variables z'_k, the decision borders will be located over the origin, similarly to what was done for the no diversity case. According to the previous definitions, the second moment of r_k is obtained as

$$E\left[|r_k|^2 \mid s\right] = 2\sigma^2_{r_k} = 2|s|^2\sigma^2_{\alpha_k} + N_0, \tag{4.98}$$

the second-order moment of r'_k is given by

$$E\left[|r'_k|^2 \mid s\right] = 2|s|^2\sigma^2_{\alpha_k} + N_0 - 2\mathbf{B}_m(l)Re\left\{E\left[r_k \cdot \hat{\alpha}_k^* \mid s\right]\right\} + \mathbf{B}_m(l)^2 E\left[\left|\hat{\alpha}_k\right|^2\right] \tag{4.99}$$

and the cross moment of r' and $\hat{\alpha}$ is

$$E\left[r'_k\hat{\alpha}^*_k \mid s\right] = E\left[r_k \cdot \hat{\alpha}^*_k \mid s\right] - \mathbf{B}_m(l)E\left[\left|\hat{\alpha}_k\right|^2\right]. \tag{4.100}$$

As for the cross-correlation coefficient, it is defined as

$$\mu'_k = \frac{E\left[r'_k\hat{\alpha}^*_k \mid s\right]}{\sqrt{E\left[|r'_k|^2 \mid s\right]E\left[|\hat{\alpha}_k|^2 \mid s\right]}} = \left|\mu'_k\right| e^{-\varepsilon'_k \cdot \mathrm{j}}. \tag{4.101}$$

Using modified decision variable $z'_{\text{re}} = \sum_{k=1}^{L} z'_{\text{re}_k}$, probability 4.95 is equivalent to the one derived for QPSK in Section 3.1.1. Therefore, the computation is performed using the same analytical expression, which we reproduce here:

$$\text{Prob}\left(z'_{\text{re}} < 0 \mid s\right) = \prod_{k=1}^{L'}\left(\frac{-2H_k}{F_k}\right)^{\theta_k}\sum_{l=1}^{L'}\sum_{i=1}^{\theta_l}(-1)^i A^1_{l,i}(G_l + H_l)^{-i}, \tag{4.102}$$

with

$$A^1_{l,i} = \frac{1}{(\theta_l - i)!}\left[\frac{\partial^{\theta_l - i}}{\partial\zeta^{\theta_l - i}}\left(\prod_{\substack{j=1\\ j\neq l}}^{L'}\frac{1}{(\zeta - G_j - H_j)^{\theta_j}}\prod_{j=1}^{L'}\frac{1}{(\zeta - G_j + H_j)^{\theta_j}}\right)\right]_{\zeta = G_l + H_l}, \tag{4.103}$$

$$F_k = 2\sigma_{r'_k}\sigma_{\hat{\alpha}_k}\sqrt{1 - \left|\mu'_k\right|^2\left(\sin\,\varepsilon'_k\right)^2}, \tag{4.104}$$

$$G_k = \frac{|\mu'_k|\cos\,\varepsilon'_k}{\sigma_{r'_k}\sigma_{\hat{\alpha}_k}\left(1 - |\mu'_k|^2\right)} \tag{4.105}$$

and

$$H_k = \frac{\sqrt{1 - |\mu'_k|^2(\sin\,\varepsilon'_k)^2}}{\sigma_{r'_k}\sigma_{\hat{\alpha}_k}\left(1 - |\mu'_k|^2\right)}. \tag{4.106}$$

Assuming the same model for the channel estimates of the different diversity branches as the one used previously for the no diversity case, that is, $\hat{\alpha}_i = a \cdot \alpha_i + \xi_i$, and applying the derived BER expressions, we present in Figure 4.16 performance curves for a 16-HQAM constellation with $k_1 = 0.4$ for several equal receiving antennas. The channel estimation error variance is $E[|\xi_i|^2] = 0.1 \cdot N_0 + 0.001 \cdot \Omega_i^2$. All the curves show irreducible error floors, which are higher for the LPB, as expected. It is visible that, even though the channel estimation error is the same for all cases, the use of multiple diversity branches can lower the error floor considerably.

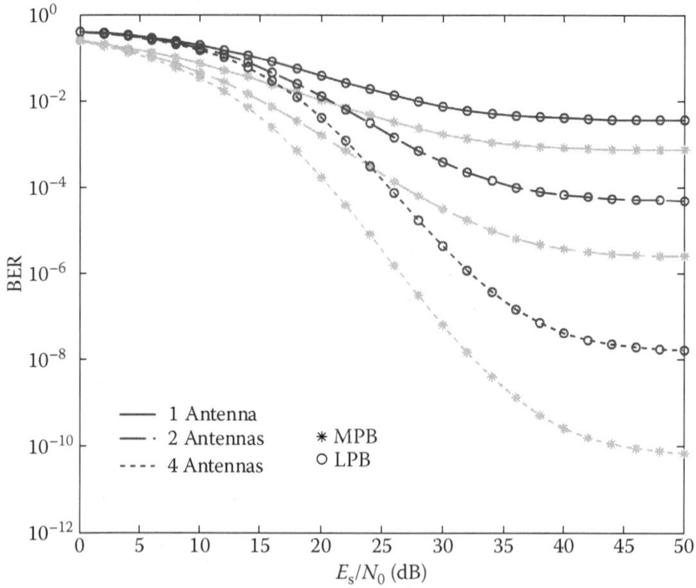

Figure 4.16 BER performance of 16-HQAM constellation, $k_1 = 0.4$, with several diversity branches for a channel estimation error variance given by $E[|\xi|^2] = 0.1 \cdot N_0 + 0.001 \cdot \Omega_i^2$.

4.3.2 Enhanced Channel Estimation for CDMA Transmissions

Starting from a basic CDMA transmitter, as presented in Section 3.4, incorporating hierarchical modulations is relatively straightforward. Figure 4.17 shows a simplified transmission chain employing M-HQAM square constellations with $m = 1/2 \cdot \log_2 M$ input bit streams with different error protection levels.

Each stream is encoded and rate matching is performed (usually puncturing) for achieving the desired bit rate. For the encoders, a powerful channel code such as a turbo

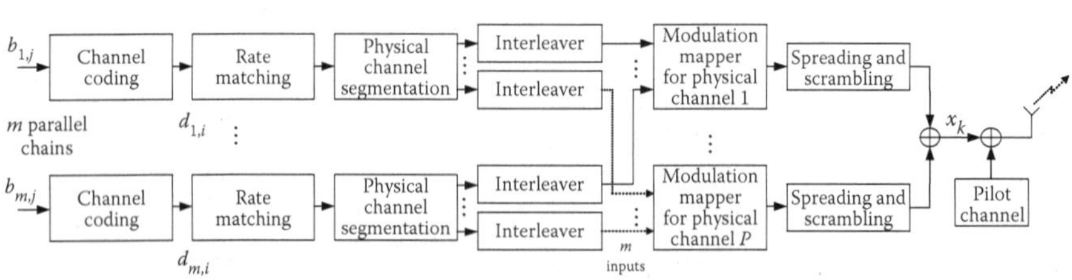

Figure 4.17 WCDMA transmitter chain incorporating hierarchical constellations.

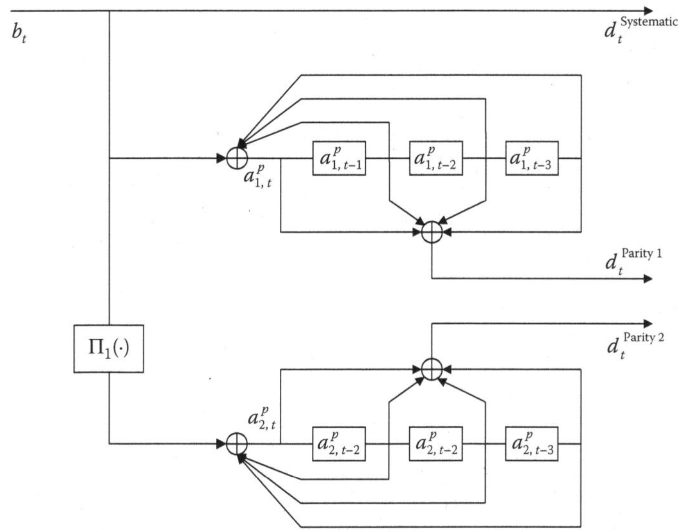

Figure 4.18 Rate 1/3 turbo encoder scheme composed of two parallel concatenated recursive systematic convolutional encoders separated by an interleaver. For each input information bit b_t, one systematic bit, $d_t^{\text{systematic}}$, and two parity bits, $d_t^{\text{parity 1}}$ and $d_t^{\text{parity 2}}$, are outputted.

code [Berrou et al. 1993], whose encoder structure is portrayed in Figure 4.18, is typically employed. After rate matching, each stream is segmented into P physical channels (each physical channel will be spread by a different OVSF), which are individually interleaved. The physical channels of the m processing chains are mapped onto the constellation symbols in the modulation mappers according to the importance attributed to the chain. The mapping is performed independently of the I and Q branches of the modulated symbols s_p (p denotes the pth physical channel) according to

$$s_p = \sum_{l=1}^{\log_2\sqrt{M}} \left[(-1)^l \frac{D_{\log_2\sqrt{M}-l+1}}{2} \prod_{m=1}^{l} d_{m,p,i} \right] + \sum_{l=1}^{\log_2\sqrt{M}} \left[(-1)^l \frac{D_{\log_2\sqrt{M}-l+1}}{2} \prod_{m=1}^{l} d_{m,p,i+1} \right] \cdot i, \tag{4.107}$$

where $d_{m,p,i}$ ($d_{m,p,i} \in \{-1, 1\}$) denotes the ith coded bit ($i = 1, \ldots, N_c$, with N_c being the number of coded bits in the frame) of the pth physical channel ($p = 1, \ldots, P$) of the mth parallel information stream ($m = 1, \ldots, 1/2 \cdot \log_2 M$). The modulated symbols are spreaded and scrambled, and the resulting physical channels are summed. Pilot symbols are added to the sequence before transmission. As discussed in Chapter 3, two methods for the transmission of pilot symbols, namely conventional data multiplexed pilots (also named PSAM) and superimposed pilots, can be employed in CDMA systems in order to achieve coherent detection at the receiver. The latter one is more typical in the downlink since the use of orthogonal spreading codes can avoid interference between superimposed pilot and data channels. Nevertheless, data multiplexed pilots can be useful for transmitting to users near the cell border, since they can be power controlled. In the following paragraphs, we will address in detail channel estimation based on the transmission of an orthogonal pilot channel.

Although the receiver could simply be the conventional RAKE receiver described in Chapters 2 and 3, for higher-order modulations it is important to employ a more robust one. In fact, mobile propagation conditions in CDMA systems usually result in frequency-selective channels that produce MPI due to superimposed signal replicas. Therefore, although in one path the physical channels are orthogonal, the presence of other multipath replicas with relative delays higher than the chip period destroys this orthogonality leading to interpath interference. Therefore, each multipath component is composed of several parallel physical channels that are not orthogonal to the other multipaths. This has an important impact on the link performance of high-order constellations (M-HQAM with $M > 4$), since they are very sensitive to interference, especially the LPBs. Moreover, large hierarchical constellations have also high channel estimation requirements as we have shown in the previous subsection. Since the usage of powerful FEC codes such as turbo codes allows the system to operate in low SNRs, the computation of accurate channel estimates becomes more complicated. Therefore, a convenient strategy to deal with both problems at the receiver has to be applied.

Regarding the MPI problem, several solutions to improve the performance have been studied in the literature for high-order constellations. In Ref. [Klein 1997], a linear chip equalizer employing ZF and MMSE followed by a code MF was proposed for suppressing multiuser interference in the downlink. This equalizer can also be employed for MPI suppression. In Refs. [Higuchi et al. 2002] and [Silva et al. 2004a], the use of an MPI canceler (MPIC) was proposed for 16-QAM HSDPA transmissions. These MPICs are based on the concept of the suboptimal interference cancelation performed by subtractive MUD schemes usually employed in the reverse link [Varanasi and Aazhang 1990; Patel and Holtzman 1994; Divsalar et al. 1998; Johansson and Svensson 1999]. The implementation of these schemes in the downlink connection takes into consideration the peculiarities of this link where all the parallel physical channels (for the same user or for different users) are orthogonal in a path and all face the same propagation channel. In Ref. [Higuchi et al. 2002], the MPIC was compared against the chip equalizer proposed in Ref. [Klein 1997]. It was shown that the MPIC is substantially less complex and although its performance is worse for high SNR values, it is able to obtain slightly better performances for low SNRs.

The presence of a turbo code (or another powerful channel code, such as LDPC codes) in the system makes it relatively easy to design an iterative receiver based on the MPIC philosophy that uses feedback information from the decoder for estimating and removing the interpath interference. This approach is based on the concept of turbo equalization [Douillard et al. 1995], where an equalizer and a channel decoder exchange information iteratively for suppressing the ISI caused by the channel. When applying the turbo equalization principle, it is common to implement the channel equalizer using soft output Viterbi algorithm (SOVA) or maximum *a posteriori* (MAP) algorithms, as was done in Refs. [Douillard et al. 1995] and [Raphaeli and Zarai 1998]. However, the required computational complexity of these algorithms for application in a WCDMA system with several physical channels makes a less complex solution more attractive.

Figure 4.19 presents a possible low-complexity solution. It corresponds to a turbo MPIC capable of providing a signal with a reduced MPI level to each finger in the RAKE similar to the one proposed in Ref. [Reynolds and Wang 2001] for convolutional coded systems. The receiver uses feedback information from the decoder to estimate the MPI and also to improve the channel estimates. This iterative channel estimation approach is called turbo estimation and was studied in Refs. [Sandell 1998], [Nefedov and Pukkila 2000], and [Valenti 2001].

For the operation of the receiver it is assumed that it has knowledge about all physical channels being transmitted. For each iteration (except in the first one), the IC block subtracts the estimated MPI, obtained in the previous iteration, from the signal received by each finger of the RAKE. The RAKE performs MRC of all despread signals processed by the fingers. The result goes into a sequence of processing blocks that perform the inverse operations of the transmitter. The demodulator computes the likelihood probabilities of the received coded bits to be used by the decoders. If turbo codes are employed, the decoder should be a typical iterative turbo decoder whose structure is presented in Figure 4.20. Two SISO decoders are employed, one for each component convolutional code. Both decoders exchange information iteratively, which can be computed using MAP algorithm [Bahl et al. 1974; Berrou et al. 1993], a simplified version of it named max-log-MAP [Robertson et al. 1995] or SOVA. Note that when using turbo codes the iterative receiver simply expands the turbo decoder iteration loops so that interference cancelation and channel estimation are also incorporated inside the loops.

Each decoder must have two outputs. One is the estimated information sequence, $\hat{\mathbf{b}}_m^{(q)}$, and the other is the sequence of LLR estimates of the code symbols, $\tilde{\boldsymbol{\lambda}}_m^{(q)}$. These LLRs are passed through the decision device, which produces either soft-decision or hard-decision estimates of the code symbols. The estimates enter the Transmitted Signal Rebuilder, which performs the same operations of the transmitter. The resulting reconstructed signal goes into the channel estimator and also into the IC, which subtracts interference from the signals fed to each RAKE finger before the next iteration. The interference subtracted from the signals provided to each RAKE finger is composed of the sum of all paths except the one that is going to be extracted by that finger. The interference signals can be weighted before the subtraction. This allows subtracting weaker interference signals in the first iterations since the confidence in the estimated transmitted signal is lower.

For the general case of a DS-CDMA system, the tth received signal sample can be expressed as

$$r_t = \sum_{l=1}^{L} \alpha_{l,t-\tau_l} \left(\sum_{p=1}^{P} s_{p,\left\lfloor \frac{t-\tau_l}{\mathrm{SF}} \right\rfloor+1} \cdot c_{p,t-\tau_l} + s_{\mathrm{pilot},\left\lfloor \frac{t-\tau_l}{\mathrm{SF}_{\mathrm{pilot}}} \right\rfloor+1} \cdot c_{\mathrm{pilot},t-\tau_l} \right) + n_t, \tag{4.108}$$

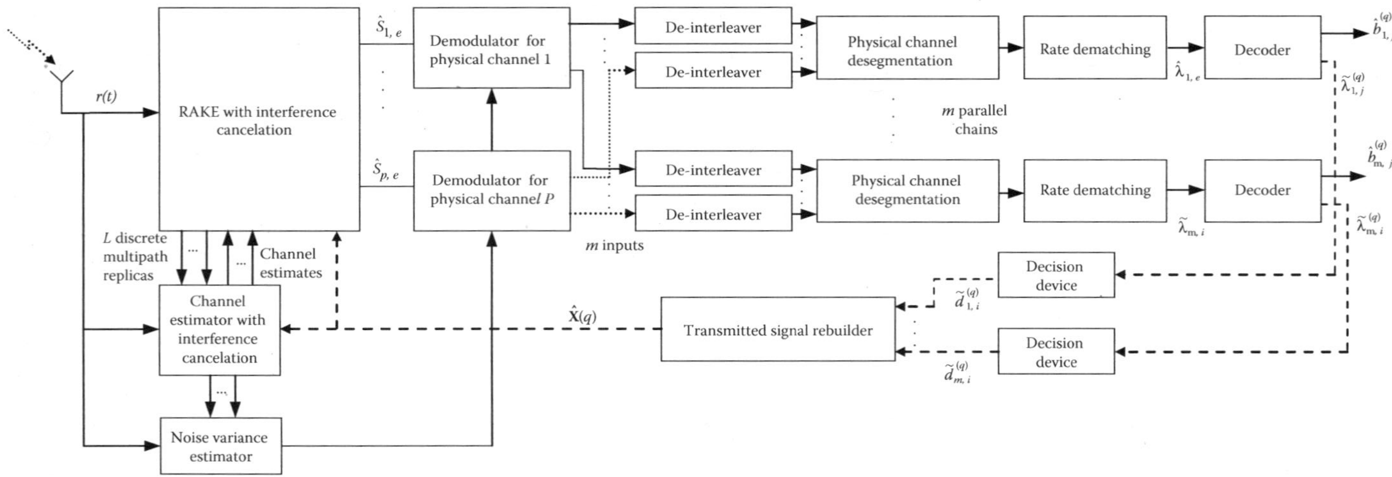

Figure 4.19 Iterative receiver structure with enhanced channel estimation for CDMA systems employing *M*-HQAM modulations.

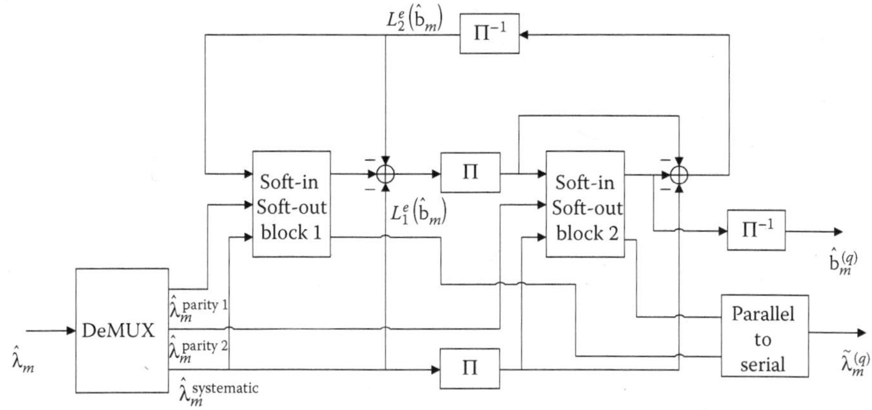

Figure 4.20 Turbo decoder structure.

where $\alpha_{l,t}$ and τ_l are the complex-valued channel gain and the time delay (in samples) of the lth path, L is the number of resolvable paths, P represents the number of physical channels, SF is the spreading factor of the data channels, SF_{pilot} is the spreading factor of the pilot channel (here a value of 256 is employed), $s_{p,t}$ and $c_{p,t}$ represent the modulated symbol and spreading signal of the pth physical channel, while $s_{\text{pilot,t}}$ and $c_{\text{pilot},t}$ correspond to the modulated symbols and spreading signal of the pilot channel. The term n_t is the AWGN noise component.

The eth dispread symbol associated with the lth finger of the pth physical channel is represented as

$$y_{p,l,e} = \frac{1}{\text{SF}} \sum_{t=(e-1)\cdot\text{SF}+1}^{e\cdot SF} r_{t+\tau_l} \cdot c_{p,t}^*. \tag{4.109}$$

Assuming perfect channel estimation, the RAKE MRC combined data sequence of the eth symbols of the pth code channel is expressed as

$$\hat{s}_{p,e} = \sum_{l=1}^{L} \hat{\alpha}_{l,e}^* \cdot y_{p,l,e}, \tag{4.110}$$

where $\hat{\alpha}_{l,e}$ is the estimated channel coefficient for path l in the eth symbol period. The estimated symbols associated with each physical channel are demodulated into LLRs and split into $1/2 \cdot \log_2 M$ different streams. After the physical channel desegmentation and rate dematching, the LLRs of the coded bits, $\tilde{\lambda}_{m,i}$ (m is the index of the information stream, with $m = 1$ corresponding to the most protected stream and $m = 1/2 \cdot \log_2 M$ being the least protected), are fed into the turbo decoder. The turbo decoder performs one decoding iteration and outputs an estimate for the jth information bit, $\hat{b}_{m,j}^{(q)}$, and also the LLR estimates of the code symbols $\tilde{\lambda}_{m,i}^{(q)}$ (q is the iteration number and i is the coded bit number). The decision device then uses these LLRs to estimate the coded bit values. It can perform either soft or hard decision. In the case of hard decision, the bit estimates are computed as

$$\tilde{d}_{m,i}^{(q)} = \begin{cases} 1, & \tilde{\lambda}_{m,i}^{(q)} \geq 0 \\ -1, & \tilde{\lambda}_{m,i}^{(q)} < 0 \end{cases}. \tag{4.111}$$

Soft decisions are obtained by computing the expected values of the bits, that is, as

$$\begin{aligned} \tilde{d}_{m,i}^{(q)} &= E\left[d_{m,i}\right] \\ &= (+1) \cdot \text{Prob}(d_{m,i} = +1) + (-1) \cdot \text{Prob}(d_{m,i} = -1) \\ &= 2 \cdot \text{Prob}(d_{m,i} = +1) - 1. \end{aligned} \tag{4.112}$$

Recalling the LLRs' definition,

$$\tilde{\lambda}_{m,i}^{(q)} = \log \frac{\text{Prob}(d_{m,i} = +1)}{\text{Prob}(d_{m,i} = -1)}, \tag{4.113}$$

we can obtain the probability of the respective bit being 1 as

$$\begin{aligned} \text{Prob}(d_{m,i} = +1) &= \mathrm{e}^{\tilde{\lambda}_{m,i}^{(q)}} \text{Prob}(d_{m,i} = -1) \\ &= \mathrm{e}^{\tilde{\lambda}_{m,i}^{(q)}} \left(1 - \text{Prob}(d_{m,i} = +1)\right) \Leftrightarrow \text{Prob}(d_{m,i} = +1) = \frac{\mathrm{e}^{\tilde{\lambda}_{m,i}^{(q)}}}{\mathrm{e}^{\tilde{\lambda}_{m,i}^{(q)}} + 1}. \end{aligned} \tag{4.114}$$

Inserting Equation 4.114 into 4.112 leads to

$$\tilde{d}_{m,i}^{(q)} = \frac{\mathrm{e}^{\tilde{\lambda}_{m,i}^{(q)}} - 1}{\mathrm{e}^{\tilde{\lambda}_{m,i}^{(q)}} + 1} = \tanh\left(\frac{\tilde{\lambda}_{m,i}^{(q)}}{2}\right). \tag{4.115}$$

These coded bit values are then modulated into symbols $\tilde{s}_p$ (p denotes the pth physical channel). Mapping the bits into M-HQAM constellation symbols can be performed independently of the I and Q branches according to Equation 4.107, where coded bit estimates are used instead of the exact ones. It is important to note that although in the proposed scheme the decision function is applied to the LLRs after the decoder and before the modulator, it could alternatively be applied directly during the remodulation process. Nevertheless, for the case of soft decision, both approaches are equivalent. To see this we deduce the expected value of the modulated symbols assuming that the bits are uncorrelated,

$$\begin{aligned}
\tilde{s}_p &= E[s_p] \\
&= \sum_{l=1}^{\log_2 \sqrt{M}} (-1)^l \frac{D_{\log_2 \sqrt{M}-l+1}}{2} \prod_{m=1}^{l} E[d_{m,i}] \\
&\quad + \sum_{l=1}^{\log_2 \sqrt{M}} (-1)^l \frac{D_{\log_2 \sqrt{M}-l+1}}{2} \prod_{m=1}^{l} E[d_{m,i+1}] \cdot \mathrm{i} \\
&= \sum_{l=1}^{\log_2 \sqrt{M}} (-1)^l \frac{D_{\log_2 \sqrt{M}-l+1}}{2} \prod_{m=1}^{l} \tilde{d}_{m,i}^{(q)} \\
&\quad + \sum_{l=1}^{\log_2 \sqrt{M}} (-1)^l \frac{D_{\log_2 \sqrt{M}-l+1}}{2} \prod_{m=1}^{l} \tilde{d}_{m,i+1}^{(q)} \cdot \mathrm{i}.
\end{aligned} \tag{4.116}$$

which is the same expression as in Equation 4.107.

These symbols can then be used to reconstruct the estimate of the transmitted signal

$$\tilde{x}_t^{(q)} = \sum_{p=1}^{P} \tilde{s}_{p,\left\lfloor \frac{t}{\mathrm{SF}} \right\rfloor+1} \cdot c_{p,t}. \tag{4.117}$$

After this, the MPI replica associated with the lth path is estimated as

$$\hat{I}_{l,t}^{(q)} = \hat{\alpha}_{l,t} \cdot \tilde{x}_t^{(q)}. \tag{4.118}$$

The interference subtracted from the signal fed to a RAKE finger is computed as the sum of all paths (with their corresponding relative delays) except the one that is going to be extracted by that finger. Hence, the input to the lth finger in the qth MPIC iteration can be represented as

$$r_{t,l}^{(q)} = r_t - w_q \sum_{\substack{j=1 \\ j \neq l}}^{L} \hat{I}_{j,t-\tau_j}^{(q-1)}, \tag{4.119}$$

where w_q is a real-valued weight factor that takes values inside the interval [0 1] and usually increases with the iteration number. This weight factor is used to reduce the impact of possible data decision errors present in the estimated MPI replicas, which are usually higher in the first iterations. After Q iterations, the $\hat{b}_{m,j}^{(Q)}$ values are used as final estimates for the information bit streams.

The transmission of a pilot channel, orthogonal to the data channels, allows a simple channel estimation processing at the receiver, which is basically the same as the one described in Chapter 3 for QPSK. However, we can take advantage of the iterative receiver structure and enhance the channel estimation process. To obtain the channel estimates for each path l, the receiver performs the following tasks:

1. Despread the received signal using

$$y_{\text{pilot},l,g} = \frac{1}{\text{SF}_{\text{pilot}}} \sum_{t=(g-1)\cdot\text{SF}_{\text{pilot}}+1}^{g\cdot\text{SF}_{\text{pilot}}} r_{t+\tau_l} \cdot c^*_{\text{pilot},t}. \tag{4.120}$$

where SF_{pilot} is the spreading factor employed in the pilot channel.

2. Obtain noisy channel estimates, $\hat{\alpha}^{\text{noisy}}_{l,g}$, in each pilot symbol position, g, by multiplying the despreaded pilot symbols, $s^*_{\text{pilot},g}$, by its conjugates, with

$$\hat{\alpha}^{\text{noisy}}_{l,g} = \frac{s^*_{\text{pilot},g}}{\left|s_{\text{pilot},g}\right|^2} \cdot y_{\text{pilot},l,g}. \tag{4.121}$$

3. The noisy channel estimates are then passed by a moving average filter with length W. The moving average filter is employed since it has low complexity, does not require knowledge of the fading rate or autocorrelation of the channel and, for slowly varying channels, it is basically equivalent to the MMSE estimate, as explained in Chapter 3. Therefore, final channel estimates are computed according to

$$\hat{\alpha}_{l,g} = \frac{1}{W} \sum_{i=g-\lfloor W/2 \rfloor}^{g+\lceil W/2 \rceil-1} \hat{\alpha}^{\text{noisy}}_{l,i}. \tag{4.122}$$

4. Since data channels can have different data rates compared to the pilot symbol rate, interpolation can be performed over the channel estimates for matching the rates. The interpolation can be implemented as a simple repeater since it is assumed that the channel is approximately stable during a pilot symbol interval.

5. After the first decoding iteration, the channel estimation can be improved by applying interference cancelation to the pilot channel as shown in Figure 4.19. The four previous processing steps are repeated but the first step is performed using the received signal with interference cancelation applied (Equation 4.119), resulting in

$$y^{(q)}_{\text{pilot},l,g} = \frac{1}{\text{SF}_{\text{pilot}}} \sum_{t=(g-1)\cdot\text{SF}_{\text{pilot}}+1}^{g\cdot\text{SF}_{\text{pilot}}} r^{(q)}_{t+\tau_l,l} \cdot c^*_{\text{pilot},t}. \tag{4.123}$$

The transmitted symbol estimates can also be used as additional pilots for improving the channel estimates. However, when considering the transmission of a pilot channel in parallel to data and with interference cancelation being applied to it, no significant improvement is achieved with that approach.

In the receiver, the channel decoder may require knowledge of noise statistics and thus it is necessary to estimate the noise variance, $\hat{\sigma}^2$. First, the noise is estimated using

$$\hat{n}_g = \frac{1}{\mathrm{SF_{pilot}}} \sum_{l=1}^{L} \sum_{t=(g-1)\cdot \mathrm{SF_{pilot}}+1}^{g\cdot \mathrm{SF_{pilot}}} r_{t+\tau_l} \cdot c^*_{\mathrm{pilot},t} - s_{\mathrm{pilot},g} \sum_{l=1}^{L} \hat{\alpha}_{l,g}, \qquad (4.124)$$

where g is the pilot symbol position. Using the resulting sequence of estimated noise samples, the respective noise variance can then be easily computed. Although we have been considering the use of orthogonal pilot channels, it is also possible to send pilot symbols time multiplexed with data in CDMA transmissions, as explained in Chapter 3. The approach used for improving the channel estimates iteratively for the case of an orthogonal pilot channel transmission can be directly applied also for this case.

As an example of the use of hierarchical constellations, we present in Figures 4.21 and 4.22 performance results obtained for a 16-HQAM ($k_1 = 0.5$) downlink transmission using a WCDMA-based Monte Carlo simulator. The CIR corresponds to the Vehicular A described in Appendix A. This environment is characterized for having several multipath replicas and thus a high level of MPI. Each tap is assumed to suffer Rayleigh fading and the velocity considered is $v = 120$ km/h. The main simulation parameters are presented in Table 4.1, corresponding to two parallel 1.8 Mbps transmissions (total of 3.6 Mbps).

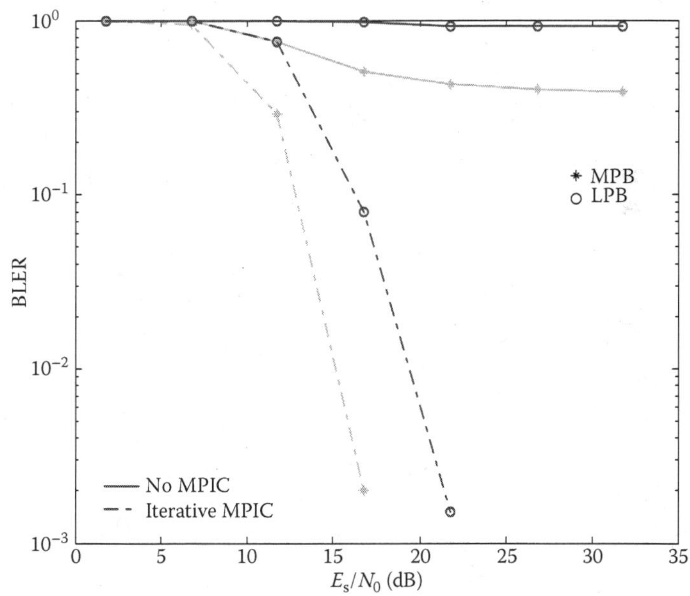

Figure 4.21 Impact of interference cancelation on a transmission rate of 2 x 1.8 Mbps (*SF* = 16, *P* = 5, and *R* = 9/12) using a 16-HQAM constellation (k_1 = 0.5) in Vehicular A environment, *v* = 120 km/h. Perfect channel estimation assumed.

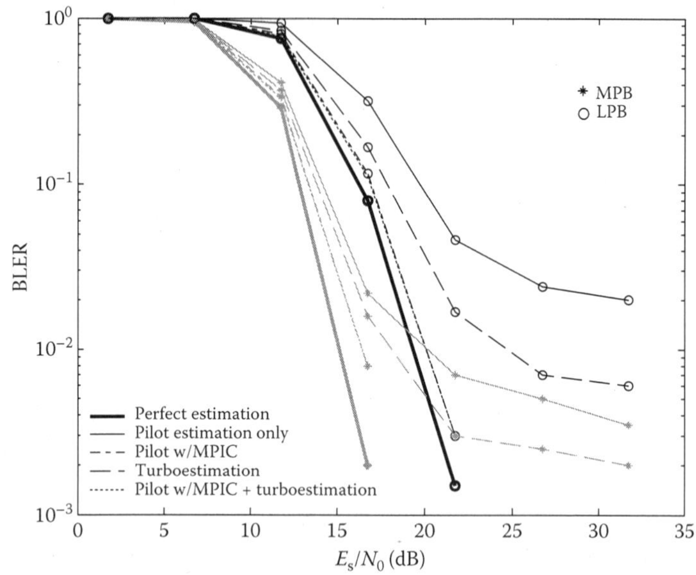

Figure 4.22 Performance results for a transmission rate of 2 × 1.8 Mbps ($SF = 16$, $P = 5$, and $R = 9/12$) using a 16-HQAM constellation ($k_1 = 0.5$) in Vehicular A environment, $v = 120$ km/h, with different channel estimation strategies.

The pilot channel uses $SF_{pilot} = 256$ with 10% of the total transmitted power. The iterative receiver described previously was employed with eight iterations, while channel estimation was performed with a moving average filter with length $W = 8$.

To show the importance of a convenient interference cancelation approach for HQAM constellations, Figure 4.21 compares the performance with and without the

Table 4.1 Simulation Parameters for 5 MHz BW

Transmission BW	5 MHz
Chip period, T_S (ns)	260
Data spreading factor, SF	16
Physical channels, P	5
Coding rate, R	9/12
Transmission rate (Mbps)	2 × 1.8
Pilot spreading factor, SF_{pilot}	256
Pilot channel power (% of total)	10

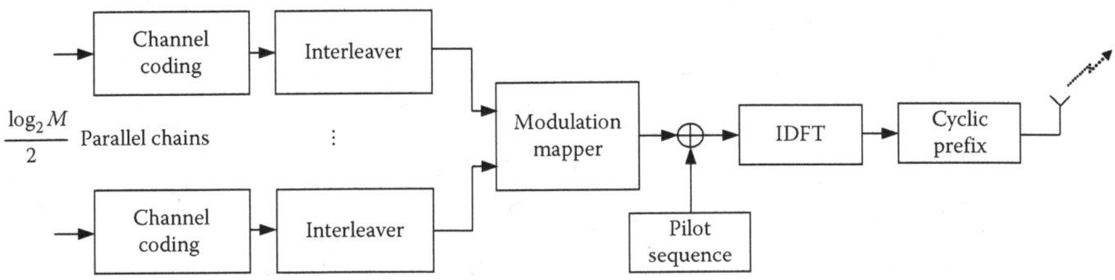

Figure 4.23 OFDM transmitter chain incorporating hierarchical constellations.

iterative MPIC. To isolate its impact, perfect channel estimation is assumed. It is visible that by using only a RAKE receiver without any sort of interference cancelation technique, it is not possible to accomplish the transmission in this environment. With the use of the turbo MPIC, severe MPI can be substantially reduced and the performance hugely improves.

Perfect channel estimation and four different nonideal channel estimation cases are compared in Figure 4.22. The first case corresponds to standard channel estimation using the pilot channel directly, the second case applies the MPIC to remove the estimated interpath interference from the pilot channel before channel estimation, the third case uses the pilot channel and the data channels (referenced as "turboestim" in the legend), and the last case uses the previous two approaches simultaneously. From the results, it seems clear that with standard estimation using only the pilot channel, the performance is severely degraded and cannot achieve a BLER target of 10^{-2}. The enhanced channel estimation technique that uses the estimated data channels as new pilots is able to improve the performance but it is still far from the case with perfect channel knowledge. When the channel is estimated using the pilot channel with interference cancelation, a substantial improvement is obtained and the results get very close to the case of perfect estimation. When using turbo estimation and interference cancelation, the results are practically overlapping and difficult to distinguish from those obtained for the case of pilot channel with interference cancelation, which means that it is not necessary to use the estimated data channels to improve the channel estimation as long as an interference canceler for the pilot channel is employed in each iteration.

4.3.3 Enhanced Channel Estimation for OFDM

After addressing CDMA systems, in this subsection we study the use of M-HQAM constellations as an efficient method for supporting MBMSs in OFDM-based systems. To achieve reliable coherent detection for all bit streams with these constellations, we will consider the usage of the two different pilot symbols transmission methods previously discussed: data multiplexed pilots and superimposed pilots.

In Figure 4.23, we show a possible configuration for an OFDM transmitter chain that incorporates M-HQAM constellations. In this scheme, there are $1/2 \cdot \log_2 M$ parallel chains for the different input bit streams. Each stream is encoded, interleaved, and mapped into the constellation symbols according to the importance attributed to it.

Pilot symbols are inserted into the modulated symbols sequence using one of two possible methods, which is then converted into the time domain using an IDFT with the CP added before transmission. Note that similar to the CDMA transmitter, each individual chain is similar to a unicast transmission scheme based on QPSK. The main difference is the addition of the modulation mapper that joins the parallel streams.

4.3.3.1 Conventional Pilots

In the first pilot transmission method that we will consider, known pilot symbols are periodically inserted into the data stream, both in time and in frequency as explained previously in Chapter 3. The respective frame is shown in Figure 4.24. According to this structure, in an OFDM system with N carriers, pilot symbols are multiplexed with data symbols using a spacing of ΔN_{T} OFDM blocks in the time domain and ΔN_{F} subcarriers in the frequency domain.

Before being transmitted, the sequence of symbols is converted into the time domain through $\{x_{i,l}, i = 0, 1, \ldots, N-1\} = \mathrm{IDFT}\{S_{k,l}, \; k = 0, 1, \ldots, N-1\}$, where $S^m_{k,l}$ is the symbol transmitted on the kth subcarrier of the lth OFDM block. The OFDM signal is then expressed as

$$x(t) = \sum_{l} \sum_{i=-N_G}^{N-1} x_{i,l} \cdot h_T(t - i \cdot T_s), \tag{4.125}$$

with T_s denoting the symbol duration, N_{G} the number of samples at the CP (with $x_{-i,l} = x_{N-i,l}, i = 1, \ldots, N_{\mathrm{G}}$), and $h_{\mathrm{T}}(t)$ the adopted pulse shaping filter.

Owing to a higher sensitivity to channel estimation errors of HQAM constellations, we can employ a receiver capable of jointly performing channel estimation and data detection through iterative processing. Figure 4.25 presents the structure of the iterative receiver for the case of data multiplexed pilots transmission.

According to the figure, the signal, which is considered to be sampled and with the CP removed, is converted to the frequency domain after an appropriate size-N DFT operation. If the CP is longer than the overall CIR, the resulting received sequence can

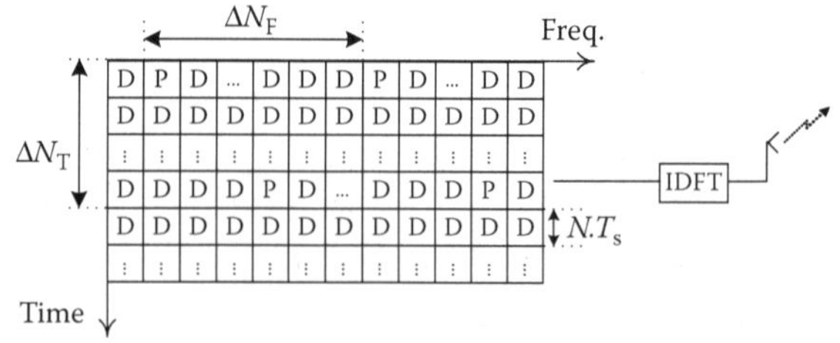

Figure 4.24 Frame structure for an OFDM transmission with data multiplexed pilots (P: pilot symbol, D: data symbol, and T_s: symbol duration).

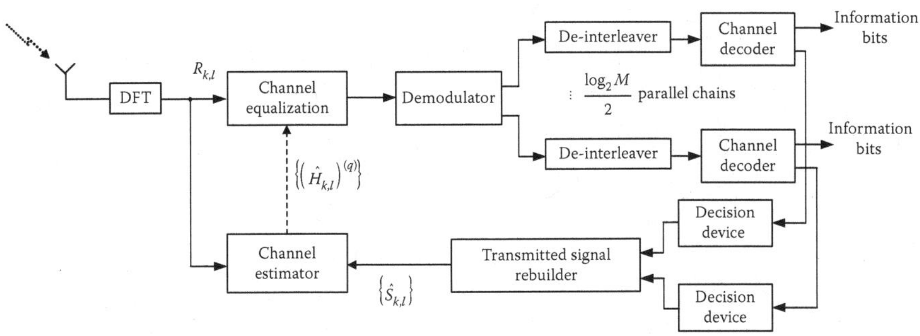

Figure 4.25 Iterative receiver structure with enhanced channel estimation for OFDM systems employing *M*-HQAM modulations and data multiplexed pilots.

be expressed as

$$R_{k,l} = S_{k,l} H_{k,l} + N_{k,l}, \tag{4.126}$$

with $H_{k,l}$ denoting the overall channel frequency response for the kth frequency of the lth time block and $N_{k,l}$ denoting the corresponding channel noise sample. The sequence of received samples enter the equalizer, which computes symbol estimates as

$$\left(\tilde{S}_{k,l}\right)^{(q)} = \frac{\left(\hat{H}_{k,l}\right)^{(q)*} (R_{k,l})^{(q)}}{\left|\left(\hat{H}_{k,l}\right)^{(q)}\right|^2}, \tag{4.127}$$

where $(\hat{H}_{k,l})^{(q)}$ represent channel frequency response estimates and q is the current iteration. After equalization, the following processing steps are similar to the ones performed in the CDMA receiver presented previously. The estimated symbol sequence passes through the demodulator that computes LLRs for the different bit streams, which are then deinterleaved and decoded in the channel decoder blocks. Each channel decoder outputs the respective estimated information sequence along with the sequence of LLR estimates for the code symbols. These LLRs enter the decision device for producing either soft-decision or hard-decision estimates of the code symbols, which are then used by the Transmitted Signal Rebuilder for reconstruction of the transmitted signal (in the frequency domain). The reconstructed symbol sequence can then be employed for a refinement of the channel estimates for the succeeding iteration.

Regarding the channel estimation process, the sequence of steps can be performed as follows:

1. Channel estimates for each pilot symbol position are simply computed as

$$\tilde{H}_{k,l} = \frac{\left(S_{k,l}^{\text{Pilot}}\right)^*}{\left|S_{k,l}^{\text{Pilot}}\right|^2} R_{k,l}, \tag{4.128}$$

where $S_{k,l}^{\text{Pilot}}$ corresponds to a pilot symbol transmitted in the kth subcarrier of the lth OFDM block. Obviously not all indexes k and l will correspond to a pilot symbol since, forcedly, we have $\Delta N_{\text{T}} > 1$ or $\Delta N_{\text{F}} > 1$.

2. Channel estimates for the same subcarrier k but in time-domain positions (index l) that do not carry a pilot symbol can be obtained through interpolation using an FIR filter with length W as follows:

$$\tilde{H}_{k,l+t} = \sum_{j=-\lfloor (W-1)/2 \rfloor}^{\lfloor W/2 \rfloor} h_t^j \tilde{H}_{k,l+j\cdot\Delta N_{\text{T}}}, \tag{4.129}$$

where t is the OFDM block index relative to the last one carrying a pilot (whose index is l) and h_t^j are the interpolation coefficients of the estimation filter that depend on the channel estimation algorithm employed. As explained in Chapter 3 there are several possible algorithms proposed in the literature such as the optimal Wiener filter interpolator [Cavers 1991] or the low-pass sinc interpolator [Kim et al. 1997].

3. Interpolation in the frequency domain can be accomplished by ensuring that the corresponding impulse response in the time domain has a length of N_G samples. This is performed through the computation of the time-domain impulse response of Equation 4.129 using $\{(\tilde{h}_{i,l})^{(q)};\ i = 0, 1, \ldots, N-1\} = \text{DFT}\{(\tilde{H}_{k,l})^{(q)};\ k = 0, 1, \ldots, N-1\}$, followed by the truncation of this sequence according to $\{(\hat{h}_{i,l})^{(q)} = w_i(\tilde{h}_{i,l})^{(q)}; i = 0, 1,\ \ldots,\ N-1\}$ with $w_i = 1$ if the ith time-domain sample is inside the CP duration and $w_i = 0$ otherwise. The final frequency response estimates are then simply computed using $\{(\hat{H}_{k,l})^{(q)};\ k = 0, 1, \ldots, N-1\} = \text{IDFT}\{(\hat{h}_{i,l})^{(q)}; i = 0, 1, \ldots, N-1\}$.

4. After the first iteration, data estimates can also be used as additional pilots for channel estimation refinement. The respective channel estimates are computed as

$$\left(\hat{H}_{k,l}\right)^{(q)} = \frac{R_{k,l}^n \left(\hat{S}_{k,l}\right)^{(q-1)*}}{\left|\left(\hat{S}_{k,l}\right)^{(q-1)}\right|^2}. \tag{4.130}$$

4.3.3.2 Superimposed Pilots

The alternative pilot transmission method that we will consider corresponds to the use of superimposed pilots. The frame structure and the pilot insertion process are shown in Figure 4.26. According to this structure, superimposed pilots are generated using a grid with a spacing of ΔN_{T} symbols in the time domain and ΔN_{F} symbols in the frequency domain.

The transmitted sequence is given as

$$X_{k,l} = S_{k,l} + S_{k,l}^{\text{Pilot}}, \tag{4.131}$$

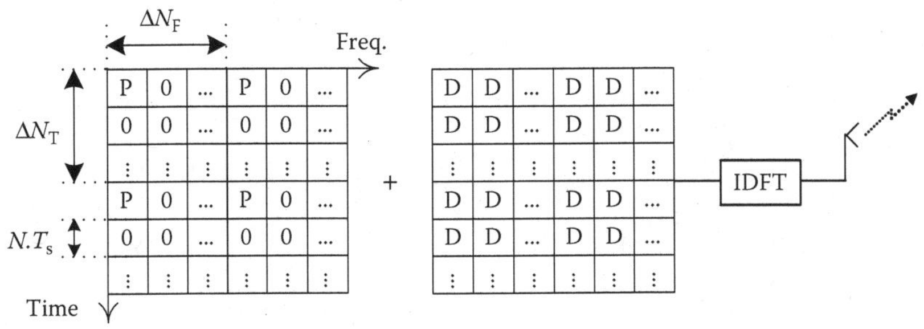

Figure 4.26 Frame structure for an OFDM transmission with superimposed pilots (P: pilot symbol, D: data symbol, and T_s: symbol duration).

where $S_{k,l}^{\text{Pilot}}$ is the implicit pilot transmitted over the kth subcarrier in the lth OFDM block and $S_{k,l}$ is the respective data symbol. The resulting sequence is converted into the time domain through the usual process, $\{x_{i,l}, i = 0, 1, \ldots, N-1\} = \text{IDFT}\{X_{k,l}, k = 0, 1, \ldots, N-1\}$, before being transmitted. OFDM signals can then be expressed also as Equation 4.125.

The transmission of pilot symbols superimposed on data will clearly result in interference between them. To reduce the mutual interference and achieve reliable channel estimation and data detection, we can employ a receiver capable of jointly performing these tasks through iterative processing, which can be a modified version of the receiver studied for conventional pilots (Figure 4.25). The structure of the modified receiver is shown in Figure 4.27. The main differences lie in the addition of the "Remove Pilots" and "Remove Data" processing blocks.

In this case, after conversion into the frequency domain, the received sampled sequence can be expressed as

$$R_{k,l} = \left(S_{k,l} + S_{k,l}^{\text{Pilot}}\right) H_{k,l} + N_{k,l}. \tag{4.132}$$

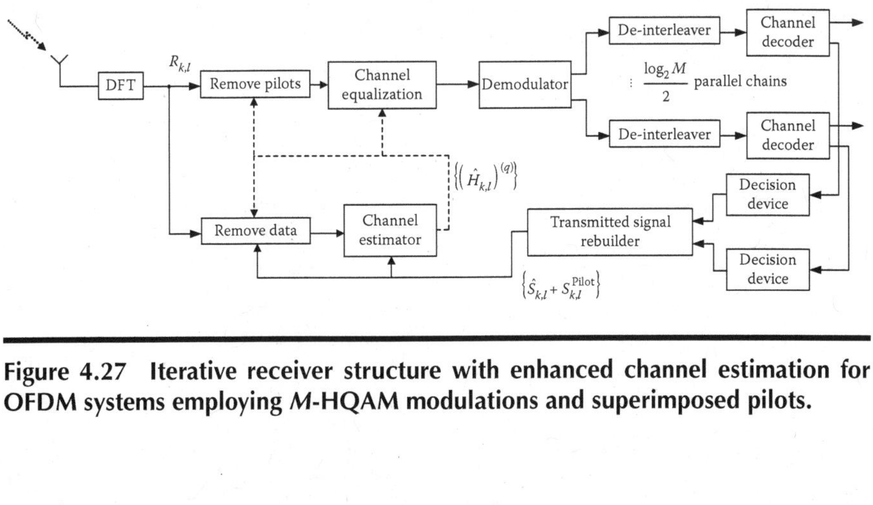

Figure 4.27 Iterative receiver structure with enhanced channel estimation for OFDM systems employing *M*-HQAM modulations and superimposed pilots.

The pilot symbols are removed from the sequence, resulting in

$$\left(Y_{k,l}\right)^{(q)} = R_{k,l} - S_{k,l}^{\text{Pilot}} \left(\hat{H}_{k,l}\right)^{(q)}, \tag{4.133}$$

where $(\hat{H}_{k,l}^{m,n})^{(q)}$ is the channel frequency response estimate and q is the current iteration. The sequences of samples then follow the same processing sequence already described for the conventional pilots method.

For accomplishing reliable channel estimation with superimposed pilots, in each iteration the receiver applies the following steps:

1. Data symbols estimates are removed from the pilots. The resulting sequence becomes

$$\left(\tilde{R}_{k,l}\right)^{(q)} = R_{k,l} - \left(\hat{S}_{k,l}\right)^{(q-1)} \left(\hat{H}_{k,l}\right)^{(q-1)}, \tag{4.134}$$

where $(\hat{S}_{k,l})^{(q-1)}$ and $(\hat{H}_{k,l})^{(q-1)}$ are the data and channel response estimates of the previous iteration. This step can only be applied after the first iteration. In the first iteration, we set $(\tilde{R}_{k,l})^{(1)} = R_{k,l}$.

2. The channel frequency response estimates is computed using a moving average with size W as follows:

$$\left(\tilde{H}_{k,l}\right)^{(q)} = \frac{1}{W} \sum_{l'=l-\lfloor W/2 \rfloor}^{l+\lceil W/2 \rceil - 1} \frac{\left(\tilde{R}_{k,l'}\right)^{(q-1)}}{S_{k,l'}^{\text{Pilot}}}. \tag{4.135}$$

3. After the first iteration, data estimates can also be used as additional pilots for channel estimation refinement. This is especially useful if a fully dense pilot distribution is not employed (i.e., $\Delta N_{\text{F}} \neq 1 \vee \Delta N_{\text{T}} \neq 1$). The respective channel estimates are computed as

$$\left(\tilde{H}_{k,l}\right)^{(q)} = \frac{\left(Y_{k,l}\right)^{(q-1)} \left(\hat{S}_{k,l}\right)^{(q-1)*}}{\left|\left(\hat{S}_{k,l}\right)^{(q-1)}\right|^2}. \tag{4.136}$$

These channel estimates can also be enhanced by ensuring that the corresponding impulse response has a length of N_{G}. This is accomplished by computing the time-domain impulse response of Equations 4.135 and 4.136 through $\{(\tilde{h}_{i,l})^{(q)}; i = 0, 1, \ldots, N-1\} = \text{DFT}\{(\tilde{H}_{k,l})^{(q)}; k = 0, 1, \ldots, N-1\}$, followed by the truncation of this sequence according to $\{(\hat{h}_{i,l})^{(q)} = w_i (\tilde{h}_{i,l})^{(q)}; i = 0, 1, \ldots, N-1\}$ with $w_i = 1$ if the ith time-domain sample is inside the CP duration and $w_i = 0$ otherwise. The final frequency response estimates are then simply computed using $\{(\hat{H}_{k,l})^{(q)}; k = 0, 1, \ldots, N-1\} = \text{IDFT}\{(\hat{h}_{i,l})^{(q)}; i = 0, 1, \ldots, N-1\}$.

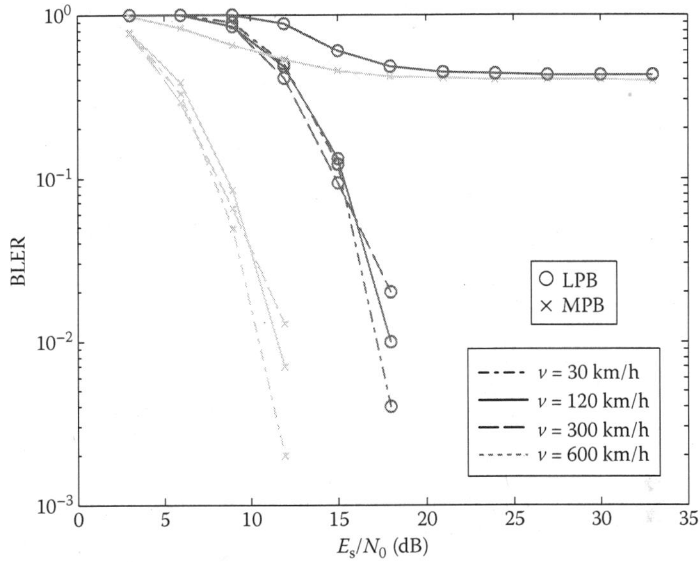

Figure 4.28 Performance of a 16-HQAM (k_1 = 0.4) transmission with data multiplexed pilots ($\Delta N_F = 4$, $\Delta N_T = 7$) for different velocities.

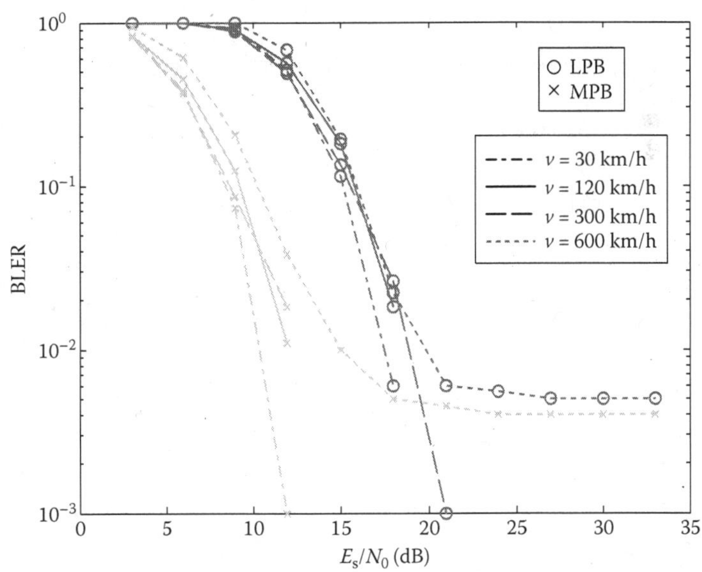

Figure 4.29 Performance of a 16-HQAM (k_1 = 0.4) transmission with superimposed pilots ($\Delta N_F = 0$, $\Delta N_T = 0$, $E\left[|S_{k,l}^{m,\text{Pilot}}|^2\right]/E\left[|S_{k,l}^{m}|^2\right] = -16$ dB) for different velocities.

Table 4.2 Simulation Parameters for 10 MHz BW

Transmission BW	10 MHz
CP length (samples)	72
FFT size	1024
Number of occupied subcarriers	600
Subframe duration (ms)	0.5
Subcarrier spacing (kHz)	15
OFDM symbols per subframe	7

One of the advantages of using superimposed pilots is that it allows us to increase significantly the density of pilots without sacrificing system capacity. In fact, we can have a pilot for each data symbol, which can be important for fast fading channels. As an example, and using an UTRA LTE-based Monte Carlo simulator, Figures 4.28 and 4.29 show the performances of a 16-HQAM ($k_1 = 0.4$) transmission for different velocities employing data multiplexed pilots ($\Delta N_\mathrm{F} = 4$, $\Delta N_\mathrm{T} = 7$) and superimposed pilots ($\Delta N_\mathrm{F} = 0$, $\Delta N_\mathrm{T} = 0$, $E[|S_{k,l}^{m,\mathrm{Pilot}}|^2]/E[|S_{k,l}^{m}|^2] = -16$ dB). A 10-MHz-BW configuration was assumed with the corresponding parameters shown in Table 4.2. The CIR is based on the Vehicular A environment (described in Appendix A) with Rayleigh fading adopted for the different paths.

All the results correspond to equivalent conditions in terms of overall percentage of transmitted energy spent on the pilots. The receivers are the iterative ones described previously (Figures 4.25 and 4.27), employing four main iterations with three inner turbo decoder iterations for each pass. Comparing both figures, we note that although both transmission techniques can attain low BLERs for velocities as high as 300 km/h, the implicit pilot transmission method shows a more robust behavior since it degrades less for very high velocities, for example 600 km/h. The reason is that we are transmitting a pilot symbol in every position and, consequently, it is easier to track the channel in the time domain, whereas the data multiplexed pilots method requires time-domain interpolation to try to estimate the channel for some positions. Note however that for lower velocities, superimposed pilots have a slightly worse performance than data multiplexed pilots due to the mutual interference between data and pilots.

Chapter 5

Multiantenna Systems

As mentioned in the previous chapters, multipath propagation is a typical phenomenon that usually plagues radio communication systems (if an MF receiver is employed). It implies that the received signal is time, frequency, and/or space dependent. Using an array of antenna elements to improve a wireless connection is an old technique, already used by Guglielmo Marconi in 1901 to increase the link gain of the trans-Atlantic transmissions of Morse codes [Berrou et al. 1993]. Marconi used four antennas arranged in a circular array to send the Morse Signal for the letter "S" at a distance of 3425 km.

We have come a long way since the pioneering days of Marconi and the first low-capacity wireless links. Today, array antennas in wireless communication systems are used to improve performance in several ways, not just for the range improvement. A very popular technology is to use smart adaptive antennas systems that consist of several antenna array elements, whose signals are processed adaptively in order to exploit the spatial dimension of the mobile radio channel. MIMO systems can be regarded as an extension of smart adaptive array antennas systems. Traditionally smart antenna systems usually employ multiple antennas on at least one side of the radio link (on the BS and/or at the mobile terminal) and combine them (beam-forming) to offer a more reliable communications link in the presence of adverse propagation conditions such as multipath fading and interference. On the other hand, multiple antennas are employed both at the transmitter and the receiver with low correlation between antennas. The use of multiple antennas combined with advanced signal processing algorithms at the transmitter and the receiver yields significant advantages over smart antennas in terms of capacity and diversity advantages.

MIMO can provide three different gains:

- Array gain, by shaping the array pattern of the antenna array
- Diversity gain, by transmitting the same signal through all the antennas
- SM gain, by sending different data streams through different antennas

In MIMO systems the data stream is demultiplexed into a number of independent streams corresponding to the number of antennas at the transmitter (N_{TX}). Each sub-stream is encoded into transmit symbols using a modulation scheme (e.g., BPSK, QPSK, M-QAM, etc.). Each one is then mapped onto one of the multiple transmit antennas and after upward frequency conversion, filtering, and amplification, the signals are launched into the wireless channel.

At the receiver, the signals are captured by the receive antennas and combined according to a specific algorithm. The demodulation and demapping operations are then performed to recover the message.

5.1 Channel Capacity for Multiantennas Systems

The capacity limit of any (SISO) telecommunications system is taken to be the resulting throughput obtained through the full usage of the allowed spectrum. However, if multiple transmit and receive antennas are employed, the capacity may be raised due to code reuse across transmit antennas. If there is a sufficient number of receive antennas, it is possible to resolve all messages, as long as the channel correlation between antennas is not too high. The pioneer work of Telatar [Telatar 1995] and more recently of Foschini [Foschini and Gans 1998] established the mathematical roots from the information theory field that with multipath propagation, multiple antennas at both the transmitter and the receiver can establish essentially multiple parallel channels that operate simultaneously on the same frequency band and at the same time.

Claude Shannon derived the following capacity formula in 1948 for an AWGN channel [Shannon 1948]:

$$C = W \log_2 \left(1 + \frac{S}{N}\right) \text{[bits/s]}, \tag{5.1}$$

where W is the BW of the channel (in Hz), S is the signal power, and N is the total noise power of the channel (both in watts). In a more practical case of a time variable and randomly fading wireless channel, the capacity is written as

$$C = W \log_2 \left[1 + \frac{S}{N} \cdot \left|\mathbf{H}^2\right|\right] \text{[bits/s]}, \tag{5.2}$$

where $\mathbf{H}^2$ is the normalized channel power transfer function (for the SISO case, $\mathbf{H}$ is a 1×1 unit power complex Gaussian amplitude of the channel).

The channel coding theorem (CCT) formulates two theorems:

- The *direct part* states that for rate $R < C$, there exists a coding system with arbitrarily low block and BERs as the code length increases $N \to \infty$.
- The *converse part* states that for $R \geq C$, the bit and block error rates are strictly higher than zero for any coding system.

The CCT therefore establishes rigid limits on the maximal supportable transmission rate of an AWGN channel in terms of power and BW.

The BW/spectral efficiency (in bits/s/Hz) characterizes how efficiently a system uses its allotted BW and is defined as

$$\eta = \frac{\text{Transmission rate}}{\text{Channel band}}. \tag{5.3}$$

From this, the Shannon limit is calculated as

$$\eta_{\text{MAX}} = \log_2\left(1 + \frac{S}{N}\right). \tag{5.4}$$

The average signal power S can be expressed as

$$S = \frac{k \cdot E_b}{T} = R \cdot E_b, \tag{5.5}$$

where E_b is the energy per bit, k is the number of bits transmitted per symbol, T is the duration of a symbol, and $R = k/T$ is the transmission rate of the system in bits/s. The total noise power is given by

$$N = N_0 W, \tag{5.6}$$

where N_0 is the one-sided noise power spectral density. The Shannon limit is obtained in terms of the bit energy and noise power spectral density, given by

$$\eta_{\text{MAX}} = \log 2\left(1 + \frac{RE}{N_0 W}\right). \tag{5.7}$$

This can be resolved to obtain the minimum bit energy required for reliable transmission, called the Shannon bound

$$\frac{E_b}{N_0} \geq \frac{2^{\eta_{\text{MAX}}} - 1}{\eta_{\text{MAX}}}. \tag{5.8}$$

The "fundamental limit" states that, for infinite amounts of BW, that is, $\eta_{\text{MAX}} \to 0$, we obtain

$$\frac{E_b}{N_0} \geq \lim_{\eta_{\text{MAX}} \to 0} \frac{2^{\eta_{\text{MAX}}} - 1}{\eta_{\text{MAX}}} = \ln(2) = -1.59\,\text{dB}. \tag{5.9}$$

This is the absolute minimum signal energy to noise power spectral density ratio required to reliably transmit one bit of information.

As more N_{RX} receive antennas are deployed, the statistics of the capacity improve (the single input, multiple output [SIMO] case):

$$C = W \log_2 \left(1 + \frac{S}{N} \sum_{rx=1}^{N_{RX}} |h_{rx}|^2 \right), \tag{5.10}$$

where h_{rx} is the gain for receive antenna rx. Note that the crucial feature of Equation 5.10 is that increasing the value of N_{RX} only results in a logarithmic increase in average capacity. Similarly, if transmit diversity is opted for, in the common case where the transmitter does not have channel knowledge, we have a MISO system with N_{TX} antennas and the capacity is given by [Foschini and Gans 1998]

$$C = W \log_2 \left(1 + \frac{S}{N \cdot N_{TX}} \sum_{tx=1}^{N_{TX}} |h_{tx}|^2 \right), \tag{5.11}$$

where the normalization by N_{TX} ensures a fixed total transmitter power and shows the absence of array gain in that case (compared to the case in Equation 5.10, where the channel energy can be combined coherently). Again, note that capacity has a logarithmic relationship with N_{TX}. Now, considering the use of diversity at both the transmitter and the receiver gives rise to a MIMO system. For N_{TX} and N_{RX} antennas, the general capacity equation [Foschini and Gans 1998; Telatar 1995] is obtained:

$$C_{EP} = W \log_2 \left(\det \left(\mathbf{I}_{N_{RX}} + \frac{S}{N \cdot N_{TX}} \mathbf{H}\mathbf{H}' \right) \right), \tag{5.12}$$

where $\mathbf{H}$ is the channel matrix and $\mathbf{H}'$ is the transpose conjugate of $\mathbf{H}$. Note that both Equations 5.11 and 5.12 are based on equal power (EP) uncorrelated sources, hence the subscript in Equation 5.12. Foschini [Foschini and Gans 1998] and Telatar [Telatar 1999] both demonstrated that the capacity in Equation 5.12 grows linearly with $m = \min(N_{TX}, N_{RX})$, rather than logarithmically (as in Equation 5.11). This result can be intuited as follows: the determinant operator yields a product of $\min(N_{TX}, N_{RX})$ nonzero eigenvalues of its (channel-dependent) matrix argument, each eigenvalue characterizing the SNR over a so-called channel eigenmode. An eigenmode corresponds to the transmission using a pair of right and left singular vectors of the channel matrix as transmit and receive antenna weights, respectively. Thanks to the properties of the log, the overall capacity is the sum of capacities of each of these modes. Note that the linear growth predicted by the theory coincides with the transmission example of Figure 5.1.

Clearly, this growth is dependent on the properties of the eigenvalues. If they decayed away rapidly, then linear growth would not occur. However (for simple channels), the eigenvalues have a known limiting distribution [Silverstein 1995] and tend to be spaced out along the range of this distribution. Hence, it is unlikely that most eigenvalues are very small and the linear growth is indeed achieved.

The fundamental consideration is that it is possible to augment the capacity/throughput, depending on the number of transmit and receive antennas. The downside of this scheme is the receiver complexity, sensitivity to interference, and correlation

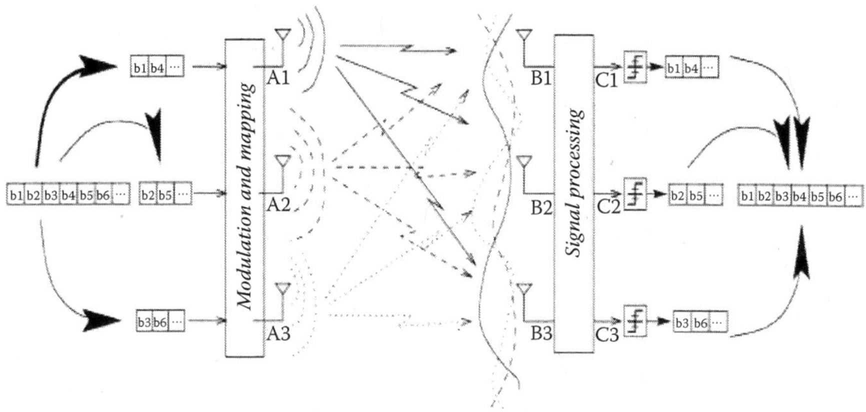

Figure 5.1 SM with three transmit and three receive antennas.

between antennas, which is more significant as the antennas are closer together. For a 3G system, for instance, it is inadequate to consider more than two or four antennas at the UE/mobile receiver since physical space is scarce, especially at the UE since it tends to be made relatively small, that is, to fit the hand of an adult person.

With the capacity defined by Equation 5.12 as a random variable, the issue arises as to how best to characterize it. Two simple methods are commonly used: the mean (or ergodic) capacity [Telatar 1995; Rapajic and Popescu 2000] and capacity outage [Foschini and Gans 1998; Lozano and Valenzuela 2001; Driessen and Foschini 1999; Shiu 1999]. Capacity outage measures (usually based on simulation) are often denoted by $C_{0.1}$ or $C_{0.01}$, that is, those capacity values supported 90% or 99% of the time, and indicate the system reliability. A full description of the capacity would require the PDF or equivalent. Some results are available in Ref. [Sengupta and Mitra 2000] but they are limited.

Some care is necessary in interpreting the previous equations. Capacity, as discussed here and in most MIMO work ([Foschini and Gans 1998] and [Telatar 1995]), is based on a "quasistatic" analysis where the channel varies randomly from burst to burst. Within a burst the channel is assumed fixed and it is also assumed that sufficient bits are transmitted for the standard infinite time horizon of information theory to be meaningful. A second note is that this discussion will concentrate on single-user MIMO systems but many results also apply to multiuser systems with receive diversity. Finally, the linear capacity growth is only valid under certain channel conditions. It was originally derived for the i.i.d. flat Rayleigh fading channel and does not hold true for all cases. For example, if large numbers of antennas are packed into small volumes, then the gains in H may become highly correlated and the linear relationship will plateau out due to the effects of antenna correlation [Chiurtu et al. 2001; Gesbert et al. 2002; Wei et al. 2002]. In contrast, other propagation effects not captured in Equation 5.12 may serve to reinforce the capacity gains of MIMO such as multipath delay spread. This was shown in particular in the case when the transmit channel is known [Raleigh and Cioffi 1998] but also in the case when it is unknown [Bölcskei et al. 2002].

More generally, the effect of the channel model is critical. Environments can easily be chosen that give channels where the MIMO capacities do not increase linearly with the numbers of antennas. However, most measurements and models available to date do give rise to channel capacities that are of the same order of magnitude as the promised theory. Also, the linear growth is usually a reasonable model for moderate numbers of antennas that are not extremely close packed.

Note that, unlike in CDMA where the user's signatures are quasiorthogonal by design, the separation of the MIMO channel relies on the presence of a rich multipath, which is needed to make the channel spatially selective. Therefore, MIMO can be said to effectively exploit multipath; this also implies that MIMO systems could not be a good radio solution for every scenario, namely for LOS-directed radio links.

5.2 Transmit and Receive Diversity Gains

This section deals with the different transmit and receive diversity gains, along with some results for WCDMA and OFDM.

5.2.1 Open/Closed Loop Techniques for Flat Fading

3GPP specifications define several transmit diversity schemes such as the standardized closed-loop modes 1 and 2 [3GPP 2003c] or the open loop space time transmit diversity (STTD) [3GPP 2003a] for two transmitting antennas. These schemes and other transmitting schemes such as open loop modes using space time block codes (STBCs) for four and eight transmitting antennas, the closed-loop selective transmit diversity (STD) and hybrid modes combining STBCs and STD were analyzed.

5.2.1.1 Open Loop Techniques

Open loop transmit diversity schemes are performed without previous knowledge of the channel state by the transmitter. Figure 5.2 shows the general scheme of an open loop transmitter scheme using M transmitting antennas.

STTD (also known as the Alamouti scheme) [Alamouti 1998] is a particular example of a scheme using two transmitting antennas and is already standardized for UMTS

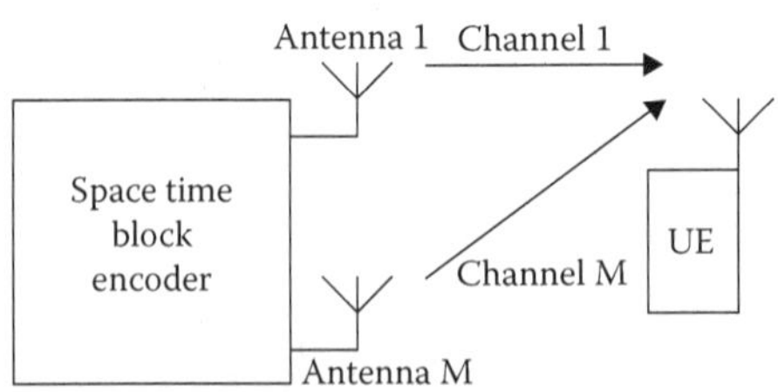

Figure 5.2 Generic block diagram of an open loop transmitter scheme.

[3GPP 2003a]. For this particular case, the transmission matrix representing the encoding process of the space time block encoder is defined as follows:

$$\mathbf{G}_2(s_0, s_1) = \begin{bmatrix} s_0 & -s_1^* \\ s_1 & s_0^* \end{bmatrix}. \tag{5.13}$$

Each column represents the sequence of symbols s_i transmitted by an antenna. Admitting that the channel coefficients for antennas 1 and 2 (h_0 and h_1) are constant during two symbol periods, the receiver performs the following processing:

$$\hat{s}_0 = h_0^* \cdot r(t_0) + h_1 \cdot r(t_1)^* = s_0 \cdot \left(h_0^2 + h_1^2\right) + n', \tag{5.14}$$

$$\hat{s}_1 = h_0^* \cdot r(t_1) - h_1 \cdot r(t_0)^* = s_1 \cdot \left(h_0^2 + h_1^2\right) + n''. \tag{5.15}$$

As it can be seen, the path diversity obtained is 2. It is possible to use other STBCs for four and eight antennas, as shown in Ref. [Tarokh et al. 1999].

For four antennas, a code with the following generator matrix can be used:

$$\mathbf{G}_4(S_0, S_1, S_2, S_3) = \begin{bmatrix} \mathbf{G}_2(S_0, S_1) & \mathbf{G}_2^*(S_2, S_3) \\ \mathbf{G}_2(S_2, S_3) & -\mathbf{G}_2^*(S_0, S_1) \end{bmatrix}. \tag{5.16}$$

The decoding is performed as

$$\begin{aligned}
\hat{s}_0 &= r_0 h_0^* + r_1^* h_1 - r_2^* h_2 - r_3 h_3^* = \sum_i |h_i|^2 \, s_0 - c_0 s_3 + \eta_0, \\
\hat{s}_1 &= r_1 h_2^* - r_0^* h_1 - r_3^* h_2 + r_2 h_3^* = \sum_i |h_i|^2 \, s_1 + c_0 s_2 + \eta_1, \\
\hat{s}_2 &= r_2 h_0^* + r_3^* h_1 + r_0^* h_2 + r_1 h_3^* = \sum_i |h_i|^2 \, s_2 + c_0 s_1 + \eta_2, \\
\hat{s}_3 &= r_3 h_0^* - r_2^* h_1 + r_1^* h_2 - r_0 h_3^* = \sum_i |h_i|^2 \, s_3 - c_0 s_0 + \eta_3, \\
c_0 &= 2\mathrm{Re}\{h_0^* h_3 - h_1 h_2^*\}.
\end{aligned} \tag{5.17}$$

In the case of eight antennas, the generator matrix is

$$\mathbf{G}_8(S_0, S_1, S_2, S_3, S_4, S_5, S_6, S_7) = \begin{bmatrix} \mathbf{G}_4(S_0, S_1, S_2, S_3) & \mathbf{G}_4^*(S_4, S_5, S_6, S_7) \\ \mathbf{G}_4(S_4, S_5, S_6, S_7) & -\mathbf{G}_4^*(S_0, S_1, S_2, S_3) \end{bmatrix}. \tag{5.18}$$

After performing a decoding processing similar to the case of four antennas, the following symbol estimates are obtained:

$$
\begin{aligned}
\hat{s}_0 &= \sum_i |h_i|^2 s_0 - c_0 s_3 - c_1 s_5 + c_2 s_6 + \eta_0, \\
\hat{s}_1 &= \sum_i |h_i|^2 s_1 + c_0 s_2 + c_1 s_4 + c_2 s_7 + \eta_1, \\
\hat{s}_2 &= \sum_i |h_i|^2 s_2 + c_0 s_1 - c_1 s_7 - c_2 s_4 + \eta_2, \\
\hat{s}_3 &= \sum_i |h_i|^2 s_3 - c_0 s_0 + c_1 s_6 - c_2 s_5 + \eta_3, \\
\hat{s}_4 &= \sum_i |h_i|^2 s_4 - c_0 s_7 + c_1 s_1 - c_2 s_2 + \eta_4, \\
\hat{s}_5 &= \sum_i |h_i|^2 s_5 + c_0 s_6 - c_1 s_0 - c_2 s_3 + \eta_5, \\
\hat{s}_6 &= \sum_i |h_i|^2 s_6 + c_0 s_5 + c_1 s_3 + c_2 s_0 + \eta_6, \\
\hat{s}_7 &= \sum_i |h_i|^2 s_7 - c_0 s_4 - c_1 s_2 + c_2 s_1 + \eta_7, \\
c_0 &= 2\mathrm{Re}\{h_0^* h_3 - h_1 h_2^*\}, \\
c_1 &= 2\mathrm{Re}\{h_0^* h_5 - h_1 h_4^*\}, \\
c_2 &= 2\mathrm{Re}\{h_0^* h_6 - h_1 h_7^*\}.
\end{aligned}
\tag{5.19}
$$

Clearly, in both the cases, the decoding does not achieve the maximum possible path diversity, because in the decoding of all symbols there is always intersymbolic interference from one symbol in the case of four antennas (c_0), and from three symbols in the case of eight antennas (c_0, c_1, c_2).

It has been shown that the STBCs constructed using an orthogonal design can achieve the maximum possible rate of one for every number of transmit antennas using any arbitrary real constellation, and for two transmit antennas using any arbitrary complex constellations (Alamouti) [Tarokh et al. 1999]. However, there does not exist any orthogonal code of rate 1 using more than two antennas [Liang and Xia 2003]—if orthogonality is essential (fully loaded systems with significant interference levels), a code with $R < 1$ should be employed for such cases.

5.2.1.2 Closed Loop Techniques

Closed loop techniques require feedback from the receiving end. For this type of techniques the transmitter has knowledge of the channel through a feedback path established between the receiver and the transmitter. In this section, the WCDMA model will be discussed. The extrapolation to the SC-FDE/OFDM model is straightforward.

3GPP specifications [3GPP 2003c] define two closed loop modes: mode 1 and mode 2. Figure 5.3 portrays the transmitter structure and the philosophy of STD. The

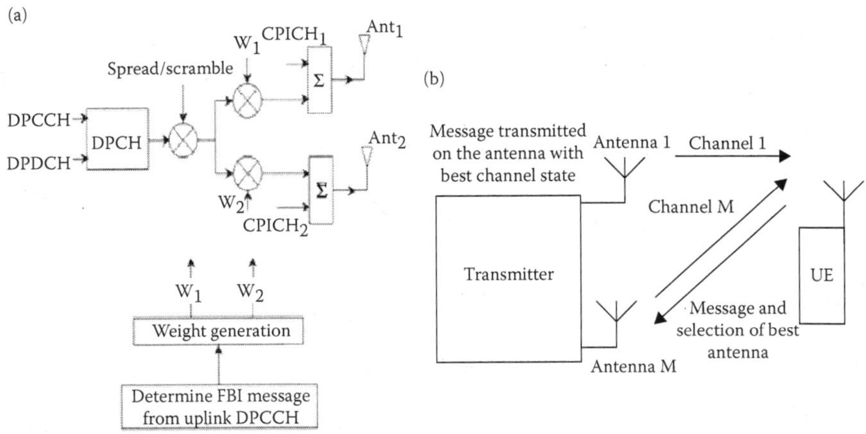

Figure 5.3 Generic DL transmitter structure for closed loop transmit diversity (a) and for the STD scheme (b).

UE uses the CPICHs transmitted through each antenna to estimate the respective channel state. Using the estimates, it computes the complex weight values w_1 and w_2 to be applied to the transmitting antennas 1 and 2 so that it maximizes

$$P = w^{H}\mathbf{H}^{H}\mathbf{H}w, \tag{5.20}$$

where $\boldsymbol{w} = [w_1 \quad w_2]^{T}$ and $\mathbf{H} = [\boldsymbol{h}_1 \quad \boldsymbol{h}_2]$ ($\boldsymbol{h}_1$ and $\boldsymbol{h}_2$ are column vectors representing the estimated CIR). The difference between 3GPP closed loop modes 1 and 2 is in the available weights constellation. Mode 1 allows the selection of four possible weight vectors w that result only in a phase difference between the signals transmitted in both antennas. Mode 2 has 16 possible weight vectors available that produce a phase and amplitude difference between the two antennas.

Figure 5.3b presents the scheme for STD mode, which is another possible closed loop transmit diversity mode. In STD, the UE is continuously measuring the reception quality of each transmit antenna and instructs the BS to transmit the message in the best antenna. The other antennas transmit pilot bits for channel estimation.

5.2.2 WCDMA

In this section, we will combine the STD technique with an STBC transmission, on a WCDMA scheme. The extrapolation to SC-FDE/OFDM is straightforward.

Combining the STD technique with an STBC transmission, a hybrid scheme is obtained, as shown in Figure 5.4. In this case, there are K groups of M antennas. In this scheme, the UE selects the best group of M antennas from the K groups available according to the best combined channel state, and uses this group for the transmission of the message using an STBC.

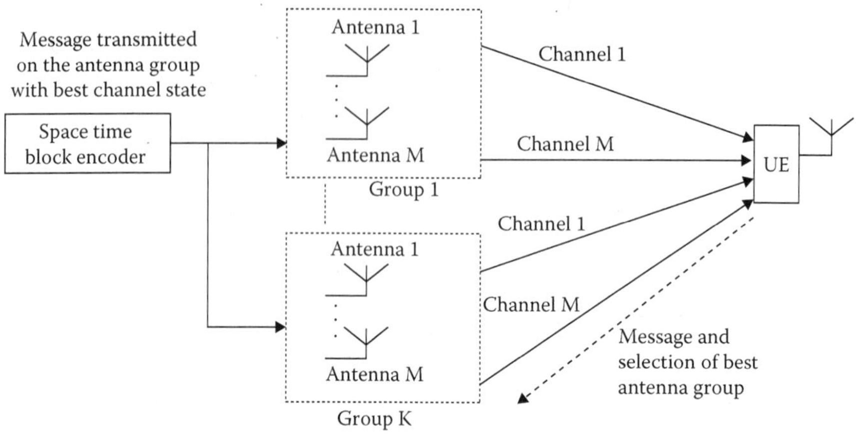

Figure 5.4 Hybrid STD/STBC transmit diversity scheme.

For the STD and the hybrid transmit diversity scheme simulations, a delay of 0.67 ms equivalent to 1 TS was used between the measured channel state and the selection of the transmitting antennas.

Different CQI modes [3GPP 2003c] (each CQI corresponds to a fixed coding rate and modulation alphabet) were simulated to observe the performance impact of the enhancements in each individual CQI. CQI values of 15 and 22 (Table 5.1) were considered.

The BLER results are displayed, which play a very important role in the study of HSDPA performance and the need for retransmissions (HARQ, with Chase combining).

Figures 5.5 and 5.6 portray several results for CQI 15 in Pedestrian A and Indoor A environments. From the results, it can be concluded that considerable performance gains can be obtained with any of the transmit diversity methods. The technique that resulted in best performance was the hybrid STBC4 + STD which, for a BLER of 10^{-2}, achieved an impressive 13 dB gain over the case with no transmit diversity. For this BLER, this technique also achieves about 5 dB improvement over the best 3GPP standardized technique (closed loop mode 2). Moreover, when comparing schemes using the same

Table 5.1 Parameters for CQI Modes Employed in Simulations

CQI Value	*Transport Block Size*	*Number of HS-PDSCH*	*Modulation*
1	137	1	QPSK
15	3319	5	QPSK
22	7168	5	16-QAM
30	25,558	15	16-QAM

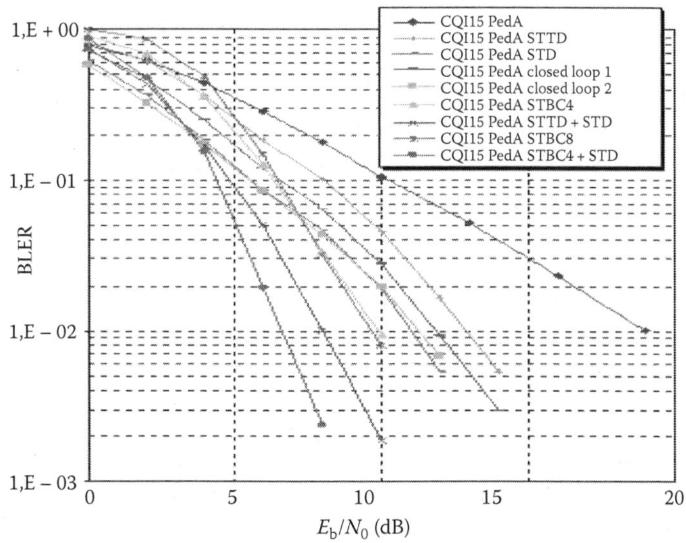

Figure 5.5 BLER performance for CQI15 mode in Pedestrian A channel.

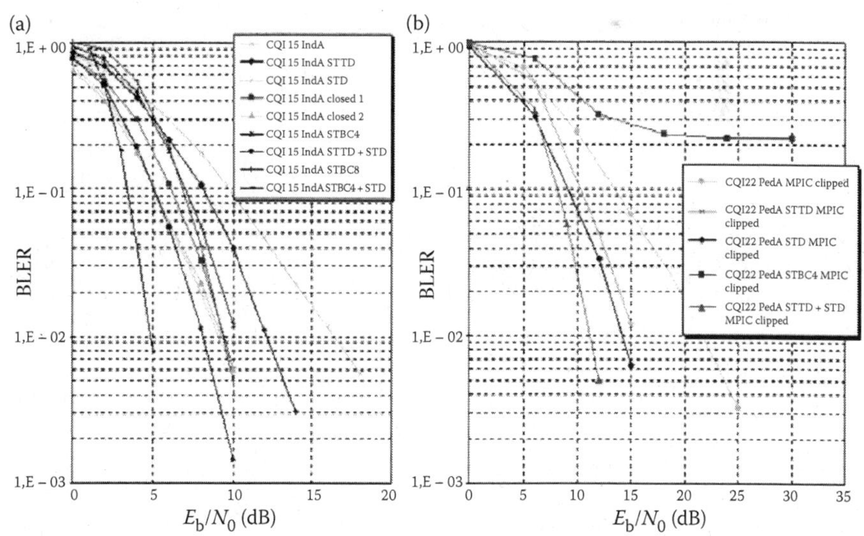

Figure 5.6 BLER performance for CQI15 mode in Indoor A channel (a) and CQI22 mode in Pedestrian A channel (b).

number of transmitting antennas, the closed loop techniques (including the hybrid) achieved better performance than the open loop techniques. In the particular case of two transmitting antennas the three different closed loop modes tested resulted in similar performances, although the closed loop mode 2 was found to be slightly better than the others, followed by the STD mode.

For open loop configurations, the performance difference between STBC4 and STBC8 is very low. Looking at the expressions 5.17 and 5.19 for the decoding of STBC4 and STBC8 transmissions, it can be seen that in both the cases each decoded symbol has interference from one (in STBC4) or three (in STBC8) other symbols. Since CQI15 uses five physical channels (five spreading codes), each symbol of an individual physical channel is, in fact, suffering from interference of 5 (in STBC4) or 15 (in STBC8) symbols from all the physical channels present. This is the reason why there is no significant improvement in the performance of CQI15 when changing from STBC4 to STBC8.

Figure 5.6b shows the performance of CQI22, which uses 16-QAM modulation, in Pedestrian A environment using an MPIC in the receiver. Due to the sensitivity of 16-QAM against interference, only a few transmit diversity schemes are presented. The STTD, STD, and the hybrid STTD + STD resulted in great performance improvement over the one transmit antenna case. As for the STBC4 scheme the performance got worse, even with the MPIC. This was as expected since, as already stated, this scheme results in interference between the decoded symbols, which, for this modulation, easily results in poor performance.

5.2.3 SC-FDE/OFDM

Block transmission techniques, with appropriate CPs and employing FDE techniques, have been shown to be suitable for high-data-rate transmission over severely time-dispersive channels [Gusmão et al. 2000; Falconer and Ariyavisitakul 2002; Falconer et al. 2002]. OFDM is the most popular modulation based on this technique.

SC modulation using FDE is an alternative approach based on this principle. As with OFDM, the data blocks are preceded by a CP, long enough to cope with the overall channel length. Due to the lower envelope fluctuations of the transmitted signals [and, implicitly a lower peak-to-mean envelope power ratio (PMEPR)], SC-FDE schemes are especially interesting for the uplink transmission (i.e., the transmission from the mobile terminal to the BS) [Gusmão et al. 2000; Falconer and Ariyavisitakul 2002].

A promising Iterative Block-Decision Feedback Equalization (IB-DFE) technique for SC-FDE was proposed in Ref. [Benvenuto et al. 2002b] and extended to other scenarios in Refs. [Dinis, et al. 2003, 2004b]. These IB-DFE receivers can be regarded as iterative DFE receivers where the feedforward and the feedback operations are implemented in the frequency domain, offering much better performance than noniterative methods [Benvenuto et al. 2002b; Dinis et al. 2003, 2004b].

In this section, we consider transmit diversity schemes for both OFDM and SC-FDE schemes, specifically the STBC with two and four antennas [Marques da Silva et al. 2009c]. For OFDM schemes we consider the conventional receiver, and for SC-FDE schemes, we consider IB-DFE receivers. For nonorthogonal codes (i.e., more than two transmit antennas), we also consider iterative receivers with cancelation of the residual interference (for SC schemes with IB-DFE receivers, this means a negligible increase in the receiver complexity).

5.2.3.1 STBC for Two Antennas

We consider block transmission schemes and the lth transmitted block has the form

$$s_l(t) = \sum_{n=-N_G}^{N-1} s_{n,l} h_T(t - nT_S), \tag{5.21}$$

with T_S denoting the symbol duration, N_G denoting the number of samples at the CP, and $h_T(t)$ is the adopted pulse shaping filter. For a single transmit antenna system, the signal $s_l(t)$ is transmitted over a time-dispersive channel and the signal at the receiver input is sampled and the CP is removed, leading to the time-domain block $\{y_{n,l}; n = 0, 1, \ldots, N-1\}$, which is then subjected to the FDE. For SC-FDE schemes, the lth time-domain block to be transmitted is $\{s_{n,l}; n = 0, 1, \ldots, N-1\}$, where $s_{n,l}$ is the nth data symbol, selected from a given constellation (e.g., a QPSK constellation) under an appropriate mapping rule (it is assumed that $s_{-n,l} = s_{N-n,l}$, $n = -N_G, -N_G + 1, \ldots, -1$); the frequency-domain blocks associated with the data are $\{S_{k,l}; k = 0, 1, \ldots, N-1\} = \text{DFT}\{s_{n,l}; n = 0, 1, \ldots, N-1\}$. For OFDM schemes, the data symbols are transmitted in the frequency domain, that is, $S_{k,l}$ are selected according to an appropriate constellation. At the output of the FDE we have the samples $\tilde{A}_{k,l} = Y_{k,l} H_{k,l}^* \big/ \left(\alpha + \left|H_{k,l}\right|^2\right)$. In the OFDM case, this equalization process is simply accomplished through $\tilde{A}_{k,l} = Y_{k,l} H_{k,l}^*$.

If we employ Alamouti's transmit diversity, we need some processing at the transmitter.

Assuming that the CP is longer than the overall CIR of each channel, the lth frequency-domain block after the FDE block (i.e., the DFT of the lth received time-domain block, after removing the CP) is $\{y_{n,l}; n = 0, 1, \ldots, N-1\} = \text{IDFT}\{Y_{k,l}; k = 0, 1, \ldots, N-1\}$, with

$$\mathbf{Y}_{k,l} = \mathbf{S}_{k,l}^{(1)} \mathbf{H}_{k,l}^{(1)} + \mathbf{S}_{k,l}^{(2)} \mathbf{H}_{k,l}^{(2)} + \mathbf{N}_{k,l}, \tag{5.22}$$

where $\left\{\mathbf{H}_k^{(m)}; k = 0, 1, \ldots, N-1\right\} = \text{DFT}\left\{h_n^{(m)}; n = 0, 1, \ldots, N-1\right\}$ denotes the channel frequency response for the kth subcarrier and the mth transmit antenna (the channel is assumed to be invariant in the frame) and $\mathbf{N}_{k,l}$ is the frequency-domain block channel noise for that subcarrier and the lth block. Assuming, for now, the conventional linear FDE for SC schemes, Alamouti's postprocessing for two antennas (STBC2) becomes

$$\begin{aligned} \tilde{\mathbf{A}}_{k,2l-1} &= \left[\mathbf{Y}_{k,2l-1} \mathbf{H}_{k,l}^{(1)*} + \mathbf{Y}_{k,2l}^* \mathbf{H}_{k,l}^{(2)}\right] \beta_k^{(2)} \\ \tilde{\mathbf{A}}_{k,2l} &= \left[\mathbf{Y}_{k,2l} \mathbf{H}_{k,l}^{(1)*} - \mathbf{Y}_{k,2l-1}^* \mathbf{H}_{k,2l}^{(2)}\right] \beta_k^{(2)}, \end{aligned} \tag{5.23}$$

where $\{\mathbf{A}_{k,m}, k=0,1,\ldots,N\} = \mathrm{DFT}\{a_{n,m}, n=0,1,\ldots,N\}$ and where $\beta_k^{(2)} = \left(\alpha + \left(\left|\mathbf{H}_{k,l}^{(1)}\right|^2 + \left|\mathbf{H}_{k,l}^{(2)}\right|^2\right)\right)^{-1}$. This leads to $\tilde{\mathbf{A}}_{k,2l-j} = \mathbf{A}_{k,2l-j} \sum_{m=1}^{M} \left|\mathbf{H}_{k,l}^{(m)}\right|^2 \beta_k^{(2)} + \mathbf{N}_{k,2l-j}^{\mathrm{eq}}$ $j=0,1$. In addition, we define $\alpha = E\left[\left|\mathbf{N}_{k,l}\right|^2\right] / E\left[\left|\mathbf{S}_{k,2l-j}\right|^2\right]$. Here, $\mathbf{N}_{k,l}^{\mathrm{eq}}$ denotes the equivalent noise for detection purposes, with $E\left[\left|\mathbf{N}_{k,l}^{\mathrm{eq}}\right|^2\right] = \left[2\sigma_N^2 \left(\sum_{m=1}^{M} \left|\mathbf{H}_{k,l}^{(m)}\right|^2\right)\right] \beta_k^{(2)}$ and $\sigma_N^2 = E\left[\left|\mathbf{N}_{k,l}\right|^2\right] / 2$.

Alamouti's postprocessing for OFDM signals is the same as defined in Equation 5.23 but without multiplying by the $\beta_k^{(2)}$ component.

5.2.3.2 STBC for Four Antennas

Using unspecified complex-valued modulation, such an improvement is possible only for the two-antenna scheme. Higher schemes with four and eight antennas with code rate one exist only in the case of binary transmission [Hochwald et al. 2001]. The proposed STBC4 scheme has $M = 4$ transmit antennas, presenting a code rate one.

The lth frequency-domain block after the FDE block (i.e., the DFT of the lth received time-domain block, after removing the CP) is $\{y_{n,l}; n = 0,1,\ldots,N-1\} = \mathrm{IDFT}\{\mathbf{Y}_{k,l}; k = 0,1,\ldots,N-1\}$, with

$$\mathbf{Y}_{k,l} = \mathbf{S}_{k,l}^{(1)}\mathbf{H}_{k,l}^{(1)} + \mathbf{S}_{k,l}^{(2)}\mathbf{H}_{k,l}^{(2)} + \mathbf{S}_{k,l}^{(3)}\mathbf{H}_{k,l}^{(3)} + \mathbf{S}_{k,l}^{(4)}\mathbf{H}_{k,l}^{(4)} + \mathbf{N}_{k,l}. \tag{5.24}$$

Assuming, for now, the conventional SC-FDE decoding (i.e., no IB-DFE receiver), the postprocessing STBC for four antennas ($M = 4$) becomes (using Equation 5.16)

$$\begin{aligned}
\tilde{\mathbf{A}}_{k,4l-3} &= \left[\mathbf{Y}_{k,4l-3}\mathbf{H}_{k,l}^{(1)*} + \mathbf{Y}_{k,4l-2}^{*}\mathbf{H}_{k,l}^{(2)} - \mathbf{Y}_{k,4l-1}^{*}\mathbf{H}_{k,l}^{(3)} - \mathbf{Y}_{k,4l}\mathbf{H}_{k,l}^{(4)*}\right]\beta_k^{(4)} \\
&= \overbrace{\mathbf{A}_{k,4l-3}\sum_{m=1}^{M}\left|\mathbf{H}_{k,l}^{(m)}\right|^2}^{\text{Desired symbol}} - \overbrace{\mathbf{C}_k\mathbf{A}_{k,4l}}^{\text{Residual interference}} + \mathbf{N}_{k,4l-3}^{\mathrm{eq}}, \\
\tilde{\mathbf{A}}_{k,4l-2} &= \left[\mathbf{Y}_{k,4l-2}\mathbf{H}_{k,l}^{(1)*} - \mathbf{Y}_{k,4l-3}^{*}\mathbf{H}_{k,l}^{(2)} - \mathbf{Y}_{k,4l}^{*}\mathbf{H}_{k,l}^{(3)} + \mathbf{Y}_{k,4l-1}\mathbf{H}_{k,l}^{(4)*}\right]\beta_k^{(4)} \\
&= \mathbf{A}_{k,4l-2}\sum_{m=1}^{M}\left|\mathbf{H}_{k,l}^{(m)}\right|^2 + \mathbf{C}_k\mathbf{A}_{k,4l-1} + \mathbf{N}_{k,4l-2}^{\mathrm{eq}}, \\
\tilde{\mathbf{A}}_{k,4l-1} &= \left[\mathbf{Y}_{k,4l-1}\mathbf{H}_{k,l}^{(1)*} + \mathbf{Y}_{k,4l}^{*}\mathbf{H}_{k,l}^{(2)} + \mathbf{Y}_{k,4l-3}^{*}\mathbf{H}_{k,l}^{(3)} + \mathbf{Y}_{k,4l-2}\mathbf{H}_{k,l}^{(4)*}\right]\beta_k^{(4)} \\
&= \mathbf{A}_{k,4l-1}\sum_{m=1}^{M}\left|\mathbf{H}_{k,l}^{(m)}\right|^2 + \mathbf{C}_k\mathbf{A}_{k,4l-2} + \mathbf{N}_{k,4l-1}^{\mathrm{eq}},
\end{aligned}$$

$$\tilde{A}_{k,4l} = \left[Y_{k,4l}H_{k,l}^{(1)*} - Y_{k,4l-1}^{*}H_{k,l}^{(2)} + Y_{k,4l-2}^{*}H_{k,l}^{(3)} - Y_{k,4l-3}H_{k,l}^{(4)*}\right]\beta_k^{(4)}$$
$$= A_{k,4l}\sum_{m=1}^{M}\left|H_{k,l}^{(m)}\right|^2 - C_k A_{k,4l-3} + N_{k,4l}^{\mathrm{eq}}, \tag{5.25}$$

with $\beta_k^{(4)} = \left(\alpha + \sum_{m=1}^{M}\left|H_{k,l}^{(m)}\right|^2\right)^{-1}$, and where $C_k = 2\mathrm{Re}\left\{H_{k,l}^{(1)*}H_{k,l}^{(4)} - H_{k,l}^{(2)}H_{k,l}^{(3)*}\right\} \Big/ \left\{\left(\sum_{m=1}^{M}\left|H_{k,l}^{(m)}\right|^2\right)\right\}$, which stands for the residual interference coefficient generated in the STBC decoding process. In the following we will show how we can remove this residual interference.

5.2.3.3 Receiver Design

We present next an IB-DFE receiver for STBC with four antennas considering SC-FDE signals. The frequency-domain block at the output of the receiver is $\{\tilde{A}_{k,4l-j}^{(i)}; k = 0, 1, \ldots, N-1\}$, with

$$\tilde{A}_{k,4l-3}^{(i)} = \overbrace{Y_{k,4l-3}F_{k,l}^{(i)(1)} + Y_{k,4l-2}^{*}F_{k,l}^{(i)(2)} - Y_{k,4l-1}^{*}F_{k,l}^{(i)(3)} - Y_{k,4l}F_{k,l}^{(i)(4)}}^{\text{STBC4 decoding plus IB}-\text{DFE feedforward}}$$
$$\overbrace{+ C_k\widehat{A}_{k,4l}^{(i)}}^{\text{Cancelation of residual interference}} \quad \overbrace{- B_{k,l}^{(i)}\bar{A}_{k,4l-3}^{(i-1)}}^{\text{IB}-\text{DFE feedback}},$$
$$\tilde{A}_{k,4l-2}^{(i)} = Y_{k,4l-2}F_{k,l}^{(i)(1)} - Y_{k,4l-3}^{*}F_{k,l}^{(i)(2)} - Y_{k,4l}^{*}F_{k,l}^{(i)(3)} + Y_{k,4l-1}F_{k,l}^{(i)(4)}$$
$$- C_k\widehat{A}_{k,4l-1}^{(i)} - B_{k,l}^{(i)}\bar{A}_{k,4l-2}^{(i-1)}, \tag{5.26}$$
$$\tilde{A}_{k,4l-1}^{(i)} = Y_{k,4l-1}F_{k,l}^{(i)(1)} + Y_{k,4l}^{*}F_{k,l}^{(i)(2)} + Y_{k,4l-3}^{*}F_{k,l}^{(i)(3)} + Y_{k,4l-2}F_{k,l}^{(i)(4)}$$
$$- C_k\widehat{A}_{k,4l-2}^{(i)} - B_{k,l}^{(i)}\bar{A}_{k,4l-1}^{(i-1)},$$
$$\widehat{A}_{k,4l}^{(i)} = Y_{k,4l}F_{k,l}^{(i)(1)} - Y_{k,4l-1}^{*}F_{k,l}^{(i)(2)} + Y_{k,4l-2}^{*}F_{k,l}^{(i)(3)} - Y_{k,4l-3}F_{k,l}^{(i)(4)}$$
$$+ C_k\widehat{A}_{k,4l-3}^{(i)} - B_{k,l}^{(i)}\bar{A}_{k,4l}^{(i-1)},$$

where C_k is as defined for Equation 5.25. The feedforward coefficients are $\left\{F_{k,l}^{(i)(m)}; k = 0, 1, \ldots, N-1; m = 1, 2, \ldots, M\right\}$ and the feedback coefficients are $\left\{B_{k,l}^{(i)}; k = 0, 1, \ldots, N-1\right\}$. The block $\left\{\widehat{A}_{n,4l-j}^{(i-1)}; n = 0, 1, \ldots, N-1\right\} = \mathrm{DFT}\left\{\widehat{a}_{n,4l-j}^{(i-1)}; n = 0, 1, \ldots, N-1\right\}$ denotes the DFT transform of the data estimates associated with the previous iteration, that is, the hard decisions associated with the time-domain block at the output of $\left\{\tilde{a}_{n,4l-j}^{(i-1)}; n = 0, 1, \ldots, N-1\right\} = \mathrm{IDFT}\left\{\tilde{A}_{k,4l-j}^{(i-1)}; k = 0, 1, \ldots, N-1\right\}$. $\left\{\bar{A}_{k,4l-j}^{(i-1)}; k = 0, 1, \ldots, N-1; j = 0, 1, 2, 3\right\}$

denotes the average signal conditioned to the FDE output for the previous iteration $\left\{\bar{a}_{n,4l-j}^{(i-1)}; n = 0, 1, \ldots, N-1\right\}$ from Equation 5.35. It is worth noting that since $\tilde{\mathbf{A}}_{k,4l-j}^{(i)}$ presents residual interference, the detection of $\mathbf{A}_{k,4l-j}^{(i)}$ should be accompanied by the detection of $\mathbf{A}_{k,4l-p}^{(i)}$ (with $p = 3 - j$) to allow the cancelation of the residual interference generated in the STBC4 decoding process.

In the case of a SISO system, Equation 5.26 takes the form $\widehat{\mathbf{A}}_{k,l}^{(i)} = \mathbf{Y}_{k,l}\mathbf{F}_{k,l}^{(i)} - \mathbf{B}_{k,l}^{(i)}\bar{\mathbf{A}}_{k,l}^{(i-1)}$, that is, there is a single branch (there is no STBC4 decoding) and there is no cancelation of the residual interference. In the case of STBC2 (two transmit antennas), there is no residual interference component.

To further improve performance with STBC4 the residual interference to be subtracted (which is a function of the estimate of the symbol that generates interference), we consider an iterative interference cancelation (IIC) that can be implemented as follows:

i. Compute $\widehat{\mathbf{A}}_{k,4l-j}^{(i)(q)}$ using Equation 5.26 without canceling the residual interference.

ii. Based on $\widehat{\mathbf{A}}_{k,4l-j}^{(i)(q)}$ from step (i), compute $\widehat{\mathbf{A}}_{k,4l-p}^{(i)(q)}$ after canceling the corresponding residual interference.

iii. Based on $\widehat{\mathbf{A}}_{k,4l-p}^{(i)(q)}$ from step (ii), compute $\widehat{\mathbf{A}}_{k,4l-j}^{(i)(q+1)}$ after canceling the residual interference $(\mathbf{C}_k\widehat{\mathbf{A}}_{k,4l-p}^{(i)})$.

iv. Repeat steps (ii) and (iii) iteratively to improve the accuracy of $\widehat{\mathbf{A}}_{k,4l-p}^{(i)}$ (cancelation of the residual interference), which will finally be used to improve the accuracy of $\widehat{\mathbf{A}}_{k,4l-j}^{(i)}$.

It can be shown that the optimum feedback coefficients are [Benvenetu 2002; Dinis et al. 2003]

$$\mathbf{B}_{k,l}^{(i)} = \sum_{m=1}^{M} \mathbf{F}_{k,l}^{(i)(m)}\mathbf{H}_{k,l}^{(m)} - 1, \tag{5.27}$$

and the feedforward coefficients are given by

$$\mathbf{F}_{k,l}^{(i)(m)} = \frac{\mathbf{Q}_{k,l}^{(m)}}{\left[\alpha + \left(1 - \left(\rho_l^{(i-1)}\right)^2\right)\sum_{m=1}^{M}\left|\mathbf{H}_{k,l}^{(m)}\right|^2\right]\gamma_l^{(i)}}, \tag{5.28}$$

with $\mathbf{Q}_{k,l}^{(m)} = \mathbf{H}_{k,l}^{(m)*}$ for $m = 1$ or 4 and $\mathbf{Q}_{k,l}^{(m)} = \mathbf{H}_{k,l}^{(m)}$ for $m = 2$ or 3. In the particular case of SISO, we only have $m = 1$ (with $M = 1$) and $\mathbf{Q}_{k,l} = \mathbf{H}_{k,l}^{*}$. In the case of STBC of order 2 (i.e., STBC2), we have $\mathbf{Q}_{k,l}^{(m)} = \mathbf{H}_{k,l}^{(m)*}$ for $m = 1$ and $\mathbf{Q}_{k,l}^{(m)} = \mathbf{H}_{k,l}^{(m)}$ for $m = 2$. The parameter $\gamma_l^{(i)}$ is defined as

$$\gamma_l^{(i)} = \frac{1}{N} \sum_{m=1}^{M} \sum_{k=0}^{N-1} \mathbf{F}_{k,l}^{(i)(m)} \mathbf{H}_{k,l}^{(m)} \tag{5.29}$$

and the correlation factor $\rho_{4l-j}^{(i-1)}$ is defined as

$$\rho_{4l-j}^{(i-1)} = \frac{E\left[\hat{a}_{n,4l-j}^{(i-1)} a_{n,4l-j}^{*}\right]}{E\left[\left|a_{n,4l-j}\right|^2\right]}. \tag{5.30}$$

It can be shown that, for the QPSK modulation, the correlation coefficient is given by [Gusmão et al. 2006]

$$\rho_{4l-j}^{(i)} = \frac{1}{2N} \sum_{n=0}^{N-1} \left(\rho_{n,4l-j}^{\mathrm{I}(i)} + \rho_{n,4l-j}^{\mathrm{Q}(i)}\right). \tag{5.31}$$

($\rho_{4l-j}^{(i)}$ is almost independent of l for large values of N, provided that $\mathbf{H}_{k,l}^{(m)}$ is constant for the frame duration), as (see Chapter 2)

$$\begin{aligned} \rho_{n,4l-j}^{\mathrm{I}(i)} &= \left|\tanh\left(\frac{L_n^{\mathrm{I}(i)}}{2}\right)\right|, \\ \rho_{n,4l-j}^{\mathrm{Q}(i)} &= \left|\tanh\left(\frac{L_n^{\mathrm{Q}(i)}}{2}\right)\right|. \end{aligned} \tag{5.32}$$

The LLRs of the "in-phase bit" and the "quadrature bit," associated with $a_{n,4l-j}^{\mathrm{I}(i)}$ and $a_{n,4l-j}^{\mathrm{Q}(i)}$, respectively, are given by

$$\begin{aligned} L_n^{\mathrm{I}(i)} &= \frac{2}{\sigma_i^2} \tilde{a}_{n,4l-j}^{\mathrm{I}(i)}, \\ L_n^{\mathrm{Q}(i)} &= \frac{2}{\sigma_i^2} \tilde{a}_{n,4l-j}^{\mathrm{Q}(i)}, \end{aligned} \tag{5.33}$$

respectively, with

$$\sigma_{i,4l-j}^2 = \frac{1}{2} E\left[\left|a_{n,4l-j} - \tilde{a}_{n,4l-j}^{(i)}\right|^2\right] \approx \frac{1}{2N} \sum_{n=0}^{N-1} \left|\hat{a}_{n,4l-j}^{(i)} - \tilde{a}_{n,4l-j}^{(i)}\right|^2, \tag{5.34}$$

(as with $\rho_{4l-j}^{(i)}$, $\sigma_{i,4l-j}^2$ is almost independent of l for large values of N, provided that $\mathbf{H}_{k,l}^{(m)}$ remains constant for the frame duration).

The conditional average values associated with the data symbols are given by

$$\bar{a}^{(i)}_{n,4l-j} = \tanh\left(\frac{L^{\mathrm{I}(i)}_{n,4l-j}}{2}\right) + \mathrm{j}\tanh\left(\frac{L^{\mathrm{Q}(i)}_{n,4l-j}}{2}\right). \tag{5.35}$$

Therefore, the several symbols of the jth order ($j = 0, 1, 2, 3$) that comprise the STBC4 block need to be decoded independently by the IB-DFE receiver, with the exception of the symbol estimates that originate the residual interference generated in the STBC4 decoding process, as shown in Equation 5.26. The *IB-DFE with soft decisions* described above does not need to perform the channel decoding in the feedback loop. As an alternative, we can define a *Turbo FDE* that employs the channel decoder outputs, instead of the uncoded "soft decisions" in the feedback loop of the IB-DFE. The main difference between *IB-DFE with soft decisions* and the *Turbo FDE* is in the decision device: in the first case the decision device is a symbol-by-symbol soft decision (for QPSK constellation this corresponds to the hyperbolic tangent, as in Equation 5.35); for the *Turbo FDE* a Soft-In, Soft-Out channel decoder is employed in the feedback loop. The Soft-In, Soft-Out block, which can be implemented as defined in Ref. [Vucetic and Yuan 2002], provides the LLRs of both the "information bits" and the "coded bits." The input of the Soft-In, Soft-Out block is LLRs of the "coded bits" at the FDE output, given by Equations 5.33 and 5.34.

The receiver for OFDM schemes with STBC2 is straightforward [Benvenuto and Tomasin 2002]. For OFDM schemes with STBC4, Equation 5.26 also applies with the difference that there is no feedback component, and the feedforward component only has the numerator of Equation 5.28. It is worth noting that these STBC schemes can easily be extended to multiple receive antennas.

5.2.3.4 Performance Results

In this section we present a set of performance results concerning the proposed receivers, for both SC-FDE and OFDM schemes with two- and four-antenna STBC schemes. We consider the BER performance, which is expressed as a function of E_b/N_0, where N_0 is the one-sided power spectral density of the noise and E_b is the energy of the transmitted bits (i.e., the degradation due to the useless power spent on the CP is not included).

Each block has $N = 256$ symbols selected from a QPSK constellation under a Gray mapping rule (similar results were observed for other values of N, provided that $N \gg 1$). Our channel has two EP paths, uncorrelated Rayleigh fading and relative delays [0 250] ns (relative to the first multipath). The channel is assumed to be invariant during the block. The duration of the useful part of the blocks (N symbols) is 1 μs and the CP has 0.125 μs duration. For SC-FDE systems we considered the *IB-DFE receiver with soft decisions* and the *Turbo FDE* with five iterations. Beyond this number the performance improvement was almost negligible.

Linear power amplification is considered at the transmitter and perfect synchronization is assumed at the receiver. The channel encoder is a convolutional code with generators $1 + D^2 + D^3 + D^5 + D^6$ and $1 + D + D^2 + D^3 + D^6$, and the coded

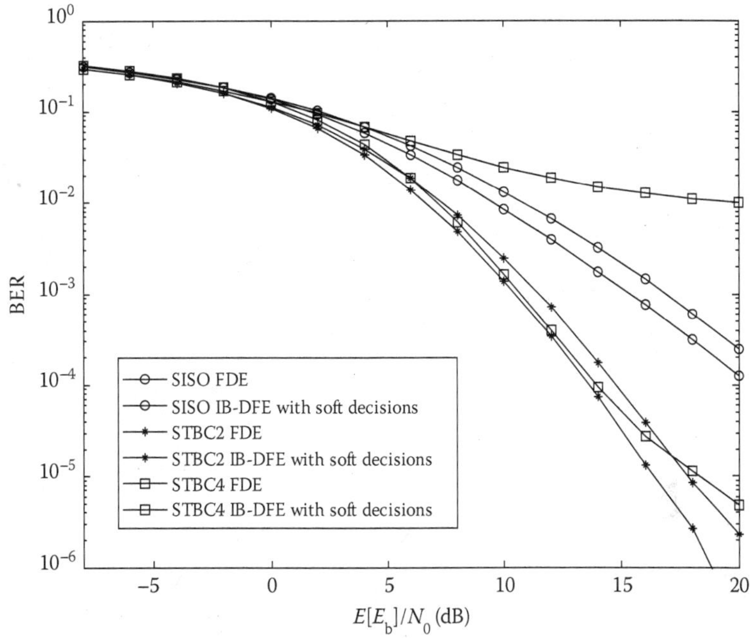

Figure 5.7 BER results for the uncoded SC-FDE.

bits associated with a given block are interleaved and mapped into the constellation points.

Figure 5.7 considers uncoded results for the SC-FDE and a linear FDE receiver (i.e., just the first iteration of the IB-DFE receiver) or the IB-DFE receiver with soft decisions (i.e., without channel decoding in the feedback loop). Clearly, the increased diversity due to STBC schemes leads to significant performance improvements when compared with the SISO case. From this figure, it is also clear that the IB-DFE performs always better than the linear FDE receiver. It can also be observed that the STBC4 with the linear FDE receiver performs very badly, due to the residual interference (generated in the STBC4 decoding process). However, when we add the IB-DFE with soft decisions to the STBC4, we have a significant performance improvement, due to the ability to mitigate the residual interference. Nevertheless, the STBC4 with an IB-DFE receiver with soft decisions is not able to outperform the STBC2.

Figure 5.8 shows a performance comparison between SC-FDE and OFDM when channel coding is considered (it is well known that uncoded performances are very poor for OFDM schemes). Note that the OFDM receiver for the STBC4 includes a residual IC, similar to the one included and described in the IB-DFE that was considered for the SC-FDE STBC4. Clearly, there is a visible advantage of SC-FDE with the Turbo FDE over the OFDM system for all schemes.

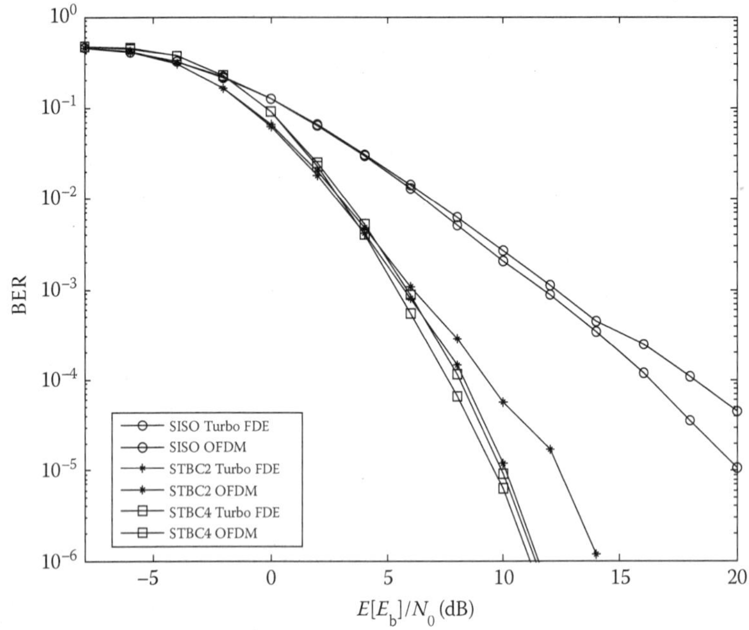

Figure 5.8 BER for coded SC-FDE and OFDM.

5.3 Spatial Multiplexing

To augment the throughput as much as possible, the SM scheme is employed. These schemes use the different transmit antennas to send out different parts of the message, which are reassembled at the receiver. The main drawback of these schemes is obviously the interference that the message sent by each transmit antenna has on the other messages.

5.3.1 Spatial Multiplexing: WCDMA

In this section, the WCDMA receivers are analyzed for SM. A focus is given on the UMTS, and its HSDPA mode.

5.3.1.1 MPIC for HSDPA

To increase the data rate, HSDPA incorporates new technologies into UMTS such as AMC and HARQ and fast cell selection. AMC relies on dynamically selecting the modulation alphabet and coding rate used according to the channel quality between the BS/Node-B and the UE. The data rate imposes a certain block size for the turbo coder, whose coding rate is changed by modifying the puncturing performed to the turbo code used. The coded message size is then split into physical channels (1–15), and afterwards modulated in order to accommodate the desired data rate. The modulation is chosen between QPSK and 16-QAM.

In WCDMA systems, the mobile propagation conditions result in frequency-selective channels that produce MPI due to the superimposed replicas. This has an important impact on the link performance, especially for the 16-QAM modulation. This modulation is very sensitive to interference and is severely degraded in this type of channels due to the dependence on both the phase and the amplitude.

3GPP specifications [3GPP 2003c] define several CQI modes for HSDPA, which represent the predefined combinations of modulation and coding rates available for use by the AMC. Most of the CQI modes require the use of several simultaneous physical channels (spreading codes) for the transmission of the data, especially the 16-QAM modes that use at least five. Each multipath component is therefore composed of several physical channels that are not orthogonal to the other multipaths, being therefore an important source of interference. The worst case arises for UEs of class 10, operating with a CQI value of 30. For this CQI, the BS uses 15 spreading codes (physical channels) for data transmission and 16-QAM modulation, thus achieving the highest available throughput: 12,779 kbps (derived from Table 5.1).

A possible solution to decrease the performance degradation, especially when using this higher-order modulation, is the implementation of an IC at the receiver. This possibility was studied for the specific case of HSDPA. So, an MPIC was developed and implemented in an HSDPA physical layer simulator for evaluating the performance gain. This MPIC estimates the interfering multipath components for each multipath and subtracts the estimates from the message, thus getting a "cleaner" signal with less interference.

To study the impact of the MPIC in the performance of the system, an HSDPA link-level simulator (LLS) was built, incorporating in the transmitter all the blocks defined in 3GPP specifications [3GPP 2003b; 3GPP 2003d], from the transport block level to the antenna. This includes cyclic redundancy check (CRC) attachment, turbo coding, interleaving, rate matching, spreading, and modulation. The channel was modeled using a TDL allowing the simulation of the environments defined in Ref. [3GPP 2003b], namely Pedestrian A and Indoor A. The receiver was built with all the blocks for performing the inverse operations and also a RAKE receiver for extracting and coherently adding the distinguishable received multipaths (using MRC). The turbo decoder used the maximum a posteriori (MAP) algorithm [Bahl et al. 1974] (working in the logarithmic domain) and performed eight iterations. For the case of 16-QAM modulation, a soft-output demodulator was built in order to obtain soft bit values to be used in the turbo decoder. The MPIC block was implemented, using clipped soft decision (CSD) and six iterations (note that each iteration is basically made of spreading and despreading operations, being therefore very fast, with the complexity of 10% of a normal turbo decoder operation), and added to the simulator.

The simulator estimated the system's BER and BLER using the Monte-Carlo method. Block errors play a very important role in the study of HSDPA performance for the study of the need for retransmissions (H-ARQ, with Chase combining). In fact, one bit could cause a faulty CRC check on a block, therefore causing the need for retransmissions, and therefore a delay and decrease in throughput. The BER values only give an estimate on how badly the blocks might arrive, and provide henceforth an estimate of the number

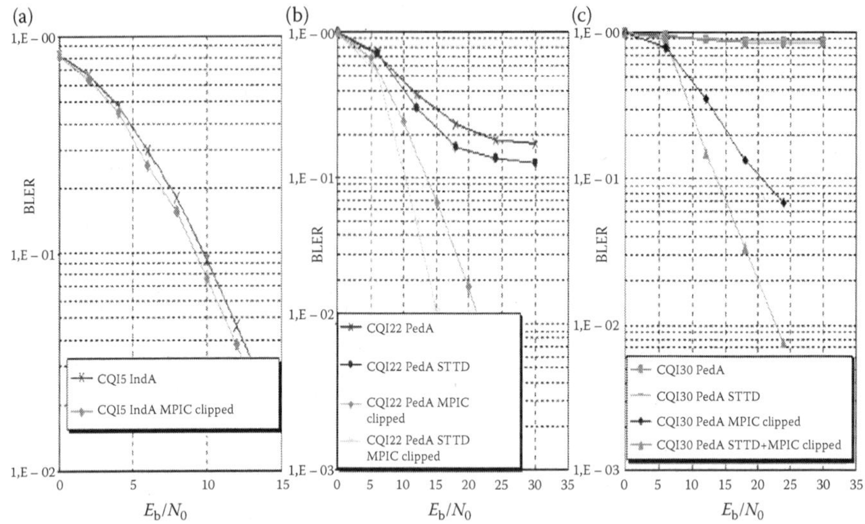

Figure 5.9 BLER performance for CQI15 in Indoor A channel (a), CQI22 in Pedestrian A channel (b), and CQI 30 in Pedestrian A channel (c).

of required H-ARQs to fix those errors. In this work, only the BLER results for the last three CQI values considered in Table 5.1 (high-throughput modes) are presented.

To increase the capacity of the system in fading channels, it is possible to introduce diversity using several transmitting antennas at the BS. This can be done using Alamouti's open loop STTD scheme [Alamouti 1998] for two transmitting antennas. In this case, STTD was implemented according to 3GPP specifications [3GPP 2003a].

Figure 5.9 portrays several BLER results. Figure 5.9a compares the results for CQI 15 with and without the MPIC in the Indoor A environment [ETSI 1998a]. Among all the CQIs using QPSK modulation, this is the one with the highest transmission rate using five physical channels (spreading codes). The performance improvement obtained with the MPIC is not very meaningful since this CQI is not very much affected by the MPI. So, for the CQIs using QPSK modulation, the MPIC is not very useful.

The graph presented in Figure 5.9b was obtained for CQI22, which uses 16-QAM modulation, for a Pedestrian A environment [ETSI 1998a]. In this case, it is clear that considerable improvement is obtained in the performance when the MPIC is used. Without the MPIC the curves for no transmit diversity and STTD had a performance "floor," which limited the minimum obtainable BLER to a value above 10%. Note that the usage of STTD in a multicode and multipath environment aids in diversity combating the fading effect, but also contributes to added interference in each multipath. Owing to this last reason, it can be seen from Figure 5.9c that the STTD scheme without MPIC for CQI 30 and Pedestrian A channel is actually worse than the scheme without STTD.

With the MPIC the "floor" was lowered substantially, making it possible to obtain much lower BLER values. This also happened for the case of CQI 30, as can be seen in Figure 5.9c, which corresponds to the higher available peak rate, using 15 physical channels and 16-QAM modulation. The STTD along with the MPIC also yielded better results than the MPIC without STTD, since a significant amount of MPI was canceled, and thus gave the STTD algorithm the chance to work effectively.

To conclude, it was shown that the CQI modes of HSDPA using 16-QAM modulation do not perform very well in multipath environments due to the low robustness of this modulation against interference. The usage of an MPI canceler using CSD for suppressing this type of interference was thus proposed. With the aid of an HSDPA LLS, it was shown that the MPIC usage resulted in a substantial performance improvement for the CQI values using the troublesome 16-QAM modulation. Furthermore, it was shown that, if STTD (using two antennas) was also used, the MPIC scheme could allow for a greater gain in performance, even though the STTD by itself performs worse in CQI 30, due to the added interference caused by the second antenna.

5.3.1.2 MMSE Receiver for WCDMA

As we have seen in the previous section, the RAKE receiver is not sufficient to cope with the interference of fully loaded systems. One of the main problems consists in the uncanceled multipath components pertaining to the same message; since the ratio of used channels per spreading factor (SF) is close to one, the cross-correlation between sequences is very significant, and left uncanceled. By employing an MPIC, substantial performance gains were observed, although still very far from the single user bound for fully loaded scenarios. Since the MPIC has its operation stemmed from the RAKE, it is only able to cancel part of the ISI (the RAKE leaves a significant noise component attached). However, for higher-order modulations than QPSK and/or MIMO systems, where interference from other transmitting antennas (ICI) significantly increase the MAI and ISI, a more powerful receiver needs to be employed.

Equalization-based receivers try to take into account all the effects that the symbols are subjected to in the transmission chain, namely the joint compensation of MAI, ICI, and ISI. In this section, linear equalizers were employed, namely the MMSE and the ZF.

5.3.1.2.1 MMSE System Matrices for a DS-CDMA System

The design of the system matrices for the MF and the ZF/MMSE schemes is presented in this section (note that the MF is also discussed since it is an integral part of both ZF and MMSE schemes). A MIMO arrangement is assumed, in which the data streams for each user are either split into different streams for each transmit antenna to increase the bit rate/capacity, or replicated (with interleaving) for each antenna, to increase the transmit diversity, reducing the necessary transmit power for nominal operations.

The transmitted signal associated with the kth spreading code and the txth transmit antenna is given by

$$x_{tx}^{(k)}(t) = A_{tx}^{(k)} \sum_{n=1}^{N\cdot\mathrm{SF}} b_{\lfloor n/\mathrm{SF}\rfloor,tx}^{(k)} c_n^{(k)} f_{\mathrm{T}}(t - nT_{\mathrm{C}}) = A_{tx}^{(k)} \sum_{n} \beta_{n,tx}^{(k)} f_{\mathrm{T}}(t - nT_{\mathrm{C}}), \quad (5.36)$$

where N is the number of data symbols to be transmitted by each antenna, $A_{tx}^{(k)} = \sqrt{E_k}$ (admitting $f_T(0) = 1$, $E\left[\left|b^{(k)}\right|^2\right] = 1$ and $E\left[\left|c^{(k)}\right|^2\right] = 1$), with E_k denoting the symbol energy, SF is the spreading factor, K is the total number of spreading codes per antenna/physical channel (or user, if each user uses only one physical channel), and $R_C = 1/T_C$ is the chip rate. The nth chip associated with the kth user and the txth antenna is $\beta_{n,tx}^{(k)} = b_{\lfloor n/\mathrm{SF}\rfloor,tx}^{(k)} c_n^{(k)}$. The data symbols to be transmitted at the txth transmit antenna are $b_{n,tx}^{(k)}$, $n = 1, 2, \ldots, N$, and the combined spreading and scrambling signature is $c_n^{(k)}$, $n = 1, 2, \ldots, N \times \mathrm{SF}$. $f_T(t)$ is the adopted pulse shape filter (a square-root raised cosine filtering is assumed).

At the receiver we have N_{RX} antennas. The signal for any given receive antenna rx, prior to the reception filter, is given by

$$y_{rx}(t) = \sum_{k=1}^{K} \sum_{n=1}^{N\cdot\mathrm{SF}} \sum_{tx=1}^{N_{TX}} \left(x_{n,tx}^{(k)}(t) * h_{n,rx,tx}^{(k)}(t) \right) + n(t), \tag{5.37}$$

with $n(t)$ denoting the channel noise, assumed Gaussian, with zero mean and the variance of the real and imaginary components denoted by σ^2. The CIR between the transmit antenna tx and the receive antenna rx is

$$h_{n,rx,tx}^{(k)}(t) = \sum_{l=1}^{L} \alpha_{n,rx,tx,l}^{(k)} \delta(t - \tau_l^{(k)}), \tag{5.38}$$

with L denoting the total number of the channel's multipath components, $\alpha_{n,tx,rx,l}^{(k)}$ denoting the complex attenuation (fading) factor for the lth path and $\tau_l^{(k)}$ is the propagation delay associated with the lth path (for the sake of simplicity, it is assumed that this delay is constant; the generalization to other cases is straightforward). The received signal associated with each antenna is submitted to a reception filter, with impulse response $f_R(t)$, which is assumed to be matched to $f_T(t)$ [i.e., $f_R(t) = f_T^*(-t)$], leading to the signal

$$r_{rx}(t) = \sum_{tx=1}^{N_{Tx}} \sum_{k=1}^{K} A_{tx}^{(k)} \sum_{l=1}^{L} \sum_{n=1}^{N\cdot\mathrm{SF}} b_{\lfloor n/K\rfloor,tx}^{(k)} c_n^{(k)} h_{n,rx,tx,l}^{(k)}\, p\left(t - nT_C - \tau_l^{(k)}\right), \tag{5.39}$$

where $p(t) = f_T(t) * f_R(t)$ [for the Nyquist pulses considered in this work $p(t) = f_T(t) * f_R(t) = f_T(t) * f_T^*(-t)$ is such that $p(nT_C) = 0$ for integer $n \neq 0$].

Using matrix algebra, the received vector is

$$r_v = \mathbf{S}\mathbf{C}\mathbf{A}\boldsymbol{b} + \boldsymbol{n}, \tag{5.40}$$

where **S**, **C**, and **A** are the spreading, channel, and amplitude matrices, respectively. The receive vector r_v encompasses the messages for all receive antennas, such that

$$r_v = \begin{bmatrix} r_{rx=1}(t) \\ r_{rx=2}(t) \\ \vdots \\ r_{rx=N_{RX}}(t) \end{bmatrix}. \tag{5.41}$$

Note that the channel matrix encompasses not only the channel coefficients, but the filter's coefficients as well. The spreading matrix accounts for the spreading and scrambling codes, as well as for the delays between users and channel replicas. The channel matrix accounts for the fading coefficients for all links between each transmit and receive antenna. For simplicity, the spreading matrix for the downlink will be described, and the assumption that the spreading and scrambling codes are the same for all the transmit antennas will be made, having a direct effect on both the spreading and the channel matrix. The other structures remain the same, for both the downlink and uplink transmissions.

5.3.1.2.2 Downlink with Equal Scrambling for All Transmit Antennas: S and C Matrices

The downlink spreading matrix **S** has dimensions $(\mathrm{SF}\cdot N\cdot N_{RX}+\psi_{\mathrm{MAX}}\cdot N_{RX})\times(K\cdot L\cdot N\cdot N_{RX})$ (ψ_{MAX} is the maximum delay of the channel's impulse response, normalized to the number of chips, $\psi_{\mathrm{MAX}}=[\tau_{\mathrm{MAX}}/T_c]$, where T_c is the chip period), and it is composed of submatrices $\mathbf{S}_{RX}$ in its diagonal for each receive antenna $\mathbf{S}=\mathrm{diag}(\mathbf{S}_{RX=1},\ldots,\mathbf{S}_{RX=\mathrm{N}_{RX}})$. Each of these submatrices has dimensions $(\mathrm{SF}\cdot N+\psi_{\mathrm{MAX}})\times(K\cdot L\cdot N)$, and they are further composed of smaller matrices $\mathbf{S}_n^L$, one for each bit position, with size $(SF+\psi_{\mathrm{MAX}})\times(K\cdot L)$. The $\mathbf{S}_{RX}$ matrix structure is made of $\mathbf{S}_{RX}=[\mathbf{S}_{\varepsilon,1},\ldots,\mathbf{S}_{\varepsilon,N}]$, with

$$\mathbf{S}_{\varepsilon,n}=\begin{bmatrix}0_{(\mathrm{SF}\cdot(n-1))\times(K\cdot L)}\\ \mathbf{S}_n^L\\ 0_{(\mathrm{SF}\cdot(N-n))\times(K\cdot L)}\end{bmatrix}. \tag{5.42}$$

The $\mathbf{S}_n^L$ matrices are made of $K\cdot L$ columns

$$\mathbf{S}_n^L=\left[s_{\mathrm{col}(k=1,l=1),n},\ldots,s_{\mathrm{col}(k=1,l=L),n},\ldots,s_{\mathrm{col}(k=K,l=L),n}\right]. \tag{5.43}$$

Each of these columns is composed of

$$s_{\mathrm{col}(kl),n}=\left[0_{(1\times\mathrm{delay}(l))},c_n(k)_{1\times\mathrm{SF}},0_{(1\times(\psi_{\mathrm{MAX}}-\mathrm{delay}(l)))}\right]^{\mathrm{T}}, \tag{5.44}$$

where $c_n(k)$ is the combined spreading and scrambling for the bit n of user k.

These $\mathbf{S}^L$ matrices are either all alike, if no long scrambling code is used, or different if the scrambling sequence is longer than the SF. The $\mathbf{S}^L$ matrices represent the combined spreading and scrambling sequences, conjugated with the channel delays. The shifted spreading vectors for the multipath components are all equal to the original sequence of the specific user

$$\mathbf{S}_n^L=\begin{bmatrix}S_{1,1,1,n} & & \cdots & \cdots & S_{K,1,1,n} & & \\ \vdots & \ddots & S_{1,1,L,n} & \cdots & \vdots & \ddots & S_{K,1,L,n}\\ S_{1,\mathrm{SF},1,n} & & \vdots & \cdots & S_{K,\mathrm{SF},1,n} & & \vdots\\ & \ddots & S_{1,\mathrm{SF},L,n} & \cdots & \cdots & \ddots & S_{K,\mathrm{SF},L,n}\end{bmatrix}. \tag{5.45}$$

Note that, to correctly model the MPI between symbols, there is an overlap between the $\mathbf{S}^L$ matrices, of ψ_{MAX}. As opposed to the SISO multipath case presented in Ref. [Latva-aho and Juntti 2000], the matrix is not trimmed for the last multipath components.

The channel matrix $\mathbf{C}$ is a $(K \cdot L \cdot N \cdot N_{RX}) \times (K \cdot N_{TX} \cdot N)$ matrix, and it is composed of N_{RX} submatrices, each one for a receive antenna

$$\mathbf{C} = \begin{bmatrix} \mathbf{C}_{RX=1} \\ \vdots \\ \mathbf{C}_{RX=N_{RX}} \end{bmatrix}. \tag{5.46}$$

Each $\mathbf{C}_{RX}$ matrix is composed of N $\mathbf{C}^{KT}$ matrices alongside its diagonals.

$$\mathbf{C}_{RX} = \begin{bmatrix} \mathbf{C}^{KT}_{1,RX} & & \\ & \ddots & \\ & & \mathbf{C}^{KT}_{N,RX} \end{bmatrix}. \tag{5.47}$$

Each $\mathbf{C}^{KT}_{n,RX}$ matrix is $(K \cdot L) \times (K \cdot N_{TX})$ and represents the fading coefficients for the current symbol of each path, user, transmit antenna, and receive antenna. The matrix structure is made up of further smaller matrices alongside the diagonal of $\mathbf{C}^{KT}_{n,RX}$, $\mathbf{C}^{KT}_{n,RX} = \mathrm{diag}\left(\mathbf{C}^{\mathrm{T}}_{n,RX,K=1}, \ldots, \mathbf{C}^{\mathrm{T}}_{n,RX,K=K}\right)$, with $\mathbf{C}^{\mathrm{T}}$ of dimensions $L \times N_{TX}$, representing the combination of fading coefficients and filters' coefficients for the user's multipath and the *tx*th antenna component.

By defining

$$c^{(k)}_{n,rx,tx,l} = \sum_{n'} \alpha^{(k)}_{n-n',rx,tx,l} p\left(n' T_{\mathrm{C}} - \left(\tau^{(k)}_l - \tau^{(k)}_0\right)\right), \tag{5.48}$$

we have

$$\mathbf{C}^{KT}_{n,RX} = \begin{bmatrix} c_{1,RX,1,1} & \cdots & c_{1,RX,N_{TX},1} & & & & \\ \vdots & & \vdots & & & & \\ c_{1,RX,1,L} & \cdots & c_{1,RX,N_{TX},L} & & & & \\ & & & \ddots & & & \\ & & & & c_{K,RX,1,1} & \cdots & c_{K,RX,N_{TX},1} \\ & & & & \vdots & & \vdots \\ & & & & c_{K,RX,1,L} & \cdots & c_{K,RX,N_{TX},L} \end{bmatrix}. \tag{5.49}$$

The $\mathbf{A}$ matrix is a diagonal of dimensions $(K \cdot N_{TX} \cdot N)$, and represents the amplitude of each user per transmission antenna and symbol, $\mathbf{A} = \mathrm{diag}\left(\mathbf{A}_{1,1,1}, \ldots, \mathbf{A}_{N_{TX},1,1}, \ldots, \mathbf{A}_{N_{TX},K,1}, \ldots, \mathbf{A}_{N_{TX},K,N}\right)$.

5.3.1.2.3 Different Scrambling for All Transmit Antennas and Uplink Modifications: **S** and **C** Matrices

The previous **S** and **C** matrices assume that all antennas use the same spreading and scrambling code. However, when not operating under full-loading conditions, the best results are obtained when different scrambling sequences are used at each transmit antenna, in order to increase diversity.

Therefore, the main changes to the matrices would be the $\mathbf{S}_n^L$ submatrix for the S matrix,

$$
\mathbf{S}_n^L = \left[\begin{matrix}
s_{1,1,1,1,n} & \cdots & s_{1,1,1,N_{TX},n} & & & & & \cdots & s_{K,1,1,1,n} & \cdots \\
\vdots & & \vdots & \ddots & s_{1,1,L,1,n} & \cdots & s_{1,1,L,N_{TX},n} & \cdots & \vdots & \\
s_{1,SF,1,1,n} & \cdots & s_{1,SF,1,N_{TX},n} & & \vdots & & \vdots & \cdots & s_{K,\mathrm{SF},1,1,n} & \cdots \\
 & & & \ddots & s_{1,\mathrm{SF},L,1,n} & \cdots & s_{1,\mathrm{SF},L,N_{TX},n} & \cdots & &
\end{matrix}\right.
$$

$$
\left.\begin{matrix}
s_{K,1,1,N_{TX},n} & & & & \\
\vdots & \ddots & s_{K,1,L,1,n} & \cdots & s_{K,1,L,N_{TX},n} \\
s_{K,\mathrm{SF},1,N_{TX},n} & & \vdots & & \vdots \\
 & \ddots & s_{K,\mathrm{SF},L,1,n} & \cdots & s_{K,\mathrm{SF},L,N_{TX},n}
\end{matrix}\right],
$$

and the $\mathbf{C}^{KT}$ submatrix for the C matrix

$$
\mathbf{C}^{KT} = \begin{bmatrix}
c_{1,1,1} & & & & & & \\
 & \ddots & & & & & \\
\vdots & & c_{N_{TX},1,1} & & & & \\
 & \vdots & & & & & \\
c_{1,L,1} & & \vdots & & & & \\
 & \ddots & & & & & \\
 & & c_{N_{TX},L,1} & & & & \\
 & & & \ddots & & & \\
 & & & & c_{1,1,K} & & \\
 & & & & & \ddots & \\
 & & & & \vdots & & c_{N_{TX},1,K} \\
 & & & & & \vdots & \\
 & & & & c_{1,L,K} & & \vdots \\
 & & & & & \ddots & \\
 & & & & & & c_{N_{TX},L,K}
\end{bmatrix}. \quad (5.50)
$$

For the uplink transmission, the **S** matrix should also portray the delays between different users (in this case, different transmit antennas). The $\mathbf{S}_n^L$ should thus be adjusted, with the columns being shifted either upward or downward, depending on the offsets between users. Note that the size (number of lines) of the $\mathbf{S}_n^L$ matrix can be increased, to account for the delays between users. Although the final **S** matrix can have a bigger number of lines than for the downlink case, its overall structure remains the same.

5.3.1.2.4 Design of the Remaining Structures

The result from the **SCA** operation (multiplication of the spreading, channel, and amplitude matrices, henceforth known as the **SCA** matrix) is depicted in Figure 5.10. It is an $N_{RX} \cdot (N \cdot \mathrm{SF} + \psi_{\mathrm{MAX}}) \times N_{TX} \cdot K \cdot N$ matrix, and it is the reference matrix for the decoding algorithms. Note that the **SCA** matrix is sparse in nature.

The resulting **SCA** matrix will have the same size as before; only the number of operations increases while constructing the **SCA** matrix (the **SC** multiplication has an increase in complexity equal to the number of transmit antennas), since values from different antennas must be treated differently.

Vector $\boldsymbol{b}$ represents the information symbols. It has length $(K \cdot N_{TX} \cdot N)$, and it has the following structure:

$$\boldsymbol{b} = \left[b_{1,1,1}, \ldots, b_{N_{TX},1,1}, \ldots, b_{1,K,1}, \ldots, b_{N_{TX},K,1}, \ldots, b_{N_{TX},K,N}\right]^{\mathrm{T}}. \quad (5.51)$$

Note that the bits of each transmit antenna are grouped together in the first level, and the bits of other interferers in the second level. This is to guarantee that the resulting matrix to be inverted has all its nonzero values as close to the diagonal as possible. Also note that there is usually a higher correlation between bits from different antennas using the same spreading code than between bits with different spreading codes.

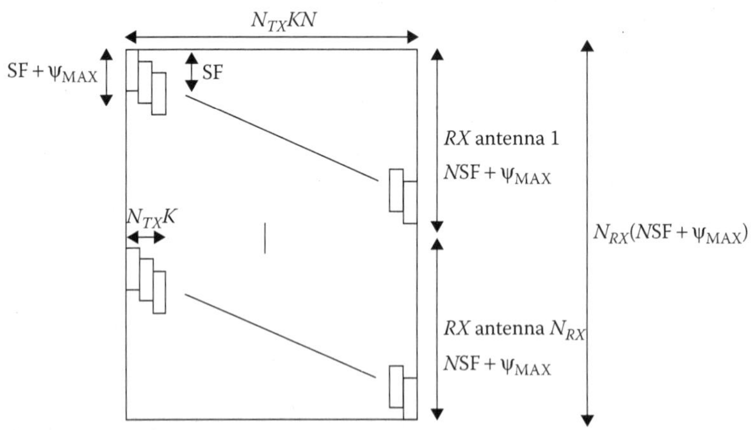

Figure 5.10 Layout of the SCA matrix.

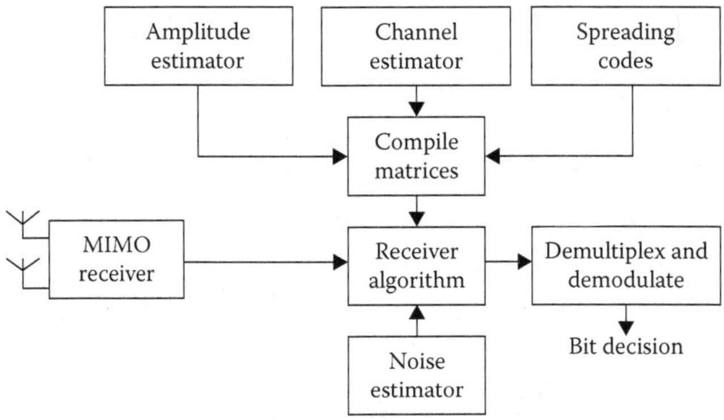

Figure 5.11 MIMO receiver for WCDMA schemes.

Finally, the $\boldsymbol{n}$ vector is a $(N \cdot \mathrm{SF} \cdot N_{RX} + N_{RX} \cdot \psi_{\mathrm{MAX}})$ vector with noise components to be added to the received vector $\boldsymbol{r}_v$, which is partitioned by N_{RX} antennas,

$$r_v = \left[r_{1,1,1}, \ldots, r_{1,SF,1}, \ldots, r_{N,1,1}, \ldots, r_{N,SF+\psi_{\mathrm{MAX}},1}, \ldots, r_{N,1,N_{RX}}, \ldots, r_{N,SF+\psi_{\mathrm{MAX}},N_{RX}}\right]^{\mathrm{T}} \tag{5.52}$$

(the delay ψ_{MAX} is only used for the final bit, although its effects are present throughout $\boldsymbol{r}_v$). Figure 5.11 illustrates the main blocks from which the receiver is compiled.

If transmit diversity is applied, the $\boldsymbol{b}$ vector should be arranged accordingly, with replicas (after multiplexing) for each antenna.

5.3.1.2.5 Receiver Schemes Using the System Matrices

Equalization-based receivers such as the MMSE and ZF take into account all the effects that the symbols are subjected to in the transmission chain, namely the MAI, ISI, and the channel effect. Using as basis the (unnormalized) MF output (obtained by applying the Hermitian to the **SCA** matrix and multiplying by the received vector),

$$y_{MF} = (SCA)^{\mathrm{H}} r_v \tag{5.53}$$

and defining **R** as

$$\mathbf{R} = \mathbf{A} \times \mathbf{C}^{\mathrm{H}} \times \mathbf{S}^{\mathrm{H}} \times \mathbf{S} \times \mathbf{C} \times \mathbf{A}, \tag{5.54}$$

the equalization matrix ($\mathbf{E}_{\mathrm{M}}$) for the MF and ZF can be written as

$$\mathbf{E}_{M,MFZF} = \mathbf{R}, \tag{5.55}$$

where the normalized MF estimate is given by

$$y_{\mathrm{NMF}} = \mathrm{diag}\left(\mathbf{E}_{M,\mathrm{MFZF}}^{-1}\right) y_{\mathrm{MF}} \tag{5.56}$$

and the ZF estimate by

$$y_{\text{ZF}} = \mathbf{E}_{M,\text{MFZF}}^{-1}\, y_{\text{MF}}, \tag{5.57}$$

which is simply applying the inverse of all effects the message was subjected to. Once again, to prevent an ill-conditioned matrix for inversion (the $\mathbf{E}_M$ might become ill-conditioned when the system is fully loaded [Divsalar et al. 1998], depending on the cross-correlations between the users' signature sequences), a small value (e.g., 1e^{-6}) should be added to all elements in the main diagonal of $\mathbf{E}_M$. In order to avoid round-off problems, the $\mathbf{E}_M$ should be rounded at a value above the minimum machine precision.

The MMSE estimate aims to minimize $E\left(\left|b - \hat{b}\right|^2\right)$. From Ref. [Kay 1993], the $\mathbf{E}_M$ includes the estimated noise power σ^2, and is represented by

$$\mathbf{E}_{M,\text{MMSE}} = \mathbf{R} + \sigma^2\mathbf{I}. \tag{5.58}$$

The MMSE estimate is thus

$$y_{\text{MMSE}} = \hat{b} = \mathbf{E}_{M,\text{MMSE}}^{-1}\, y_{\text{MF}}. \tag{5.59}$$

Both the ZF- and MMSE-based receivers are seldom used due to their perceived complexity, especially for wideband MIMO systems (with frequency-selective fading channels). Due to the multipath causing ISI, the whole information block is usually decoded at once (although there are some decoding variants in which the block is divided into smaller blocks [Silva et al. 2005c; Shoumin and Zhi 2004], requiring some overlapping between symbols, in order to provide best results), requiring the use of a significant amount of memory and computing power for the algebraic operations.

However, if the sparseness of the matrices is taken into account, only a fraction of the memory and computing power is required. As can be inferred from the previously described matrices, all system matrices are sparse in nature and consist of submatrices that are sparse themselves.

The most troublesome matrix to deal with is the $\mathbf{E}_M$, due to its inversion (more precisely, the resolution of the equation system leading to the final estimate). Fortunately, the $\mathbf{E}_M$ is also sparse by itself, it being possible to handle it with simplicity. For instance, considering a maximum-loading simulation case using 16-QAM modulation, SF = 16 (16 physical channels), $L = 2$ multipaths (the second multipath with a 1 chip delay, resembling the Indoor A or Pedestrian A channel), a MIMO of 2-transmit-antenna/2-receive-antenna system, and a block size of 1024 bits ($N = 256$ symbols, with 4 bits per symbol) per physical channel of each transmit antenna, the matrix's diagonal width (MDW), which in this case is $(K \cdot L \cdot N_{TX}) \cdot 3/2 = 96$, is roughly 1.2% of the matrix's width (MW) of $(K = 16) \cdot (N_{TX} = 2) \cdot (N = 256) = 8192$, MDW = 1.2% MW.

Another aspect of the $\mathbf{E}_M$ is that it is Hermitian positive definite, and thus it can be decomposed using the Cholesky decomposition. Since it is a banded matrix (with all elements concentrated on its diagonal), there is no Cholesky fill-in since the band is dense [cases with small chip delays (Figure 5.12)], and thus it presents itself as if the sparse reverse Cuthill–McKee ordering algorithm [George and Joseph Liu 1981] had

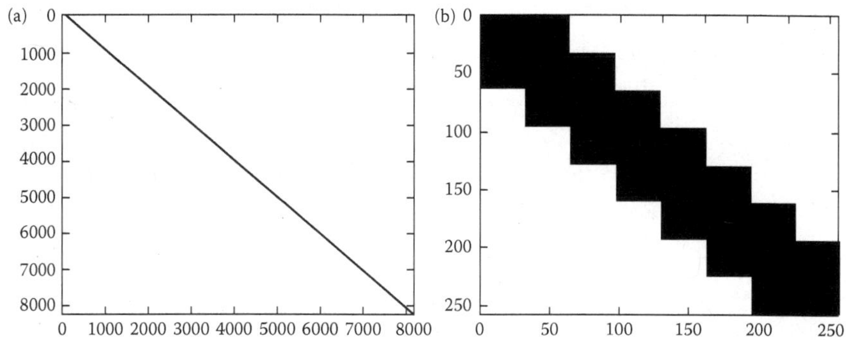

Figure 5.12 Equalization matrix for a two-tap channel, $K = 16$, $N_{TX} = 2$ (a) and diagonal close-up for the same case (b).

been applied to it (a good preordering for LU or Cholesky factorization of matrices that come from long, skinny problems). The next section introduces a detailed description of the $\mathbf{E}_M$, and enhanced algorithms to solve it.

5.3.1.2.6 Performance Results of the Equalization Techniques

Both the MF and the ZF/MMSE schemes were simulated. Since the ZF/MMSE results are based on the equalization of the MF results, it is interesting to compare the three schemes. Simulations were run considering block sizes of 1024 bits, and the reference UMTS channels. A MIMO setting with an equal number of transmitters and receivers was implemented. An SF of 16 was used (hence 16 physical channels per transmit antenna). Minimum (usage of 1 physical channel per transmit antenna) and maximum (usage of 16 physical channels per transmit antenna) loading conditions were considered.

5.3.1.2.6.1 MF Results The MF results (Figure 5.13) are important for the MMSE receiver, since it is a crucial part of the algorithm. It can be seen that, for the minimum loading case (Figure 5.13a), the results of Vehicular A are best and of Pedestrian A are worse, due to multipath diversity. The diagonal of the $\mathbf{E}_M$ was used for normalization of y_{MF}, instead of using just the estimated channel coefficients as is usually done for standard RAKE receivers. The extra information from the $\mathbf{E}_M$ allows minimizing the correlation effect, and thus the multipath diversity can be exploited for higher-order modulations when there is little interference, contrary to the normal RAKE.

Due to interference from other antennas, and the fact that the simple MF algorithm does not perform any type of interference canceling or equalization, the lowest MIMO orders provide the best results. For the fully loaded case (Figure 5.13b), the results for Pedestrian A are better than those for the Vehicular A channel, due to the high amount of MPI. The lowest MIMO orders still provide the best results, due to the reduced interference.

The uplink transmission of the MF receiver (Figure 5.14) yields worse results, due to the increased interference caused by the correlation of the spreading codes, which are

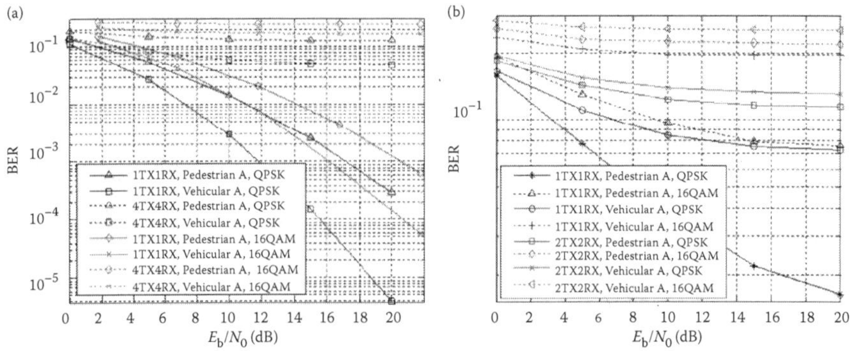

Figure 5.13 DL BER performance for MF, using $k = 1$ (a) and $k =$ SF (b).

no longer orthogonal. Note that the error floor is now over 10%, making it virtually impossible to operate at full loading with this scheme.

5.3.1.2.6.2 MMSE and ZF Results Figure 5.15 portrays some BER results for the MMSE receivers. As expected, the best results were obtained for the minimum loading case (Figure 5.15a) of the highest MIMO orders (highest receive diversity). Since the Vehicular A channel has the greater number of taps, the best results are obtained for this channel (note that perfect channel estimation is assumed). Indoor A is the second best, since it has a second tap of greater power than the pedestrian A channel, which is predominantly a 1 tap channel.

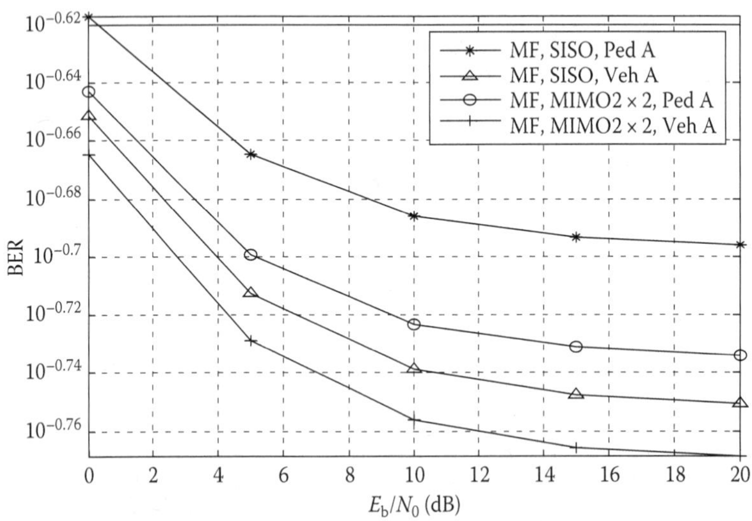

Figure 5.14 UL BER performance for MF, using $k =$ SF.

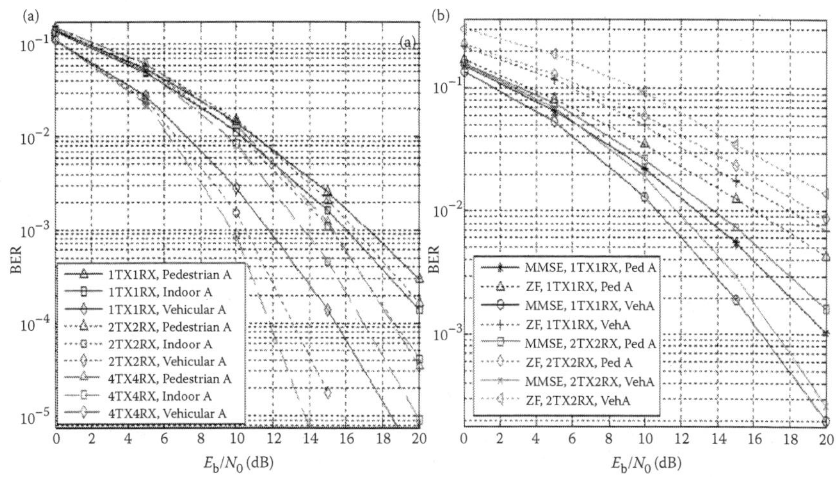

Figure 5.15 BER performance for MMSE, using k = 1 and SF = 16 (a) and BER performance for MMSE and ZF, using k = SF (b).

For the fully loaded system (Figure 5.15b), it can be seen that the situation is quite different, with the lowest MIMO orders yielding the best results, due to the reduction of interference. Thus, the best and the worst channel's performances are still Vehicular A and Pedestrian A, for high values of E_b/N_0.

Regarding the ZF results, it can be seen that these are significantly worse than those of the MMSE algorithm, with differences of performance over 10 dB. Noise estimation plays a determining role in the MMSE algorithm, especially in fully loaded systems, where the interference is high.

Some uplink results for both the MMSE and ZF are presented in Figure 5.16. The performance is also worse than for the downlink, due to the added interference between spreading sequences. The ZF results are disastrous (above 10%), augmenting the importance of the estimation of the noise component for the MMSE.

5.3.2 Spatial Multiplexing: SC-FDE

Future wireless systems are required to support high QoS at high data rates. Moreover, due to power and BW constraints, these systems are also supposed to operate with small transmit powers, especially at the MTs and to have high spectral efficiencies.

It is well known that, by employing multiple antennas at both the transmitter and the receiver, we can increase substantially the capacity of a given system [Foschini 1996]. To exploit this potential, several Bell Laboratory Layered Space-Time (BLAST) architectures have been proposed for flat fading MIMO channels [Foschini and Gans 1998; Wolniasky et al. 1998]. The BLAST techniques were extended to frequency-selective channels through the use of time-domain MIMO-DFEs [Al-Dhahir and Sayed 2000; Lozano and Papadias 2002].

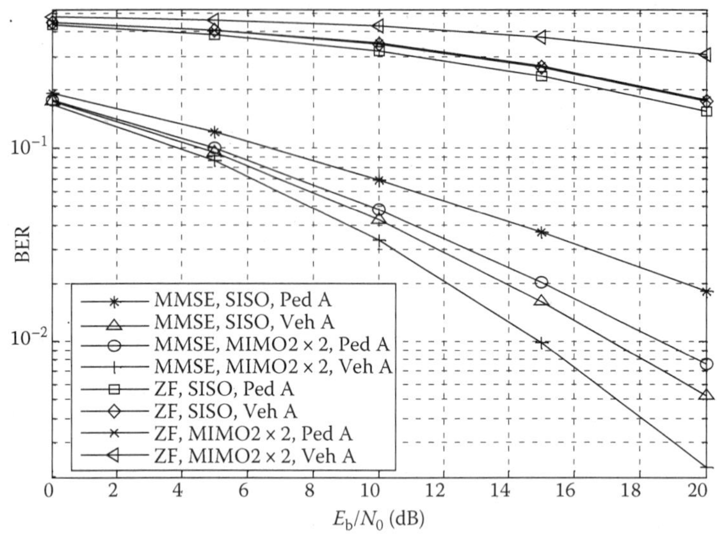

Figure 5.16 UL BER performance for MMSE and ZF, using k = SF.

This concept can be extended to space division multiple access (SDMA) techniques where we employ multiple antennas at the BS to increase the number of simultaneous users in a given cell, allowing a significant increase in the system spectral efficiency, while reducing the transmit power requirements for the MTs [Winters et al. 1992; Tidestav et al. 1999; Roy and Falconer 2003; Sfar et al. 2003].

However, for the high data rates of broadband wireless systems, we can have severe time-dispersion effects associated with the multipath propagation. In this case, conventional time-domain equalization schemes are not practical, since the number of operations per symbol is proportional to ISI span. This can be more serious when conventional time-domain equalization methods are employed in high-data-rate MIMO systems.

As mentioned before in this chapter, block transmission techniques, with appropriate cyclic extensions and employing FDE techniques, have been shown to be suitable for high-data-rate transmission over severe time-dispersive channels, since the number of operations per symbol grows logarithmically with the block duration (and therefore the ISI span), due to the FFT implementation [Gusmão et al. 2000; Wang and Giannakis 2000; Falconer 2002].

The OFDM schemes [Cimini 1985; Bingham 1990] are the most popular modulations based on this technique. Since the OFDM transmission over time-dispersive channels can be regarded as a parallel transmission over a set of nondispersive channels, one for each subcarrier, the extension of MIMO/BLAST techniques to OFDM schemes is straightforward [Raleigh and Cioffi 1998; Raleigh and Jones 1999], eventually with some preprocessing at the transmitter and/or employing adaptive loading schemes. However, the OFDM signals have high envelope fluctuations and a high PMEPR leading to amplification difficulties. For this reason, several techniques have been proposed for reducing the envelope fluctuations of OFDM signals [Jones and Wilkinson 1996; Müller et al.

1997; Dinis and Gusmão 2004]. However, these PMPER-reducing techniques require an increased signal processing effort, especially at the transmitter side, and, possibly, some signal distortion when a nonlinear signal processing is employed. Moreover, even for the most sophisticated techniques, the transmitted signals still have PMEPRs much higher than the PMEPRs for SC signals based on similar constellations.

The SC modulations, using block transmission techniques and FDE [Sari et al. 1994], are an alternative approach for broadband wireless systems. As with OFDM modulations, the data blocks are preceded by a CP, long enough to cope with the channel length. The received signal is transformed to the frequency domain, equalized in the frequency domain, and then transformed back to the time domain. The achievable performances, as well as the overall implementation complexities, are similar for SC schemes with FDE and OFDM schemes [Gusmão et al. 2000; Montezuma and Gusmão 2001; Falconer 2002]. However, the signal processing effort is concentrated at the receiver for the SC case. This, combined with the lower envelope fluctuations of SC signals, makes them preferable for the uplink transmission (i.e., the transmission from the MT to the BS), while the OFDM schemes might be a better choice for the downlink transmission (i.e., the transmission from the BS to the MT). For this reason, a mixed OFDM plus SC mode air interface was proposed, employing an SC scheme with FDE in the uplink and an OFDM scheme in the downlink [Gusmão et al. 2000; Falconer 2002].

It is well known that nonlinear equalizers can significantly outperform linear equalizers [Proakis 1995]. Among nonlinear equalizers, the DFEs are especially popular, due to their performance/complexity trade-offs. For this reason, a hybrid time-frequency SC-FDE was proposed in Ref. [Benvenuto and Tomasin 2002], employing a frequency-domain feedforward filter and a time-domain feedback filter. Therefore, the hybrid time-frequency-domain DFE has a better performance than a linear FDE. This hybrid time-frequency-domain DFE was extended to MIMO scenarios in Ref. [Tubbax et al. 2003]. However, as with conventional, time-domain DFEs, it can suffer from error propagation, especially for long feedback filters. A promising IB-DFE approach for SC transmission was proposed in Ref. [Benvenuto and Tomasin 2002] where both the feedforward and the feedback parts are implemented in the frequency domain. Since the feedback loop takes into account not just the hard decisions for each block, but also the overall block reliability, the error propagation problem is significantly reduced. Consequently, the IB-DFE techniques offer much better performances when compared with the noniterative methods [Benvenuto and Tomasin 2002; Dinis et al. 2003]. These IB-DFE techniques can be regarded as low-complexity turbo equalization schemes [Tüchler et al. 2002], since the feedback loop uses the equalizer outputs instead of the channel decoder outputs.

A promising frequency-domain receiver structure for SC-based MIMO systems was recently proposed [Dinis and Gusmão 2004]. This receiver combines the IB-DFE concept with BLAST principles to define an iterative LST (layered space-time architecture) receiver suitable for broadband wireless communications systems. As with IB-DFE schemes, the channel decoder outputs are not required in the feedback loop. Therefore, the receive structure proposed in Ref. [Dinis and Gusmão 2004] can be regarded as a low-complexity turbo equalization for MIMO channels [Abe and Matsumoto 2003].

In this section we consider the use of SDMA techniques in broadband wireless systems. We focus on the receiver design for the uplink (i.e., the BS) when an SC block transmission with FDE is assumed. The BS employs multiple antennas, allowing the detection of several MTs transmitting simultaneously (i.e., using the same TSs and the same frequency bands). Following the detection scheme proposed in Ref. [Dinis and Gusmão 2004], we present and evaluate a class of frequency-domain receivers for MIMO channels with an iterative multiuser detection. Our approach is to cancel the ISI and the multiuser interference, in an iterative way. At each iteration, we use the users' estimated blocks in the previous iteration, as well as the corresponding blockwise reliability.

5.3.2.1 Iterative Frequency-Domain Detection for MIMO-SC Systems

We consider the uplink of a broadband wireless system where an SC modulation is employed by each MT. We have an SDMA architecture corresponding to a MIMO system with P users (MTs), transmitting independent data blocks, and N receive antennas at the BS, as depicted in Figure 5.17. It is assumed that each MT has a single transmit antenna (the generalization to the case where we have multiple antennas at the MTs is straightforward). The time-domain block transmitted by the pth user is $\{s_{m,p}; m = 0, 1, \ldots, M-1\}$, with $s_{m,p}$ denoting the mth data symbol of the pth user, which is selected from a given constellation (e.g., a QPSK constellation), under an appropriate mapping rule. A CP, preceding each block, is used to avoid interblock interference and to make the linear convolution associated with the channel equivalent to a cyclic convolution with respect to the useful, M-length, part of the block. At the receiver, the CP is discarded.

The time-domain block at the nth receive antenna is $\left\{y_m^{(n)}; m = 0, 1, \ldots, JM-1\right\}$, where an oversampling factor J is assumed. The corresponding frequency-domain block, obtained after an appropriate size-JM DFT operation, is $\left\{Y_k^{(n)}; k = 0, 1, \ldots, JM-1\right\}$, where

$$\mathbf{Y}_k^{(n)} = \sum_{p=1}^{P} \mathbf{S}_{k,p}\mathbf{H}_{k,p}^{(n)} + \mathbf{N}_k^{(n)}, \tag{5.60}$$

with $\mathbf{H}_{k,p}^{(n)}$ denoting the overall channel frequency response from the pth transmitting antenna to the nth receiver antenna, for the kth frequency, and $N_k^{(n)}$ denoting the corresponding channel noise. The block $\{\mathbf{S}_{k,p}; k = 0, 1, \ldots, M-1\}$ is the size-M DFT of the pth user's data block $\{s_{m,p}; m = 0, 1, \ldots, M-1\}$, and it is assumed that $\mathbf{S}_{k,p}$ is periodic, with period M (i.e., $\mathbf{S}_{k,p} = \mathbf{S}_{k+M,p}$ for any k).

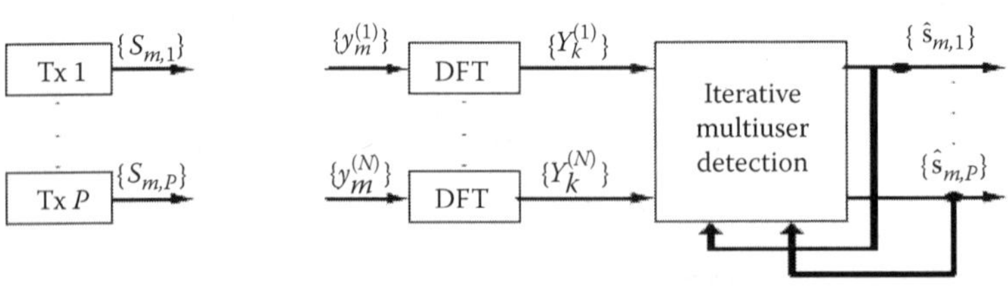

Figure 5.17 System model.

5.3.2.1.1 Receiver Structures

We consider a frequency-domain iterative multiuser detection that combines IB-DFE principles with LST interference cancelation. Each iteration consists of P detection stages, one for each user.

When detecting a given user, the interference from previously detected users is canceled, as with conventional LST receivers. However, contrarily to conventional LST receivers, we also cancel the residual ISI from the user that is being detected. Moreover, these interference and residual ISI cancelations take into account the reliability of each of the previously detected users. For a given iteration, the receiver structure for the detection of the pth user is illustrated in Figure 5.18. We have N frequency-domain feedforward filters (one for each receive antenna) and P frequency-domain feedback filters (one for each user). The feedforward filters are designed to minimize both the ISI and the multiuser interference that cannot be canceled by the feedback filters, due to decision errors in the previous detection steps. This structure can be regarded as an equalizer with multiuser interference suppression properties. After an IDFT operation, the corresponding time-domain outputs are passed through a hard-decision device so as to provide an estimate of the data block transmitted by the pth user.

We consider in this section two alternative detection schemes:

- Iterative detection with successive interference cancelation (SIC)

  The SIC architecture is closely related to the LST receivers proposed in Ref. [Dinis and Gusmão 2004]: for the first iteration and the detection of a given user, the interference from previously detected users is canceled; for the remaining iterations, we cancel the interference from all users (using the most updated version of each user), as well as the residual ISI for the user that is being detected. The detection procedure can be summarized as follows:

  – First iteration:

    (1) Detect user 1; (2) detect user 2 by removing the interference from user 1; (3) detect user 3 by removing the interference from users 1 and 2; and (4) proceed until the detection of user P.

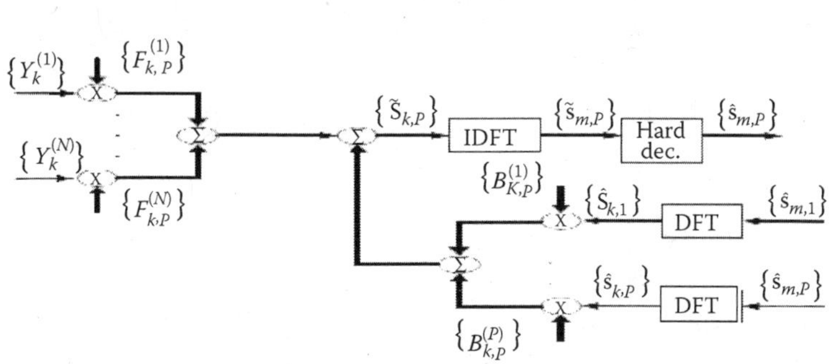

Figure 5.18 Detection of the *p*th user, for a given iteration.

- Remaining iterations:

  (1) Repeat the detection of user 1, now removing the interference from all users ($p \neq 1$) and the residual ISI ($p = 1$); (2) repeat the detection of user 2, now removing the interference from all users ($p \neq 2$) and the residual ISI ($p = 2$); and (3) proceed until the detection of user P.

- Iterative detection with PIC

  This second architecture combines IB-FDE principles with PIC. Each iteration consists of P parallel detection stages. At the first iteration, all of the users are detected simultaneously using linear processing. For the remaining iterations, we repeat the detection of each user ($p = 1, 2, \ldots, P$), now using the users' estimate from the previous iteration to remove the interference from the other users ($p' \neq p$) and the residual ISI ($p' = p$). Although the complexity of this structure is almost the same as the previous one, a parallel design can be advantageous from the implementation point of view.

5.3.2.1.2 Derivation of the Equalizer Coefficients

Throughout this section, it is assumed that the average received powers for the different users are sorted in a descending order, that is,

$$\sum_{n=0}^{N-1}\sum_{k} E\left[|\mathbf{H}_{k,p}^{(n)}|^2\right] E_{S,p} < \sum_{n=0}^{N-1}\sum_{k} E\left[|\mathbf{H}_{k,p-1}^{(n)}|^2\right] E_{S,p-1}, \quad p = 1, 2, \ldots, P, \tag{5.61}$$

with

$$E_{S,p} = E\left[|s_{m,p}|^2\right] = \frac{1}{M} E\left[|\mathrm{S}_{k,p}|^2\right], \tag{5.62}$$

denoting the average symbol energy for the pth user.

Let us first assume that there is no oversampling at the receiver (i.e., $J = 1$). In this case, the frequency-domain samples associated with the pth user at the output of the equalizer/MUD are given by [Dinis 2004]

$$\tilde{\mathrm{S}}_{k,p} = \sum_{n=1}^{N} \mathbf{F}_{k,p}^{(n)} \mathbf{Y}_k^{(n)} - \sum_{p'=1}^{P} \mathbf{B}_{k,p}^{(p')} \hat{\mathrm{S}}_{k,p'} = \sum_{n=1}^{N} \mathbf{F}_{k,p}^{(n)} \mathbf{Y}_k^{(n)} - \mathbf{B}_{k,p}^{(p)} \hat{\mathrm{S}}_{k,p} - \sum_{p' \neq p} \mathbf{B}_{k,p}^{p'} \hat{\mathrm{S}}_{k,p'}, \tag{5.63}$$

where $\mathbf{F}_{k,p}^{(n)}$ ($k = 0, 1, \ldots, M-1; n = 1, 2, \ldots, N$) denotes the feedforward coefficients and $\mathbf{B}_{k,p}^{(p')}$ ($k = 0, 1, \ldots, M-1; p = 1, 2, \ldots, P$) denotes the feedback coefficients. The coefficients $\{\mathbf{B}_{k,p}^{(p)}; k = 0, 1, \ldots, M-1\}$ are used for residual ISI cancelation and the coefficients $\{\mathbf{B}_{k,p}^{(p')}; k = 0, 1, \ldots, N-1\}(p' \neq p)$ are used for interference cancelation.

The block $\{\hat{\mathbf{S}}_{k,p'}; k = 0, 1, \ldots, M-1\}$ is the DFT of the block $\{\hat{s}_{m,p'}; m = 0, 1, \ldots, M-1\}$, where the time-domain samples $\hat{s}_{m,p'}$, $m = 0, 1, \ldots, M-1$, are the latest estimates for the p'th user transmitted symbols, that is, the hard decisions associated with the block of time-domain samples $\{\tilde{s}_{m,p'}; m = 0, 1, \ldots, M-1\} = \text{IDFT}\{\tilde{\mathbf{S}}_{k,p'}; k = 0, 1, \ldots, N-1\}$. For the ith iteration of an SIC receiver, $\hat{s}_{m,p'}$ is associated with the ith iteration for $p' < p$ and the $(i-1)$th iteration for $p' \geq p$ (in the first iteration, we do not have any information for $p' \geq p$ and $\hat{s}_{m,p'} = 0$); for the PIC receiver, $\hat{s}_{m,p'}$ is always associated with the previous iteration (for the first iteration $\hat{s}_{m,p'} = 0$).

The forward and backward coefficients, $\mathbf{F}_{k,p}^{(n)}$, and $\mathbf{B}_{k,p}^{(p')}$, respectively, are chosen so as to maximize the SNIR for the pth user, defined as

$$\text{SNIR}_p = \frac{|\gamma_p|^2 E_S}{E\left[\left|\varepsilon_{m,p}^{\text{eq}}\right|^2\right]}, \tag{5.64}$$

with

$$\gamma_p = \frac{1}{M} \sum_{k=0}^{M-1} \sum_{n=1}^{N} \mathbf{F}_{k,p}^{(n)} \mathbf{H}_{k,p}^{(n)} \tag{5.65}$$

being regarded as the average overall channel frequency response for the pth user, after combining the outputs of the N feedforward filters; and with

$$\varepsilon_m^{\text{eq}} = \tilde{s}_{m,p} - \gamma_p s_{m,p} \tag{5.66}$$

denoting the overall noise.

From Ref. [Dinis 2004], the optimum values of $\mathbf{B}_{k,p}^{(p')}$ are

$$\mathbf{B}_{k,p}^{(p')} = \begin{cases} \rho_{p'} \sum_{n'=1}^{N} \mathbf{F}_{k,p}^{(n',i)} \mathbf{H}_{k,p'}^{(n')} & \Leftarrow \; p' \neq p, \\ \rho_p \left(\sum_{n'=1}^{N} \mathbf{F}_{k,p}^{(n',i)} \mathbf{H}_{k,p}^{(n')} - 1 \right) & \Leftarrow \; p' = p. \end{cases} \tag{5.67}$$

The feedforward coefficients can be written in the form

$$\mathbf{F}_{k,p}^{(n)} = \sum_{p'=1}^{P} \mathbf{H}_{k,p'}^{(n)*} \mathbf{C}_{k,p}^{(p')}, \tag{5.68}$$

with the set of coefficients $\{\mathbf{C}_k^{(p')}; p' = 1, 2, \ldots, \mathbf{P}\}$ satisfying the set of P equations

$$\sum_{p''=1}^{P} \mathbf{C}_{k,p}^{(p'')} \cdot \left(\left(1 - \rho_{p'}^2\right) \sum_{n'=1}^{N} \mathbf{H}_{k,p''}^{(n')*} \mathbf{H}_{k,p'}^{(n')} + \alpha_p \delta_{p',p''} \right) = \delta_{p,p'}, \quad p' = 1, 2, \ldots, P, \tag{5.69}$$

($\delta_{p,p'} = 1$ if $p = p'$ and 0 otherwise), and

$$\alpha_p = \mathrm{SNR}_p = \frac{E_{S,p}}{2\sigma_N^2}, \tag{5.70}$$

$$\sigma_N^2 = \frac{1}{2M} E\left[\left|\mathbf{N}_k^{(n)}\right|^2\right]. \tag{5.71}$$

For the special case where $P = 1$ (and $p = 1$), it can be shown that

$$\mathbf{F}_{k,1}^{(n)} = \frac{\mathbf{H}_{k,1}^{(n)*}}{\alpha_1\left(1-\rho_1^2\right)\sum_{n'=1}^{N}\left|\mathbf{H}_{k,1}^{(n')}\right|^2}, \quad l = 1, 2, \ldots, N, \tag{5.72}$$

which corresponds to feedforward coefficients of an IB-DFE with N-branch space diversity [Dinis et al. 2003]. For the first iteration,

$$\mathbf{F}_{k,1}^{(n)} = \frac{\mathbf{H}_{k,1}^{(n)*}}{\alpha_1\sum_{n'=1}^{N}\left|\mathbf{H}_{k,1}^{(n')}\right|^2}, \quad l = 1, 2, \ldots, N, \tag{5.73}$$

corresponding to a linear FDE with N-branch space diversity [Gusmão et al. 2003].

It should be noted that, for the first iteration ($i = 0$), we do not have any information about $\mathbf{S}_{k,p'}$ for $p' \leq p$, for the SIC receiver, or all $\mathbf{S}_{k,p'}$, for the PIC receiver. Therefore, the corresponding correlation coefficients are zero, leading to $\mathbf{B}_{k,p}^{(p')} = 0$. After the first iteration, and if the residual BER is not too high, $\hat{s}_{m,p'} = s_{m,p'}$ for most of the data symbols, and $\hat{\mathbf{S}}_{m,p'} \approx \mathbf{S}_{k,p'}$; this means that we can use the feedback coefficients to eliminate a significant part of the residual ISI, as well as the residual multiuser interference. Naturally, when $\hat{s}_{m,p'} = s_{m,p'}$ for all data symbols, $\hat{\mathbf{S}}_{k,p'} = \mathbf{S}_{k,p'}$ and $\rho_{p'} = 1$, leading to

$$\mathbf{F}_{k,p}^{(n)} = \mathbf{H}_{k,p}^{(n)*}, \quad n = 1, 2, \ldots, N, \tag{5.74}$$

$$\mathbf{B}_{k,p}^{(p)} = \mathrm{SNR}_p\left(\sum_{n'=1}^{N}\left|\mathbf{H}_{k,p}^{(n')}\right|^2 - \frac{1}{M}\sum_{k=0}^{M-1}\sum_{n'=1}^{N}\left|\mathbf{H}_{k,p}^{(n')}\right|^2\right), \tag{5.75}$$

and

$$\mathbf{B}_{k,p}^{(p')} = \mathrm{SNR}_p\sum_{n'=1}^{N}\left|\mathbf{H}_{k,p'}^{(n')}\right|^2, \tag{5.76}$$

which corresponds to the total elimination of the ISI and total multiuser interference cancelation.

5.3.2.1.3 Complexity Analysis

Both SIC and PIC receivers require N size-JM DFT operations, one for each receiver antenna, and a pair of DFT/IDFT operations (with size-M) for the detection of each user, at each iteration. For the equalization/interference cancelation, we need JNPM multiplications for the first iteration of a PIC receiver and $JNPM + P(P-1)M/2$ multiplications for the first iteration of an SIC receiver; for the remaining iterations, we need $(JN + P)PM$ multiplications for both SIC and PIC receivers.

The most complex part of the algorithm is the computation of the feedforward coefficients, since we need to solve the M systems of P Equation 5.68 (one for each of the M frequencies), for each iteration and each user. Naturally, for slow varying channels, this operation is not required for all blocks.

5.3.2.1.4 Performance Results

In this section we consider the use of the proposed receiver in an SDMA system where each user has one transmit antenna and the BS has N receive antennas. The data block consists of $M = 256$ QPSK data symbols, plus an appropriate CP (similar results were obtained for other values of M, provided that $M \gg 1$). We consider uncoded BER performances under perfect synchronization and channel estimation conditions. A square-root-raised-cosine filtering with roll-off zero is assumed (i.e., $J = 1$). A similar behavior was observed for a milder filtering, although a slight performance improvement was observed for $J > 1$, due to the diversity effect inherent to the transmitted signals.

First, we evaluate the performance of the proposed receivers for a channel that has independent, Rayleigh-distributed fading at each frequency. This channel model can be regarded as a severely time-dispersive scenario with rich multipath propagation. Figure 5.19 shows the BER performances for different users and different iterations, when we have $P = 4$ users with the same average receive power and $N = 4$ receive antennas at the BS. An SIC receiver is assumed. For the sake of comparisons, we also include the MFB performance, defined in Chapter 2. From this figure, we can observe that, for the first iteration, the users have very different performances: more than 6 dB from user 1 to user 4, at BER $= 10^{-4}$. This difference decreases as we increase the number of iterations,

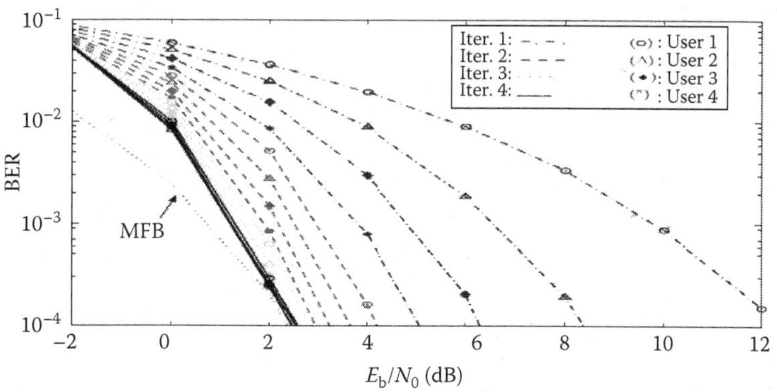

Figure 5.19 BER for the different users and iterations, along with corresponding MFB values for $n = p = 4$.

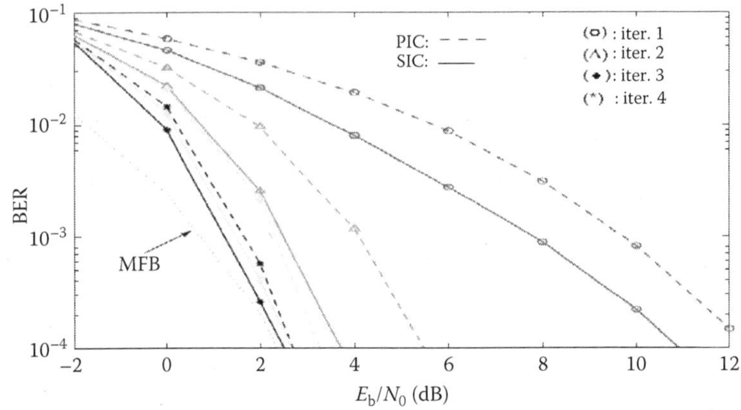

Figure 5.20 Average BER of an SIC/PIC receiver, along with MFB for $n = p = 4$.

with all users having almost the same performance after three iterations. Moreover, the resulting performance can be very close to the MFB after four iterations: the required E_b/N_0 for an average BER $= 10^{-4}$ are 13.8, 6.8, 5.7, 5.7, and 5.4 dB, for iterations 1–4 and for the MFB, respectively. This shows that the proposed receiver is able to eliminate a significant part of the ISI and multiuser interference.

Figure 5.20 shows the average BER performances (averaged over all the users) for SIC and PIC receivers, for the different iterations. After the first iteration, the performance of the PIC receiver is almost 2 dB worse than the performance of the SIC receiver. After iteration 3, this gap decreases to less than 1 dB, and after iteration 4, decreases to 0.2 dB. Once again, the BER performances after four iterations are very close to the corresponding MFB for both structures. It should be mentioned that, for the PIC receiver, all users have the same average BER.

5.3.3 Spatial Multiplexing: OFDM

In Figure 5.21, we show a possible configuration for a MIMO–OFDM transmitter chain that incorporates HQAM constellations. In this scheme, there are $1/2\log_2 M$ parallel chains for the different input bit streams. For example, with a 16-HQAM we can use two parallel chains, while for 64-QAM we can employ three parallel chains. Each stream is encoded, interleaved, and mapped into the constellation symbols according to the importance attributed to it. Pilot symbols are inserted into the modulated symbols sequence, which is then converted to the time domain using an IDFT. Finally the resulting stream is split into several smaller streams that are transmitted simultaneously by M_{tx} antennas. Note that each individual chain is similar to a unicast transmission scheme based on QPSK. The main difference is the addition of the modulation mapper that joins the parallel streams.

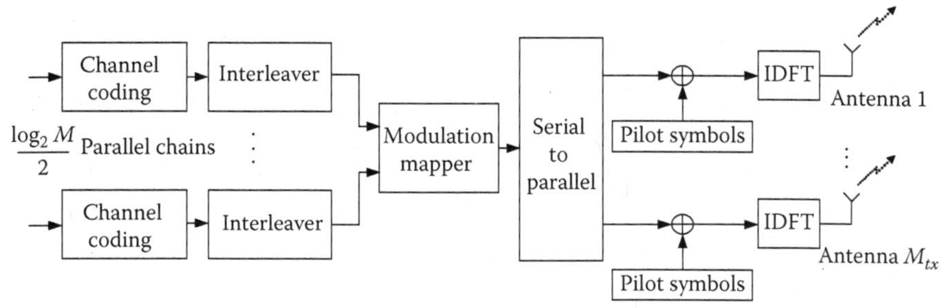

Figure 5.21 Transmitter chain.

5.3.3.1 Frame Structure

Before being transmitted, the sequences of symbols are converted to the time domain through $\{x_{i,l}^m, i = 0, 1, \ldots, N - 1\} = \text{IDFT}\ \{\mathbf{S}_{k,l}^m, k = 0, 1, \ldots, N - 1\}$, where $\mathbf{S}_{k,l}^m$ is the symbol transmitted by the kth subcarrier of the lth OFDM block using antenna m. The transmitted OFDM signals are then expressed as

$$x^m(t) = \sum_l \sum_{i=-N_G}^{N-1} x_{i,l}^m \cdot h_T(t - i \cdot T_s), \tag{5.77}$$

with T_s denoting the symbol duration, N_G the number of samples at the CP (with $x_{-i,l}^m = x_{N-i,l}^m$, $i = 1, \ldots, N_G$), and $h_T(t)$ the adopted pulse shaping filter.

Figure 5.22 presents the structure of the iterative OFDM receiver. N_{rx} receiving antennas are assumed to be used. According to the figure, the signal, which is considered to be sampled and with the CP removed, is converted to the frequency domain after an appropriate size-N DFT operation. If the CP is longer than the overall CIR, the resulting sequence received in antenna n can be expressed as

$$\mathbf{R}_{k,l}^n = \sum_{m=1}^{M_{tx}} \mathbf{S}_{k,l}^m \mathbf{H}_{k,l}^{m,n} + \mathbf{N}_{k,l}^n, \tag{5.78}$$

with $\mathbf{H}_{k,l}^{m,n}$ denoting the overall channel frequency response between transmit antenna m and receiving antenna n for the kth frequency of the lth time block and $\mathbf{N}_{k,l}^m$ denoting the corresponding channel noise sample.

The sequences of samples (Equation 5.78) enter the MIMO equalizer (spatial demultiplexer block) that separates the streams transmitted simultaneously by the multiple antennas. This can be accomplished with several methods available in the literature. Here we used an MMSE equalizer [Kay 1993], a maximum likelihood soft output criterion (MLSO), and an interference canceler (IC) [Souto et al. 2006]. It is possible to perform some of the receiver iterations using one spatial demultiplexing technique, such as the MMSE, and the others using a different one, such as the IC, as was studied in Ref. [Souto et al. 2006]. In any case, after MIMO equalization the demultiplexed symbol

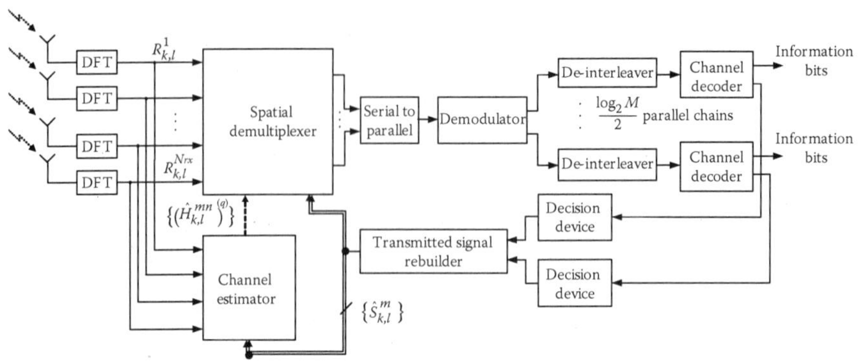

Figure 5.22 Iterative OFDM receiver structure.

sequences are serialized and pass through the demodulator, de-interleaver, and channel decoder blocks. This channel decoder has two outputs. One is the estimated information sequence and the other is the sequence of LLR estimates of the code symbols. These LLRs go through the decision device, which outputs either soft-decision or hard-decision estimates of the code symbols, and enter the transmitted signal rebuilder that performs the same operations of the transmitter (interleaving, modulation, and conversion of serial streams to parallel streams). The reconstructed symbol sequences are subsequently used for a refinement of the channel estimates and also for possible improvement of the spatial demultiplexing task (in case of employing an IC) for the succeeding iteration.

The possible MIMO equalization techniques are now going to be briefly described. Using matrix notation, the MMSE estimates of the transmitted symbols in subcarrier k and OFDM block l are given by [Kay 1993]

$$\hat{\mathbf{S}}_{k,l} = \mathbf{H}^H_{k,l} \cdot \left(\mathbf{H}_{k,l}\mathbf{H}^H_{k,l} + \sigma^2\mathbf{I}\right)^{-1} \mathbf{R}_{k,l}, \tag{5.79}$$

where $\hat{\mathbf{S}}_{k,l}$ is the $M_{tx} \times 1$ estimated transmitted signal vector with a different transmit antenna in each position, $\mathbf{H}_{k,l}$ is the $M_{rx} \times M_{tx}$ channel matrix (perfect estimation is assumed) with each column representing a different transmit antenna and each line representing a different receive antenna, $\mathbf{R}_{k,l}$ is the $M_{rx} \times 1$ received signal vector with a different receive antenna in each position, and σ^2 is the noise variance.

In the MLSO criterion, we use the following estimate for each symbol:

$$\begin{aligned}\hat{S}^m_{k,l} &= E\left[S^m_{k,l} \mid \mathbf{R}_{k,l}\right] = \sum_{s_i \in \Lambda} s_i \cdot P\left(S^m_{k,l} = s_i \mid \mathbf{R}_{k,l}\right) \\ &= \sum_{s_i \in \Lambda} s_i \cdot \frac{P\left(S^m_{k,l} = s_i\right)}{p\left(\mathbf{R}_{k,l}\right)} p\left(\mathbf{R}_{k,l} \mid S^m_{k,l} = s_i\right),\end{aligned} \tag{5.80}$$

where s_i corresponds to a constellation symbol from the modulation alphabet Λ, $E[\cdot]$ is the expected value, $P(\cdot)$ represents a probability, and $p(\cdot)$ represents PDF. Considering equiprobable symbols we have $P\left(S_{k,l}^{m}=s_i\right)=1/M$, where M is the constellation size. The PDF values required in Equation 5.80 can be computed as

$$p\left(\mathbf{R}_{k,l}\mid \mathbf{S}_{k,l}^{m}=s_i\right)=\frac{1}{M^{M_{tx}-1}}\sum_{\mathbf{S}_{k,l}^{\text{interf}}\in\Lambda^{M_{tx}-1}} p\left(\mathbf{R}_{k,l}\middle| \mathbf{S}_{k,l}^{m}=s_i,\mathbf{S}_{k,l}^{\text{interf}}\right), \tag{5.81}$$

with

$$p\left(\mathbf{R}_{k,l}\middle| \mathbf{S}_{k,l}^{m}=s_i,\mathbf{S}_{k,l}^{\text{interf}}\right)=\frac{1}{\left(2\pi\sigma^2\right)^{N_{rx}}}\exp\left[\sum_{n=1}^{N_{rx}}-\frac{\left|\mathbf{R}_{k,l}^{n}-\mathbf{H}_{k,l}(n,:)\cdot\mathbf{s}\right|^2}{2\sigma^2}\right], \tag{5.82}$$

where $\mathbf{S}_{k,l}^{\text{interf}}$ is a $(M_{tx}-1)\times 1$ vector representing a possible combination of symbols transmitted simultaneously by all antennas except antenna m, $\mathbf{s}$ is a $M_{tx}\times 1$ vector composed by $\mathbf{S}_{k,l}^{\text{interf}}$ and s_i, and $\mathbf{H}_{k,l}(n,:)$ is the nth line of channel matrix $\mathbf{H}_{k,l}$.

An IC can also be used inside the spatial demultiplexer block, but usually it is only recommendable after the first receiver iteration [Souto et al. 2006]. In this case, in iteration q, for each transmit antenna m and receive antenna n, the IC subtracts the interference caused by all the other antennas. This can be represented as

$$\left(\mathbf{R}_{k,l}^{n,m}\right)^{(q)}=\mathbf{R}_{k,l}^{n}-\sum_{\substack{m'=1\\ m'\neq m}}^{M_{tx}}\left(\hat{\mathbf{S}}_{k,l}^{m'}\right)^{(q-1)}\mathbf{H}_{k,l}^{m',n}, \tag{5.83}$$

where $\left(\hat{\mathbf{S}}_{k,l}^{m}\right)^{(q-1)}$ represents the data symbols estimates of the previous iteration for transmit antenna m, subcarrier k, and OFDM block l.

5.3.3.2 Results

From the results in Figure 5.23, we can see that the schemes using the MLSO provide the best performance, followed by the MMSE coupled with IC, MMSE, and finally ZF.

5.4 Antenna Correlation

The antenna correlation effect is a serious aspect to be taken into account by the system designer. In fact, the spacing of the antennas at both the transmitter and the receiver is essential to guarantee a low correlation value between antennas. A low correlation value implies that the channel associated with each antenna is independent of the other antennas, thus providing a greater degree of diversity.

A good antenna spacing is easy to obtain at the BS, but it is much harder to achieve a sufficient spacing at the UE. This section analyzes the potential performance decrease associated with antenna correlation, for values taken as realistic in the UMTS standard.

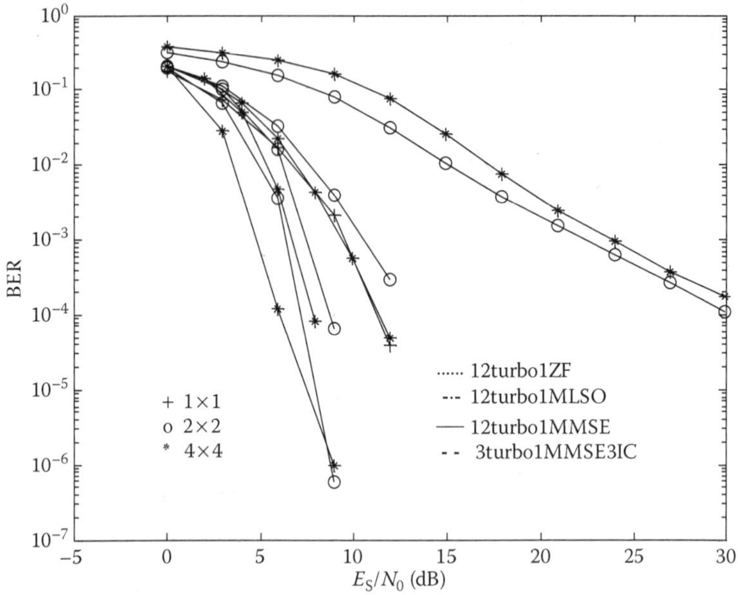

Figure 5.23 OFDM SM results (QPSK chains).

Antenna correlation deals with the correlation between channels arriving/departing at adjacent antennas, according to various parameters such as antenna spacing, angle spread, and angle of arrival. The correlation between antennas is commonly defined as the envelope correlation coefficient between signals received at two antenna elements. The received baseband signals are modeled as two complex random processes X and Y with an envelope correlation coefficient of

$$\rho_{\text{env}} = \left| \frac{E\{(X - E(X))(Y - E(Y))^*\}}{\sqrt{E\{|X - E(X)|^2\} E\{|Y - E(Y)|^2\}}} \right| \tag{5.84}$$

[note that this is not equal to the correlation coefficient of the envelopes (magnitude) of two signals, a measure commonly used in cases where no complex data are available].

Since the correlation coefficient is independent of the mean, only the random parts of the channel are of interest. In order to simplify notation, only the random channel components will be considered until the remainder of this section. Considering the general case of frequency-selective propagation, the channel is modeled as a TDL:

$$g(t, \tau) = \sum_{l=1}^{L} g_l(t)\delta(\tau - \tau_l), \tag{5.85}$$

where L is the number of taps, $g_l(t)$ are the time-varying tap coefficients, and τ_l are the tap delays. Considering two channels with CIRs $g_1(t, \tau)$ and $g_2(t, \tau)$ and assuming that equivalent taps in both channels have EP ($\sigma_{1,l}^2 = \sigma_{2,l}^2 = \sigma_l^2$), and that taps with different delays are uncorrelated within a channel as well as between channels,

$$E\left\{g_{ik}(t)g_{jl}^*(t)\right\} = 0, \quad \forall k \neq l, \tag{5.86}$$

the antenna correlation coefficient becomes

$$\rho_{\text{env}} = \left|\frac{\rho_1\sigma_1^2 + \rho_2\sigma_2^2 + \cdots + \rho_L\sigma_L^2}{\sigma_1^2 + \sigma_2^2 + \cdots + \sigma_L^2}\right|, \tag{5.87}$$

where ρ_l are the correlation coefficients between each pair of taps $g_{1l}(t)$ and $g_{2l}(t)$:

$$\rho_l = \frac{E\left\{g_{1l}(t)g_{2l}^*(t)\right\}}{\sigma_{1l}^2\sigma_{2l}^2}. \tag{5.88}$$

From Equation 5.87, it can be seen that the antenna correlation can be related to the individual tap correlations, where σ_l^2 are the individual tap gains. To obtain a simple solution for setting these tap correlations depending on the required antenna correlation, it can be imposed that all tap correlations should be equal. Therefore, Equation 5.87 simply states that all tap correlations have to be set to the antenna correlation.

To generate a sequence of random state vectors with specified first-order statistics (mean vector μ and correlation matrix $\mathbf{R}$), the following transformation can be used:

$$\tilde{\boldsymbol{v}} = \sqrt{\mathbf{R}} \cdot \boldsymbol{v} + \mu, \tag{5.89}$$

where $\boldsymbol{v}$ is a vector of independent sequences of circularly symmetric complex Gaussian-distributed random numbers with zero mean and unit variance. The correlation matrix $\mathbf{R}$ is defined and factored as

$$\mathbf{R} = E\left\{\tilde{\boldsymbol{v}}\tilde{\boldsymbol{v}}^H\right\} = \begin{bmatrix} 1 & \rho_{12} & .. \\ \rho_{12}^* & 1 & .. \\ .. & .. & \ddots \end{bmatrix} = \mathbf{R}^{1/2}\mathbf{R}^{H/2} = \mathbf{R}^{1/2}\mathbf{R}^{1/2}, \tag{5.90}$$

with ρ_{12} being the correlation between channels 1 and 2. In Ref. [Da-Shan Shiu et al. 2000], the correlation between different MIMO channel elements is modeled under the assumption that the correlation among receive antennas is independent of the correlation between transmit antennas (and vice versa). The underlying justification for this approach is to assume that only immediate surroundings of the antenna array impose the correlation between array elements and have no impact on correlations observed between the elements of the array at the other end of the link, which is a reasonable assumption for indoor [Kay Yu et al. 2001], city center environments (such as Pedestrian A), and Vehicular A. A way of including this type of antenna signal correlation into the MIMO channel model for Rayleigh flat-fading-like channels, using real-valued correlation coefficients, is given by

$$\mathbf{R} = \mathbf{R}_{RX} \otimes \mathbf{R}_{TX}, \tag{5.91}$$

where $\otimes$ denotes the Kronecker product and $\mathbf{R}_{TX}(M_{TX} \times M_{TX})$ and $\mathbf{R}_{RX}(N_{RX} \times N_{RX})$ denote the correlation observed on the transmitter and the receiver, respectively. For complex correlation coefficients, the lower diagonal part of matrix R must have its coefficients turned into the respective conjugate values. A simple model for deriving the correlation values, an approximation for the fading correlation between two adjacent antenna elements averaged over all possible orientations of the two antennas in a given wave field [Durgin and Rappaport 1999], will be used as a basis, for simplicity

$$\rho(d) = \mathrm{e}^{-23\Lambda^2 d^2}. \tag{5.92}$$

In this equation, d is the distance in wavelength between two antennas and Λ is the angular spread according to Ref. [Durgin and Rappaport 1999]. Note that this definition of Λ is very general, that is, it is defined for any distribution of power in the azimuth plan, and values close to 0.0 denote completely directional scenarios, whereas those at 1.0 represent more uniform spreading of energies in space. Based on this approximation, a correlation model is proposed for linear arrays at both the transmitter and the receiver with equidistant antenna spacing d_{TX} and d_{RX}, respectively, at the transmitter and the receiver, resulting in the following Toeplitz structure correlation matrices

$$\mathbf{R}_{RX} = \begin{bmatrix} 1 & \rho_r & \rho_r^4 & \cdots & \rho_r^{(n_r-1)^2} \\ \rho_r & 1 & \rho_r & \ddots & \vdots \\ (\rho_r^4) & \rho_r & 1 & \ddots & \rho_r^4 \\ \vdots & \ddots & \ddots & \ddots & \rho_r \\ \left(\rho_r^{(n_r-1)^2}\right) & \cdots & (\rho_r^4) & \rho_r & 1 \end{bmatrix},$$
$$\mathbf{R}_{TX} = \begin{bmatrix} 1 & \rho_t & \rho_t^4 & \cdots & \rho_t^{(n_t-1)^2} \\ \rho_t & 1 & \rho_t & \ddots & \vdots \\ (\rho_t^4) & \rho_t & 1 & \ddots & \rho_t^4 \\ \vdots & \ddots & \ddots & \ddots & \rho_t \\ \left(\rho_t^{(n_t-1)^2}\right) & \cdots & (\rho_t^4) & \rho_t & 1 \end{bmatrix}, \tag{5.93}$$

where ρ_r and ρ_t represent $\rho(d_{RX})$ and $\rho(d_{TX})$, respectively. Note that this model can range from the totally uncorrelated scenario ($\rho_r = \rho_t = 0$) to the fully correlated scenario ($\rho_r = \rho_t = 1$). For small ρ_r and ρ_t (much smaller than 1), the exponents higher than 4 can be discarded, which translates the Toeplitz matrices to a tridiagonal structure. This tridiagonal correlation model can be viewed as approximating a scenario in which it is assumed that the fading correlation has a certain given value for any pair of adjacent antenna elements, and that any other pair of antenna elements exhibits independent fading, on both sides of the communication link. For further simplification of the spatial correlation, the following relation can be set:

$$\rho_\gamma = \rho_t, \tag{5.94}$$

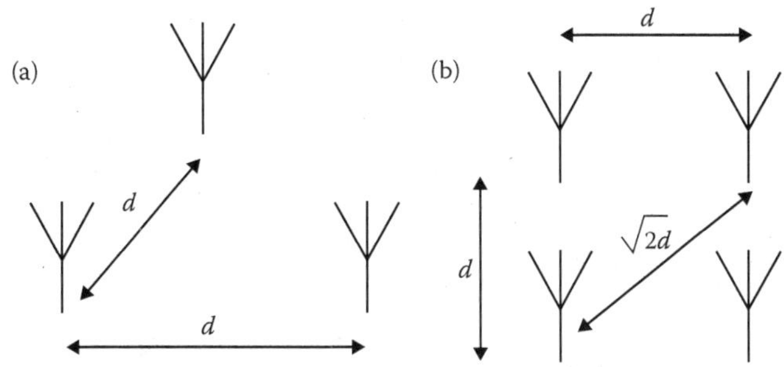

Figure 5.24 Triangular (a) and square (b) antenna dispositions.

which leads to a single-parameter MIMO correlation model. For nonlinear antenna arrangements, such as triangular or squared dispositions (Figure 5.24), different matrix structures are obtained, according to the dependence on the distance between antennas.

The square disposition was assumed for the 4 × 4 setting, since it seems the most obvious in terms of space savings. Therefore, for the simulated cases of 2 × 2 and 4 × 4, the correlation matrices are as follows:

$$2 \times 2: \quad \mathbf{R}_{RX} = \begin{bmatrix} 1 & \rho_r \\ \rho_r & 1 \end{bmatrix}, \quad \mathbf{R}_{TX} = \begin{bmatrix} 1 & \rho_t \\ \rho_t & 1 \end{bmatrix}, \tag{5.95}$$

$$4 \times 4: \quad \mathbf{R}_{RX} = \begin{bmatrix} 1 & \rho_r & \rho_r & \rho_r^2 \\ \rho_r & 1 & \rho_r^2 & \rho_r \\ \rho_r & \rho_r^2 & 1 & \rho_r \\ \rho_r^2 & \rho_r & \rho_r & 1 \end{bmatrix}, \quad \mathbf{R}_{TX} = \begin{bmatrix} 1 & \rho_t & \rho_t & \rho_t^2 \\ \rho_t & 1 & \rho_t^2 & \rho_t \\ \rho_t & \rho_t^2 & 1 & \rho_t \\ \rho_t^2 & \rho_t & \rho_t & 1 \end{bmatrix}. \tag{5.96}$$

5.5 Channel Estimation

MIMO schemes have emerged as one of the most promising methods for capacity increase in a communication system [Telatar 1995; Foschini and Gans 1998]. Although in SISO schemes a reliable channel estimation processing is required for accomplishing coherent detection at the receiver, in MIMO schemes it plays an even more crucial role since the performance of the spatial signal processing depends on the accuracy of the channel estimates. Typically, these estimates are obtained with the help of training symbols that are multiplexed with the data, either in the time domain or in the frequency domain [Hoher et al. 1997; Sanzi and Speidel 2000] as is proposed for UTRA LTE [3GPP, 25.814 2006]. This approach can result in inefficient use of the available BW, which is of crucial importance for any communication system. Therefore it is desirable to reduce the overheads required for channel estimation purposes. An alternative method, which was first proposed in Ref. [Farhang-Boroujeny 1995] and has been receiving recent attention [Hoeher and Tufvesson 1999; Mazzenga 2000; Zhu et al. 2003; Orozco-Lugo et al. 2004; Coldrey and Bohlin 2005; Ghogho et al. 2005], relies on the idea of pilot embedding where a pilot sequence is summed to the data sequence and transmitted simultaneously. This approach, usually referred to as "implicit pilots" transmission (also named superimposed pilots, embedded pilots, or hidden pilots), demands that some of the power be spent on the pilot sequence but allows us to increase significantly the pilots' density without sacrificing system capacity. In fact, we can have a pilot for each data symbol. One of the problems in this type of pilot transmission techniques relies on the interference levels between data and pilots, which might be high, especially when employing multiple antennas since each pilot symbol will be affected by interference of several data symbols simultaneously. Moreover, there is also interference to the data symbols due to the pilots. As a consequence, this can lead to irreducible noise floors and substantial performance degradation unless the mutual interference is somehow dealt with by the receiver.

Comparisons between conventional and implicit pilot approaches were made in Ref. [Coldrey and Bohlin 2005] from a theoretical point of view. In this section we also perform a comparison between implicit pilots and data multiplexed pilots through the simulation of a complete OFDM system (work on SC-FDE can also be found in Ref. [Marques da Silva et al. 2009b]) based on UTRA LTE with the addition of HQAM signal constellations since the objective is the efficient support of multicast/broadcast transmissions (only one signal transmitted to several users). The study considers SISO and MIMO schemes, with the later one being employed with the aim of increasing the transmission bit rate through SM. To deal with the problems of mutual interference between pilots and data, several iterative receivers are considered.

5.5.1 Data Multiplexed Pilots

These are the most common type of pilots, which are multiplexed alongside the data bits. The transmitter structure is the same as presented in Figure 5.21.

5.5.1.1 Frame Structure

The first pilot transmission method evaluated in this section corresponds to the one being considered for the future UTRA LTE [3GPP, 25.814 2006] where known pilot symbols are periodically inserted into the data stream. The frame structure adopted is shown in Figure 5.25. According to this structure, in a MIMO–OFDM system with N carriers, pilot symbols are multiplexed with data symbols using a spacing of ΔN_{T} OFDM blocks in the time domain and ΔN_{F} subcarriers in the frequency domain. To avoid interference between pilots of different transmitting antennas, FDM is employed for the pilots, which means that pilot symbols cannot be transmitted over the same subcarrier in different antennas. Data symbols are not transmitted on subcarriers reserved for pilots in any antenna; therefore, the minimum allowed spacing in the frequency domain is $(\Delta N_{\mathrm{F}})_{\min} = M_{tx}$.

5.5.1.2 Receiver Structure

Due to a higher sensitivity to interference and channel estimation errors of HQAM constellations, we employ receivers capable of jointly performing channel estimation and data detection through iterative processing for both pilot transmission methods.

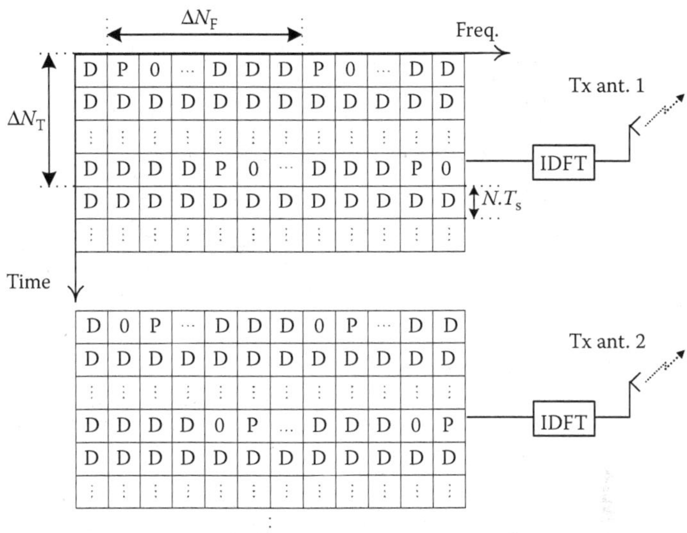

Figure 5.25 Frame structure for a MIMO–OFDM transmission with data multiplexed pilots (P: pilot symbol, D: data symbol, *Ts*: symbol duration).

Figure 5.22 presents the structure of the iterative receiver for the case of data multiplexed pilots transmission. It is assumed that N_{rx} receiving antennas are used.

The receiver structure is similar to the one described in Section 5.3.3, with the exception that the channel is not known, but estimated ($\hat{\mathbf{H}}$ instead of $\mathbf{H}$).

5.5.1.3 Channel Estimation

To obtain the frequency channel response estimates for each transmitting/receiving antenna pair, the receiver applies the following steps in each iteration:

1. The channel estimate between transmit antenna m and receive antenna n for each pilot symbol position is simply computed as

$$\tilde{\mathrm{H}}_{k,l}^{m,n} = \frac{\left(\mathrm{S}_{k,l}^{m,\mathrm{Pilot}}\right)^*}{\left|\mathrm{S}_{k,l}^{m,\mathrm{Pilot}}\right|^2}\mathbf{R}_{k,l}^{n}, \tag{5.97}$$

   where $\mathrm{S}_{k,l}^{m,\mathrm{Pilot}}$ corresponds to a pilot symbol transmitted in the kth subcarrier of the lth OFDM block using antenna m. Obviously not all indexes k an l will correspond to a pilot symbol since $\Delta N_\mathrm{T} > 1$ and $\Delta N_\mathrm{F} > 1$.

2. Channel estimates for the same subcarrier k, transmit antenna m and receive antenna n but in time-domain positions (index l) that do not carry a pilot symbol can be obtained through interpolation using an FIR filter with length W as

follows:

$$\hat{\mathrm{H}}^{m,n}_{k,l+t} = \sum_{j=-\lfloor (W-1)/2 \rfloor}^{\lfloor W/2 \rfloor} h_t^j \tilde{\mathrm{H}}^{m,n}_{k,l+j\cdot\Delta N_\mathrm{T}}, \tag{5.98}$$

where t is the OFDM block index relative to the last one carrying a pilot (which is the block with index l) and h_t^j are the interpolation coefficients of the estimation filter that depend on the channel estimation algorithm employed. There are several proposed algorithms in the literature such as the optimal Wiener filter interpolator [Cavers 1991] or the low-pass sinc interpolator [Kim 1991].

3. After the first iteration the data estimates can also be used as pilots for channel estimation refinement [Valenti 2001]. The respective channel estimates are computed as

$$\left(\hat{\mathrm{H}}^{m,n}_{k,l}\right)^{(q)} = \frac{\mathrm{R}^n_{k,l}\left(\hat{\mathrm{S}}^m_{k,l}\right)^{(q-1)*}}{\left|\left(\hat{\mathrm{S}}^m_{k,l}\right)^{(q-1)}\right|^2}. \tag{5.99}$$

5.5.2 Implicit Pilots

5.5.2.1 Frame Structure

The alternative pilot transmission method studied in this section corresponds to the use of implicit pilots. The frame structure and the pilot insertion process are shown in Figure 5.26. According to this structure, the implicit pilots are generated using a grid with a spacing of ΔN_T symbols in the time domain and ΔN_F symbols in the frequency domain. Similarly to the other method, to avoid interference between implicit pilots of different transmitting antennas FDM is employed for the pilots, so that pilot symbols in different antennas are not transmitted over the same subcarrier. In this case the minimum allowed spacing in the frequency domain is $(\Delta N_\mathrm{F})_{\min} = M_{tx}$ and the transmitted sequences are given by

$$\mathrm{X}^m_{k,l} = \mathrm{S}^m_{k,l} + \mathrm{S}^{m,\mathrm{Pilot}}_{k,l}, \tag{5.100}$$

where $\mathrm{S}^{m,\mathrm{Pilot}}_{k,l}$ is the implicit pilot transmitted over the kth subcarrier, in the lth OFDM block using antenna m. The resulting sequences are converted to the time domain through the usual process, $\{x^m_{i,l}, i = 0, 1, \ldots, N-1\} = \mathrm{IDFT}\{X^m_{k,l}, k = 0, 1, \ldots, N-1\}$ before being transmitted.

5.5.2.2 Receiver Structure

To reduce the mutual interference between pilots and data, which will be particularly high in MIMO systems with implicit pilots, and to achieve reliable channel estimation and data detection, we propose the use of a modified version of the receiver of Figure 5.26,

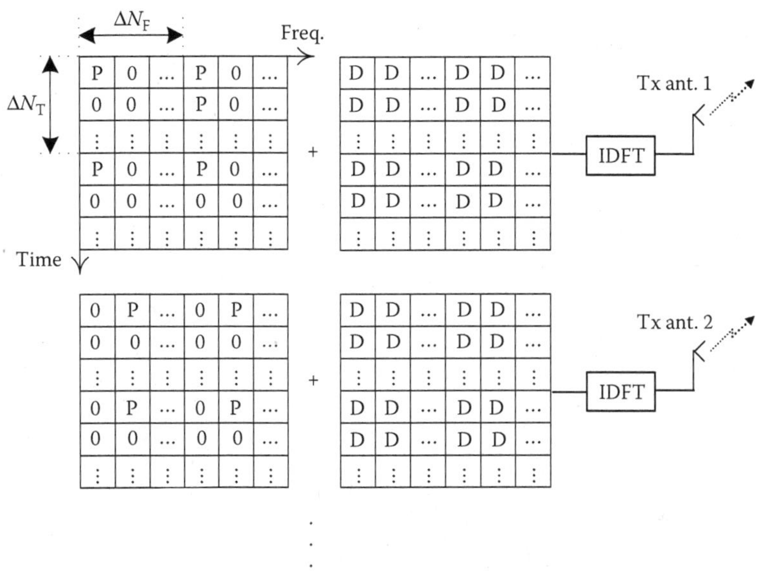

Figure 5.26 Proposed frame structure for MIMO–OFDM transmission with implicit pilots (P: pilot symbol, D: data symbol).

whose structure is shown in Figure 5.27. The main differences compared with the previous receiver lie in the addition of the "remove pilots" and "remove data" processing blocks.

Therefore, in this case, after conversion to the frequency domain the received sampled sequence can be expressed as

$$\mathbf{R}_{k,l}^{n} = \sum_{m=1}^{M_{tx}} \left(\mathbf{S}_{k,l}^{m} + \mathbf{S}_{k,l}^{m,\text{Pilot}} \right) \mathbf{H}_{k,l}^{m,n} + \mathbf{N}_{k,l}^{n}. \tag{5.101}$$

The pilot symbols are then removed from the sequence resulting in

$$\left(\mathbf{Y}_{k,l}^{n}\right)^{(q)} = \mathbf{R}_{k,l}^{n} - \sum_{m=1}^{M_{tx}} \mathbf{S}_{k,l}^{m,\text{Pilot}} \left(\hat{\mathbf{H}}_{k,l}^{m,n} \right)^{(q)}, \tag{5.102}$$

where $\left(\hat{\mathbf{H}}_{k,l}^{m,n} \right)^{(q)}$ is the channel frequency response estimate and q is the current iteration. Since only one of the antennas can transmit a pilot over each carrier, the summation in Equation 5.102 has only one term. The sequences of samples (Equation 5.102) enter the spatial demultiplexer block that can apply any of the equalization methods discussed in the receiver structure, and follow the processing sequence already described in the previous sections.

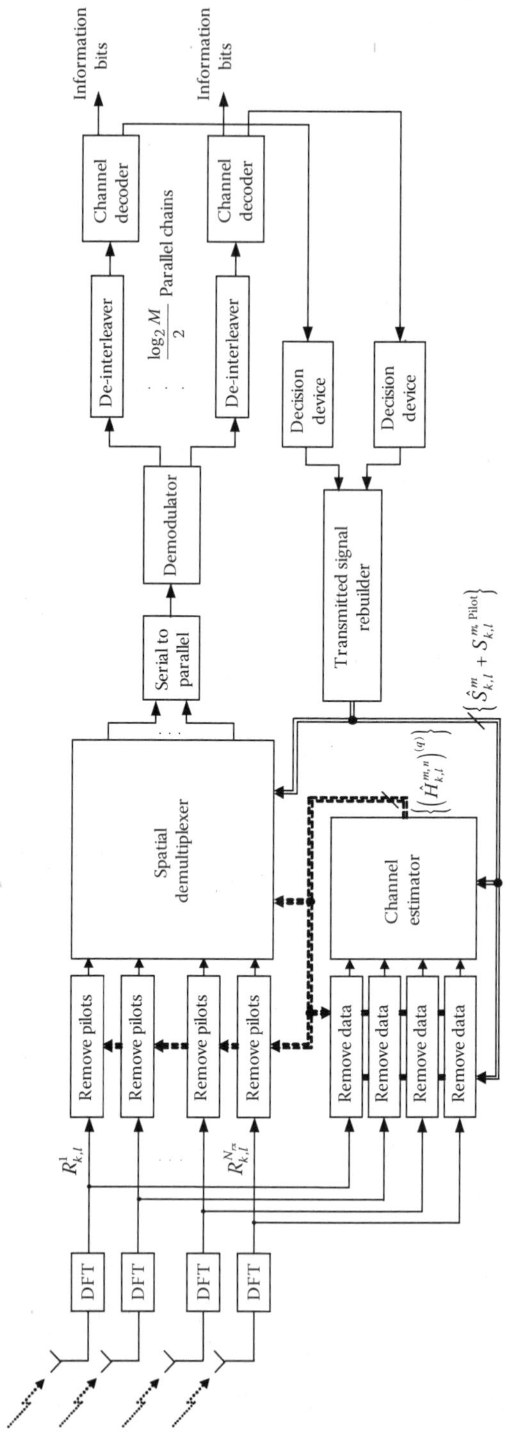

Figure 5.27 Iterative receiver structure for implicit pilots transmission.

5.5.2.3 Channel Estimation

In this case, with the existence of implicit pilots, in each iteration, the receiver applies the following steps to obtain the frequency channel response estimates for each transmitting/receiving antenna pair:

1. Data symbols estimates are removed from the pilots using

$$\left(\tilde{\mathrm{R}}_{k,l}^{n}\right)^{(q)} = \mathrm{R}_{k,l}^{n} - \sum_{m'=1}^{M_{tx}} \left(\hat{\mathrm{S}}_{k,l}^{m'}\right)^{(q-1)} \left(\hat{\mathrm{H}}_{k,l}^{m,n}\right)^{(q-1)}, \tag{5.103}$$

   where $\left(\hat{\mathrm{S}}_{k,l}^{m'}\right)^{(q-1)}$ and $\left(\hat{\mathrm{H}}_{k,l}^{m,n}\right)^{(q-1)}$ are the data and channel response estimates of the previous iteration. This step can only be applied after the first iteration. In the first iteration, we set $\left(\tilde{\mathrm{R}}_{k,l}^{n}\right)^{(1)} = \mathrm{R}_{k,l}^{n}$.

2. The channel frequency response estimates are computed using a moving average with size W as follows:

$$\left(\hat{\mathrm{H}}_{k,l}^{m,n}\right)^{(q)} = \frac{1}{W} \sum_{l'=l-\lfloor W/2 \rfloor}^{l+\lceil W/2 \rceil - 1} \frac{\left(\tilde{\mathrm{R}}_{k,l'}^{n}\right)^{(q-1)}}{\mathrm{S}_{k,l'}^{m,\mathrm{Pilot}}}. \tag{5.104}$$

3. After the first iteration if a fully dense pilot distribution is not employed (i.e., $\Delta N_{\mathrm{F}} \neq 1 \vee \Delta N_{\mathrm{T}} \neq 1$), then the data estimates can also be used as pilots for channel estimation refinement.

5.5.3 Simulation and Numerical Results

To study and compare the behavior of the proposed schemes, several Monte Carlo simulations were performed for 16-HQAM ($k_1 = 0.4$) and 64-HQAM ($k_1 = k_2 = 0.4$) constellations (other values of k_1 and k_2 can be used but the conclusions would be similar). Two streams of bits with different error protection levels were used in the case of 16-HQAM and three for 64-HQAM. Although this study is valid for any OFDM system, we performed the simulations using the parameters from UTRA LTE documents [3GPP, 25.905 2006; 3GPP 2007b] for a 10 MHz BW, which were already described in Chapter 4. The CIR is based on the Vehicular A environment from Ref. [ETSI 1998b]. The channel encoders were rate-1/2 turbo codes based on two identical recursive convolutional codes characterized by $G(D) = [1(1+D^2+D^3)/(1+D+D^3)]$ [3GPP 2003b]. A random interleaver was used within the turbo encoders. The results presented next will be shown as a function of E_S/N_0, where E_S is the average symbol energy and N_0 is the single-sided noise power spectral density. When using data multiplexed pilots, the spacing employed was $\Delta N_{\mathrm{F}} = 5 + M_{tx}$ and $\Delta N_{\mathrm{T}} = 4$ or 7 (the two possible configurations proposed in Ref. [3GPP 2007b]) and a sinc filter interpolation with length $W = 2$ was applied at the receiver.

In the graphs' legends, IPB means intermediate protected bits. Regarding the pilots powers, we will denote it as β_P, which is defined as $\beta_P = E\left[\left|S_{k,l}^{m,\text{Pilot}}\right|^2\right] \Big/ E\left[\left|S_{k,l}^{m}\right|^2\right]$, where $E[\cdot]$ represents the expected value computed over all positions (k, l) containing pilot symbols in the case of the numerator and overall positions containing data symbols in the case of the denominator. For data multiplexed pilots, β_P will always be 0 dB.

5.5.3.1 Results

The first evaluation phase consisted in studying the impact of both pilot transmission methods in SISO transmissions, whose results were presented in Chapter 4. In the second part of the study, we extended the evaluation to a MIMO 2 × 2 transmission. Table 5.2 shows the individual block sizes employed for these simulations, which once again admits the use of a subframe composed of seven OFDM blocks.

First, it was important to evaluate the impact of the use of an iterative receiver approach when multiple antennas are being employed. Figure 5.28 compares the performance of the different receiver methods of Table 5.3 for a MIMO 2 × 2 transmission employing a 16-HQAM ($k_1 = 0.4$) constellation with data multiplexed pilots (for $\Delta N_T = 7$). It can be seen that, although the receiver with the MMSE equalizer alone performs worse than the receiver with the MLSO equalizer, the performance can be substantially improved and one can achieve lower BLERs when an IC is applied in the last iterations. This is due to the fact that the information feedback by the receiver in the first iterations affects the MMSE equalizer through the improvement of the channel estimates. After a certain number of iterations, the use of the MMSE may not be able to provide any more substantial improvements (if the channel estimates are already sufficiently accurate). Therefore, in the last iterations, it is preferable to employ an IC since the feedback information allows the subtraction of better estimates of interference and thus there is always the possibility of improving the performance with a high number of iterations even if the channel estimates are already accurate. Similar conclusions could be achieved for implicit pilots, as was done in Ref. [Souto 2007]. For the remainder of this section the receiver configuration considered will be method 3.

When using multiple transmit antennas, the implicit pilot will be affected by higher levels of interference since there will be M_{tx} data symbols being transmitted at the same time as the pilots. Therefore, before comparing the two pilot transmission methods, it was important to study how the overall performance of the implicit pilot method

Table 5.2 Individual Block Sizes (in bits) Employed for MIMO 2 x 2 Simulations

Pilot Method	ΔN_T	*Block Size*
Data multiplexed	4	3852
Data multiplexed	7	4024
Implicit	1	4196

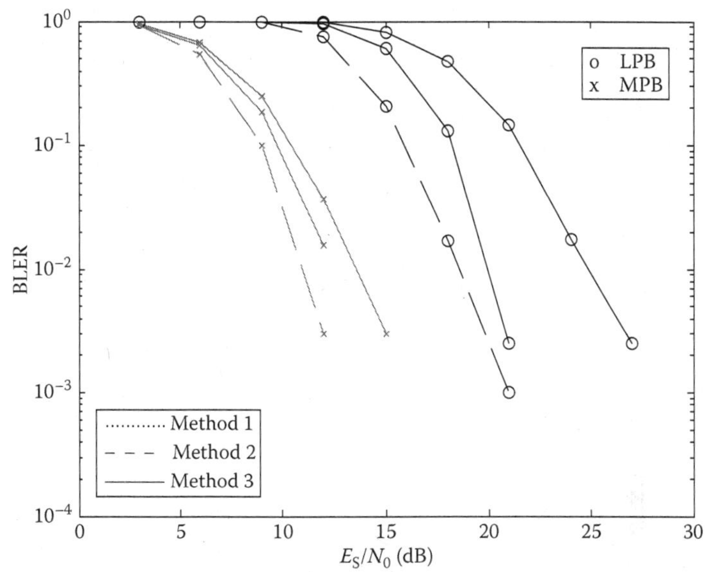

Figure 5.28 Performance of a 16-HQAM ($k_1 = 0.4$) hierarchical constellation with data multiplexed pilots ($\Delta N_T = 7$) using several receiver methods, with a MIMO 2 x 2 setting.

would be affected for different power levels applied for the pilots. Figure 5.29 shows some performance results. The first conclusion is that the curves for $\beta_P = -10.5$ dB, which correspond to a case where the percentage of power spent on the pilots is the same as in the data multiplexed pilot transmission with $\Delta N_T = 7$, have considerably high irreducible BLER floors. This means that the implicit pilots are being severely affected by the interference of the multiple data streams. Only when the power level is raised to -4.5 dB we start obtaining acceptable performances. If we keep increasing the power spent on the pilots the improvement in the channel estimation quality does not result in a better performance as can be seen from the $\beta_P = -3$ dB curve. In fact, increasing β_P

Table 5.3 Different Receiver Methods Applied for the MIMO 2×2 Simulations

Receiver Method	*Inner Turbo Decoder Iterations*	*Receiver Iterations*
1	12	1 with MMSE
2	3	2 with MMSE + 2 with IC
3	12	1 with MLSO

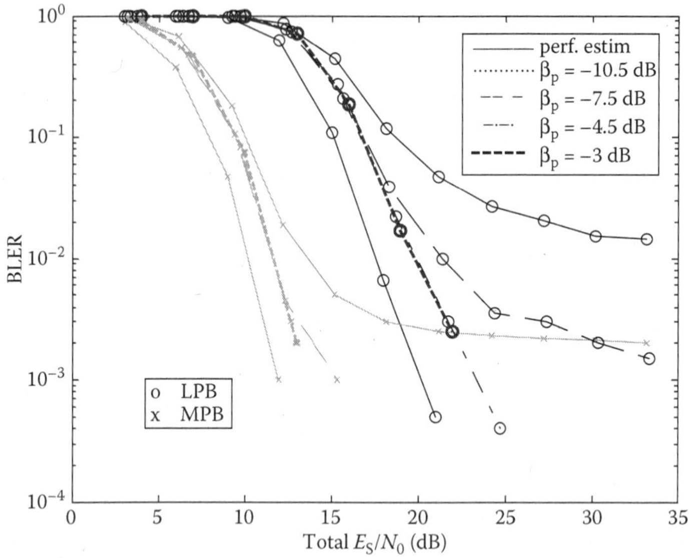

Figure 5.29 Performance of a 16-HQAM (k_1 = 0.4) hierarchical constellation for different β_P values. MIMO 2 × 2.

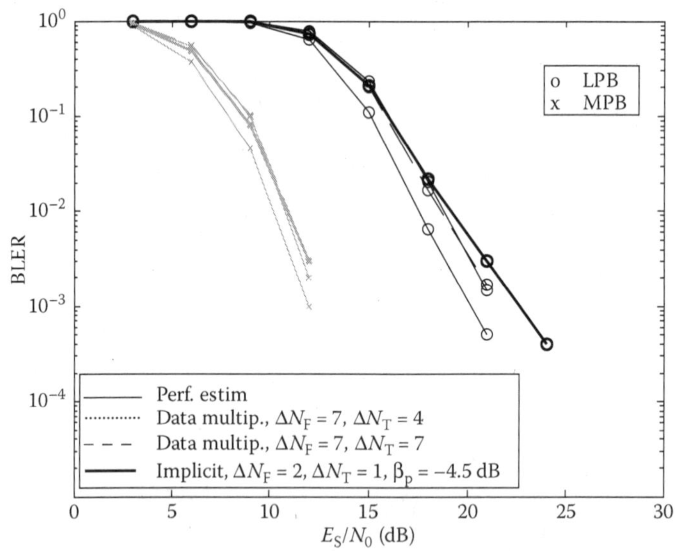

Figure 5.30 Data multiplexed pilots versus implicit pilots. 16-HQAM (k_1 = 0.4), MIMO 2 × 2.

even further would just result in similar curves dislocated to the right (worse performances) due to the higher power required for the pilots.

Therefore, for comparison of the implicit pilots against the data multiplexed pilots, we used $\beta_P = -4.5$ dB. The results are presented in Figure 5.30. Once again both pilot spacing considered for data multiplexed pilots have similar performances, which means that the longest time-domain spacing can be used also for MIMO 2×2 transmissions. As for the implicit pilot results, they are very close to the data multiplexed pilots performance for the MPB and only show a small degradation in the LPB.

Chapter 6

Multiresolution Techniques for MBMS

3GPP has launched the study item Evolved UTRA and UTRAN, which aimed at studying means to achieve further substantial leaps in terms of service provisioning and cost reduction. LTE constitutes a key standard that paves the way for the future mobile broadband landscape envisioned by 4G wireless networks. Actually, LTE addresses the emerging trend for the mass provision of rich multimedia services, such as mobile TV, in a more power and spectral efficient way than its predecessor UMTS. To this direction, the Evolved-MBMS (E-MBMS) framework is envisaged to play an instrumental role for the LTE proliferation in the mobile TV [3GPP 2007a]. E-MBMS constitutes the evolutionary successor of MBMS, which was introduced in the Release 6 of UMTS. In MBMS, efficient power allocation techniques should be implemented so as to ensure the mass provision of multimedia applications to mobile users. This chapter presents several multiresolution techniques and mechanisms for MBMS based on the WCDMA technology and for E-MBMS based on the OFDMA technology for next-generation cellular networks. The basic role of these techniques/mechanisms is to optimize the power and the resources during MBMS transmissions.

6.1 Introduction to Multiresolution Using WCDMA and OFDMA

Orthogonal frequency division multiplexing/OFDMA OFDM/OFDMA [Bingham 1990; Sari et al. 1994; Koffman and Roman 2002], used in the physical layer (downlink connection) of LTE, is an attractive choice to meet requirements for high data rates, with correspondingly large transmission BWs and flexible spectrum allocation.

OFDM also allows for a smooth migration from earlier radio access technologies and is known for achieving high performance in frequency-selective channels. Furthermore it enables frequency-domain adaptation, provides benefits in broadcast scenarios and is well suited for MIMO processing. The possibility to operate in vastly different spectrum allocations is essential. Different BWs are realized by varying the number of subcarriers used for transmission, while the subcarrier spacing remains unchanged. In this way, operation in spectrum allocations of 1.4, 3, 5, 10, 15, and 20 MHz can be supported.

For MBMS support within a certain cell coverage area for a given coverage target, the modulation and coding scheme (MCS) of the MBMS transport channel typically has to be designed under worst-case assumptions. Except for cell-edge users who typically experience large intercell interference, users with better channel conditions (closer to the BS) could receive the same service with a better quality (e.g., video resolution), as their receiving SNR would allow usage of a higher-rate MCS. Hierarchical modulations [Cover 1972; Ramchandran et al. 1993; Jiang and Wilford 2005; Wang et al. 2008], which have been specified for broadcast systems such as DVB-T or MediaFLO, is one way of accounting for unequal receiving conditions. Here, a signal constellation such as 16QAM, with each symbol being represented by four bits, is interpreted in a sense that the first two bits belong to an underlying QPSK alphabet. This enables the use of two independent data streams with different sensitivity requirements. The so-called high-priority stream employs QPSK modulation and is designed to cover the whole service area. The low-priority stream requires the constellation to be demodulated as 16QAM, and provides an additional or refined service via the two additional bits. These may transport an additional MBMS channel with a different type of service, or an enhancement stream that, for example, leads to enhancing the resolution of the base stream. A design parameter that determines the constellation layout allows the control of the amount of distortion that the enhancement symbols add to the baseline constellation, and can be used to control the ratio of coverage areas or service data rates. Theoretical evaluation of this type of modulations where the dependence of the individual bit streams performance on the constellation design parameter is explicitly shown has been previously presented in Refs. [Vitthaladevuni and Alouini 2004; Souto et al. 2007a].

Specifically for broadcast and multicast transmissions in a mobile cellular network, depending on the communication link conditions, some receivers will have better SNR than others and thus the capacity of the communication link for these users is higher. Hierarchical constellations and MIMO (SM [Foschini 1996; Foschini and Gans 1998]) are methods that are able to offer multiresolution and take advantage of the different link capacities. These two forms of multiresolution were considered for the WCDMA technology in Refs. [Soares et al. 2007a; Soares et al. 2007b; Correia et al. 2008]. In OFDMA-based networks, the transmission of different fractions of the total set of subcarriers [physical resource blocks (PRBs)] depending on the position of the mobiles is another way to offer multiresolution. In WCDMA-based networks, the transmission of different spreading codes (multicode), depending on the position of the mobiles, assures multiresolution. Any of these methods is able to provide unequal bit error protection. In any case there are two or more classes of bits with different error protection levels, to which different streams of information can be mapped. Regardless of the channel conditions, a given user always attempts to demodulate both types of bits, the most protected and the ones carrying additional resolution. Depending on its position inside the cell, more or less blocks with additional resolution will be correctly received by the mobile user. However, the basic quality will be always correctly received independent of the position of any user, within the 95% target of coverage.

For increasing distance between terminals and BSs, decreasing bit rates are correctly received due to the decrease in SNR. AMC is a technique that maximizes the total throughput for unicast transmissions. The decrease of SNR with distance is common to unicast or broadcast/multicast transmissions. However, for broadcast/multicast, the same video content is transmitted and AMC is not possible without personal uplink feedback. With the introduction of multiresolution techniques, the maximization of the total throughput is the goal to be achieved.

There is support for MBMS in the first version of LTE specifications. However, specifications for E-MBMS are in early stages. Two important scenarios have been identified for E-MBMS: one is single-cell PMP (SC-PMP) broadcast; the second is Multimedia Broadcast SFN (MBSFN). The scenario SC-PMP corresponds to the PMP mode of MBMS already specified in the Release 6 for WCDMA, operating on a single cell scenario. MBSFN is a new feature that is being introduced in the LTE specification. MBSFN is envisaged for delivering services such as the mobile TV using the LTE infrastructure, and is expected to be a competitor to DVB-for Handhelds (DVB-H)-based TV broadcast. In MBSFN, the transmission happens from a time-synchronized set of enhanced-Node Bs (eNBs) using the same resource block. This enables over-the-air combining, thus improving the SINR significantly compared to MFN operation. The CP used for MBSFN is slightly longer, and this enables the UE to combine transmissions from different eNBs, thus somewhat negating some of the advantages of SFN operation. There will be six symbols in a slot of 0.5 ms for MBSFN operation versus seven symbols in a slot of 0.5 ms for SC operation.

System-level simulations for broadcast/multicast with multiresolution and different fractional frequency reuse for LTE are necessary to evaluate achievable capacity and intersite gains compared to single-resolution systems, that is, WCDMA based. Taking the 95% coverage as reference, the evaluation of the achievable capacity gain (number of transmitted mobile TV channels for WCDMA and LTE) was done in Ref. [Correia et al. 2007]. The ISD gain is also evaluated allowing for a substantial reduction in the number of cell sites when LTE replaces WCDMA. The scenario based on the use of SFN with the MBSFN channel will be also evaluated for 16QAM/64QAM hierarchical modulations and compared with the present MBMS network WCDMA based without and with hierarchical modulations.

By allocating the proper TSs and carrier frequencies, LTE provides intracell orthogonality between users in both uplink and downlink. Nevertheless, intercell interference is a major problem, as compared to WCDMA/HSPA, especially for users at the edge of the cell. Several measures can be implemented in order to minimize this problem. The most important is fractional frequency reuse. Two different fractional frequency reuses will be evaluated in this chapter.

6.2 Objectives and Requirements

In the 3GPP LTE specification, two types of E-MBMS transmission scenarios exist:

- Multicell transmission (MBSFN) on a dedicated frequency layer or on a shared frequency layer.
- Single-cell transmission (SC-PMP: Single-cell PMP) on a shared frequency layer.

Contrarily to LTE, in 3GPP Release 6 (UMTS), the only specified transmission scenario for the MBMS transmission is the SC-PMP. However, soft-diversity combining is possible as long as the delay between different BS transmissions of the same content allows macrodiversity.

Multicell transmission in an SFN area is a way to improve the spectral efficiency: since all E-MBMS cells transmit the same E-MBMS session data, the signals can be combined for a UE located at a cell boundary. Furthermore, the multicell transmission may be provided over a cell group that comprises cells that transmit the same service.

In contrast, the single-cell transmission covers only one cell or one eNB. In addition, the concept of dynamic MBSFN area is introduced where the E-MBMS transmission is switched off in some cells of the MBSFN area when a certain E-MBMS is not needed there. In some cases, the released resource can be reused for other E-MBMS or unicast services. The decision to turn off the MBSFN transmission in a cell is based on two factors:

- Local existence, which refers to the number of UEs that are both interested in the current E-MBMS and located in this cell.
- Contribution of neighboring cells, which refers to the number of UEs that are both interested in the current E-MBMS and located in neighboring cells having the same MBSFN transmission.

The cells of an MBSFN area contribute to the MBSFN transmission only if there are UEs that are interested in the particular service in this MBSFN area.

The introduction of hierarchical modulation in a broadcast cellular service such as E-MBMS requires a scalable video coded as presented in Chapter 1. The base layer transmission provides the minimum quality, and one or more enhancement layers offer improved quality at increasing bit/frame rates and resolutions. Besides being a potential solution for content adaptation, scalable video schemes may also allow an efficient usage of radio resources in E-MBMS.

According to Release 6 of 3GPP the single-resolution scheme corresponds to the transmission of QPSK with at least 95% coverage. The assignment of the fraction of the total transmission power reserved for E-MBMS has implications in the coverage and average throughput of the multiresolution based on the hierarchical 16-QAM/64-QAM schemes. The multicell interference distribution has also strong impact in the coverage and throughput. An interesting design parameter is the channel bit rate (and its coding rate) associated to the multiresolution scheme. An optimization of this parameter has also strong impact on the achievable coverage and average throughputs.

Regardless of the channel conditions and user location, a given user always attempts to demodulate both the base layer and the enhancement layers carrying additional resolution. For good multiresolution design, the basic information will be always correctly received independent of the position of any user, within the 95% coverage target. However, depending on its position inside the cell, more or less blocks with additional resolution will be correctly received by the mobile user.

The objective of this work is the design of multiresolution schemes for both technologies, WCDMA and OFDMA, in the two different scenarios, namely MBSFN and SC-PMP with intercell interference without and with macrodiversity support. In addition, it is intended to measure the corresponding multiresolution gain of total throughput compared to the reference total throughput of the single-resolution scheme based on the QPSK transmission.

6.3 Evaluation Methodology and Simulation Assumptions

Typically, radio network simulations can be classified as either link level [radio link (RL) between the BS and the user terminal] or system level (several BSs with large number of mobile users). A single approach would be preferable, but the complexity of such simulators (including everything from transmitted waveforms to multicell network) is far too high for the required simulation resolutions and simulation time. Therefore, separate but interconnected link- and system-level approaches are needed.

The LLS is needed for the system simulator to build a receiver model that can predict the receiver Block Error Rate (BLER)//BER performance, taking into account channel estimation, interleaving, modulation, receiver structure, and decoding. The system-level simulator (SLS) is needed to model a system with a large number of mobiles and BSs and also to evaluate algorithms operating in such a system.

As the simulation is divided into two parts, an approach for linking the two simulators must be defined. Conventionally, the information obtained from the LLS is inserted into the SLS through the utilization of a specific performance parameter (BLER) corresponding to a specific SNR estimated in the terminal or BS. In Figure 6.1, the interaction between the simulators is shown.

6.3.1 LLS Design

LLSs for WCDMA and OFDMA were developed in Matlab and take into account the specifications of 3GPP MBMS Releases 6 and 7 [3GPP 2006d], respectively, regarding the signal processing of transport and physical channels of WCDMA and OFDMA. It satisfied two essential requirements:

- Serve as reference for all the link-level simulations with multiresolution and parameters estimation
- Serve as a platform to the different multiresolution improvements tested and quantified

Typical time interval of each link-level simulation is 0.5 s. The entire WCDMA or OFDMA signal processing at the transmitter was included in the LLS as well as several

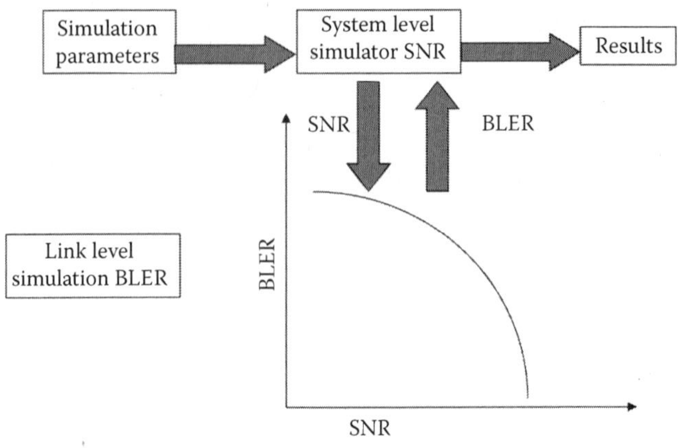

Figure 6.1 Interaction between LLS and SLS.

different receiver structures. To achieve reliable channel estimation and data detection, we employ a receiver capable of jointly performing these tasks through iterative processing. The structure of the different iterative receivers was presented in Chapters 4 and 5 (see also Refs. [3GPP 2006c; Souto et al. 2008]). Clearly, the receiver structure for AWGN channels is less complex (only a few turbo decoder iterations and no channel estimation or channel equalization is required).

Multipath Rayleigh fading channels were considered in the simulator since it comprises a more realistic scenario for evaluating hierarchical high-order QAM modulations due to their sensitivity to the channel parameters estimation. As indicated in the previous chapters, the receiver structures are nonlinear, iterative, and include channel parameters estimation for the analyzed multipath Rayleigh fading channels. Table 6.1 presents the main link-level simulation parameters considered in the LLs.

6.3.2 Radio Access Network (RAN) SLS

For the purpose of validating the work presented in this section, SLSs were developed in Java, using a discrete event-based philosophy, which captures the dynamic behavior of the RAN system. This dynamic behavior includes the user (e.g., mobility and variable traffic demands), radio interface, and RAN with some level of abstraction. The SLSs work at slot duration or TTI rate and typical time interval of each simulation is 600 s.

The channel model used in the SLS considers three types of losses: distance loss, shadowing loss, and multipath fading loss (one value per 0.5 ms). The model parameters depend on the environment. For the distance loss, the Okumura–Hata model from the COST 231 project was used [Damosso 1999]. Shadowing is due to the existence of large obstacles such as buildings and the movement of UEs in and out of the shadows. This is modeled through a process with a log-normal distribution and a correlation distance.

Table 6.1 Link-Level Simulation Parameters

WCDMA Parameters	*Value*
S-CCPCH slot format	12, 14 (128 kbps, 256 kbps)
Transport block size and number of transport blocks per TTI	Varied according to information bit rate (128 or 256 kbps) and TTI value
TTI	20, 40, 80 ms
CPICH Ec/Ior	−10 dB (10%)
P-SCH Ec/Ior	−15 dB (3%)
S-SCH Ec/Ior	−15 dB (3%)
OCNS orthogonal channel noise simulator	Used to sum the total Tx. Ec/Ior to 0 dB (100%)
Channel estimation	Enabled
Power control	Disabled
Channels	Pedestrian B, 3 km/h and vehicular A, 30 km/h
OFDMA parameters	Value
Transmission BW	10 MHz
CP size (samples)	256
FFT size	1024
Carriers space (kHz)	15
Available BW	9 MHz
Sample time (ns)	65
Slot duration (ms)	0.5
Number of used subcarriers/sector	200, 600
Channels	Vehicular A 30 km/h and MBSFN 30 km/h

The considered interference is the sum of intracell and intercell interferences. Both have a noise-like character. This is mainly due to the large number of sources adding to the signal, which are similar in signal strength.

Link performance results are used as input by the SLS where several estimates for coverage and throughput purposes can be made by populating the scenario topology uniformly and giving users a random mobility. The estimates are made for every TTI being the packets that are received with a BLER over 1% considered to be well received. The estimate for coverage purposes are made for an average of five consecutive received packets. If the average received BLER of these packets is above 1%, the mobile user is considered to be within coverage area. For the throughput calculation, the estimation is made based on each individual packet received with a BLER higher than 1%.

The multipath fading employed in the SLS corresponds to the 3GPP channel models, where the ITU Vehicular A (3, 30 km/h) and Pedestrian B (3 km/h) (see Ref. [3GPP 2006c] Annex B) and both SC-PMP and MBSFN environments were chosen as references. The latter models were also used in the LLS but at much higher rate. The vehicular A channel model was chosen for OFDMA with SC-PMP because it is an important test channel in 3GPP specifications. Furthermore, it allows direct comparison against previous system-level simulations implemented for the WCDMA [Correia et al. 2007].

In OFDM systems, an important parameter is the maximum delay of the multipath profile and its relation with the duration of the time guard between OFDM symbols to avoid ISI. 3GPP has specified a short time guard with about 4.75 μs and a long one with 16.67 μs. The long time guard was considered in this chapter, making the performance less sensitive to the chosen propagation channel. However, there is a reduction of the transmitted bit rates.

A uniform distribution of mobile users is generated at the beginning of each simulation. Typical number of users chosen for each simulation run was 20 per sector. Each mobile has random mobility with the specified 3 or 30 km/h. Dynamic SLSs such as the ones presented in this chapter are very accurate, the main limitation is the hypothetical urban macrocellular test scenario that is different from any real one.

Table 6.2 shows the system-level simulation parameters for OFDMA and WCDMA. It presents the parameters used in the system-level simulations based on 3GPP documents [3GPP 2006b; 3GPP 2007b; 3GPP 2008a].

Table 6.2 System-Level Simulation Parameters for Urban Macrocellular Scenarios

OFDMA Parameters	*Value*
Max Tx. power (dBm)/sector	46
Frequency reuse	1/3, 1
Interfering cells transmit with % of maximum power	90
Cellular layout	Hexagonal
Sectors	3 sectors/cell
Number of cell sites	19
Site-to-site distance	Variable

continued

Table 6.2 (continued) System-Level Simulation Parameters for Urban Macrocellular Scenarios

OFDMA Parameters	*Value*
Antenna gain of the BS + cable loss	14.5 dBi
Width of the beam of the antenna at −3 dB	70°
Propagation model	Okumura–Hata
Downlink thermal noise	−100 dB
Fade out standard deviation due to shadowing	10 dB
WCDMA parameters	Value
Node B total transmit power (sector)	43 dBm
Hexagonal cellular layout	19 cells
Sectorization	Yes, 3 sectors/cell
Site-to-site distance	1000 m
Interfering cells transmit with % of maximum power	80
BS antenna gain + cable losses	14.5 dBi
Antenna beamwidth, −3 dB	70°
Propagation models	Okumura–Hata
Thermal noise DL	−100.3 dB
Orthogonal factor	0.4, 0.5
Std. of shadow fading	10 dB
Correlation between sites forslow fading	0.5

Figure 6.2 illustrates the cellular layout (trisectorial antenna pattern) indicating the fractional frequency reuse of 1/3 considered in the system-level simulations of OFDMA. One-third of the available BW was used in each sector to reduce the multicell interference. As depicted in the figure, the identification of the sources of multicell interference, that is, the use of the same adjacent subcarriers (named PRBs) is given by the sectors with the same gray shadow/number, namely dark gray/1, light gray/2, or average gray/3. WCDMA networks have frequency reuse of 1 (every sector has the same gray shadow/number),

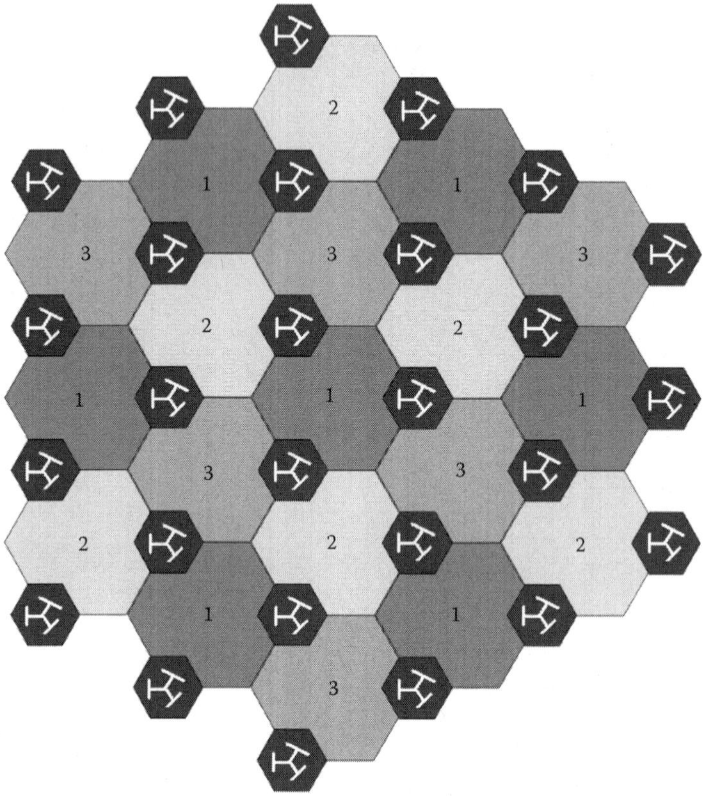

Figure 6.2 Cellular layout including the frequency reuse of 1/3 (colors/numbers of the cells).

which is another important frequency reuse scenario also considered and evaluated with OFDMA networks.

For 16-QAM hierarchical constellations two classes of bits with different error protection are used. The darkest gray around the antennas only indicates the minimum coverage of the weak bits blocks, while the other gray shadow indicate the coverage of the strong bits blocks.

In the analysis of the scenario SC-PMP, there is one RL between the mobile and the closest BS. Synchronism between the transmissions from different BSs with the same gray shadow resulting in interference from all but one cell with the same gray shadow is not assumed at any time. However, in the scenario SC-PMP with macrodiversity combining the two or three best RLs, it is assumed that there is time synchronization between the two or three closest BS sites with the same gray shadow. In this case, the multicell interference is reduced because only the other BS sites with the same gray shadow remain unsynchronous and capable to interfere.

In the MBSFN scenario there are, at least, three RLs between the mobile and the three closest BSs. It is assumed that the existence of time synchronism between the transmissions from the closest BSs with the same gray shadow resulting in much less interference from the cellular environment. In this scenario there is macrodiversity combining at least the three best RLs. In addition, the interfering BSs must be at least 5 km away from the reference BS because the long CP, with 16.67 μs, and the frequency of 2 GHz was taken into account. Therefore, only distant BS sites are capable of interfering in the MBSFN scenario.

6.4 System-Level Performance Results

6.4.1 WCDMA Broadcast Multiresolution Results

For MBMS PMP mode transmissions, the available channel is the FACH, which is mapped onto the Secondary Common Control Physical Channel (S-CCPCH). Results for different multiresolution systems will be next analyzed associated to this specific transport channel of UMTS.

6.4.1.1 Multicode Results

Figure 6.3 presents the link-level S-CCPCH performance in terms of fraction of the total transmitted power from the BS (equal to the rate of the energy transmitted per chip to the total power spectral density transmitted from the BS, denoted as E_c/I_{or} in dB) necessary to achieve the corresponding BLER graduated on the vertical axis. For the reference BLER $= 10^{-2}$, the use of a single spreading code with spreading factor $SF = 8$ (bit rate of 256 kbps) imposes to have a geometry factor of 0 dB in order to achieve E_c/I_{or} less than 80% (-1 dB), considering the Vehicular A propagation channel. This means that we can only offer such a high bit rate for users located in the middle of the cell, not near the border. The use of two spreading codes each with $SF = 16$ (bit rate of 128 kbps) transmitted with different power allows an increase in coverage ($G = -3$ dB) and throughput. This will be next confirmed with system-level results. Remind that for multiresolution using multicode, the base layer is transmitted with higher power ($G = -3$ dB) and the enhancement layer with lower power ($G = 0$ dB). Here, we will assume that $E_c/I_{or} = 80\%$ is available for the transmission of MBMS in the SC-PMP scenario, without any macrodiversity combining.

In Figure 6.4, the average coverage versus E_c/I_{or} is presented. The introduction of multicode allows multiresolution and an increase in average coverage for the same total transmitted power. With multicode, the aggregate bit rate of 256 kbps is achievable with two data streams of 128 kbps. For Pedestrian B channel model, the enhancement data stream is transmitted with $E_c/I_{or1} = 30\%$ assuring 62% coverage and the base data stream is transmitted with $E_c/I_{or2} = 50\%$ offering 85% coverage. For the same total transmitted power $E_c/I_{or} = 80\%$ a single stream of 256 kbps allows a coverage of 66%. Similar coverage values (58% and 81% for multiresolution and 64% for single resolution) are obtained for the Vehicular A propagation channel considering the same E_c/I_{or} values.

In Figure 6.5, the average throughput versus E_c/I_{or} is presented. Recall that blocks with errors are not retransmitted in MBMS PMP mode. Taking as reference the same E_c/I_{or} values used for comparison of coverage, namely $30\% + 50\% = 80\%$ and the

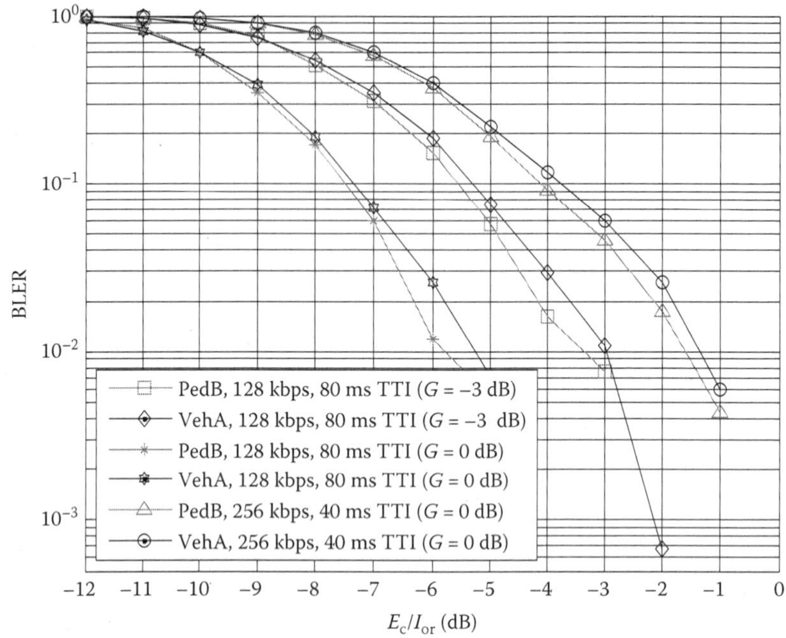

Figure 6.3 BLER versus Tx. power for S-CCPCH different geometries and propagation channels.

Pedestrian B channel, we will check that multicode allows an increase in throughput. For the enhancement data stream, $E_c/I_{or1} = 30\%$ assures average throughput 80 kbps. With $E_c/I_{or2} = 50\%$, the throughput of 110 kbps is achieved for the base data stream. The total throughput is 190 kbps for multiresolution. For the single-resolution stream with $E_c/I_{or} = 80\%$, the achieved throughput is 172 kbps. The throughput gain is around 10%.

Figure 6.6 presents the macrodiversity gains when soft combining three simultaneous S-CCPCH channels. Comparing Figure 6.4 for one RL (1RL) with Figure 6.6 for three RLs we can observe, with the reference coverage 95%, a decrease in the required E_c/I_{or} to 18% for 256 kbps.

As shown in Figure 6.5 the throughput of 256 kbps could not be achieved using a single S-CCPCH RL. Figure 6.7 presents the throughput gains for soft combining three simultaneous S-CCPCH channels. In Figure 6.7 combining the three best RLs provides a decrease in the required E_c/I_{or} to 20% for 256 kbps.

6.4.1.2 Hierarchical Modulation Results

Figure 6.8 presents an alternative way of offering multiresolution where the bit rate is 256 kbps using hierarchical 16-QAM modulation and a single spreading code with

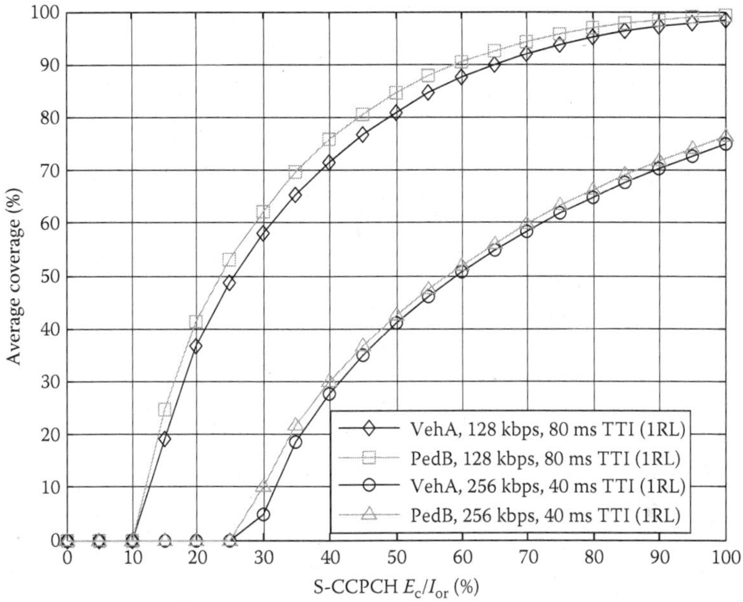

Figure 6.4 S-CCPCH average coverage versus Tx. power for (1RL).

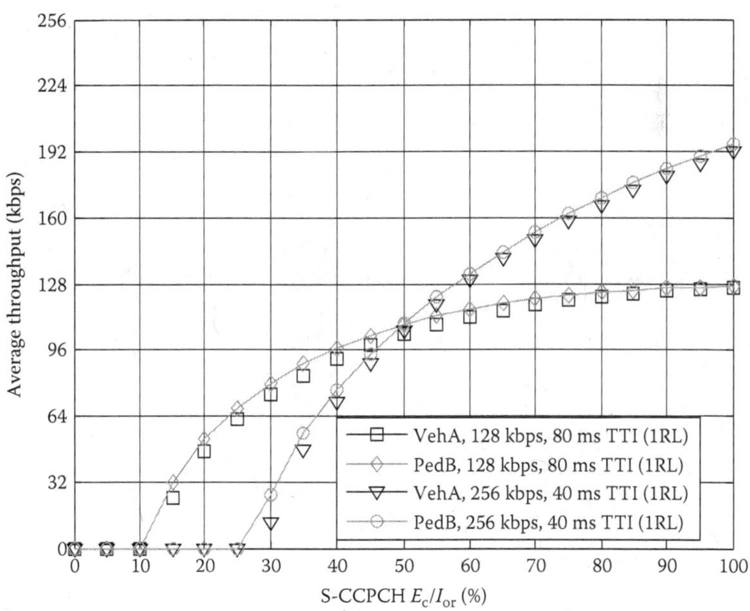

Figure 6.5 S-CCPCH average throughput versus Tx. power (1RL).

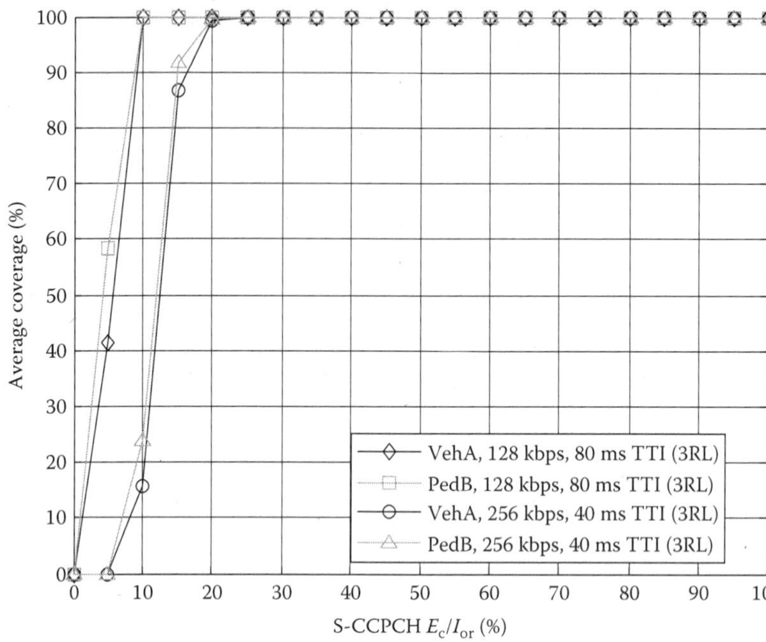

Figure 6.6 S-CCPCH average coverage versus Tx. power (soft combining 3RL).

$SF = 16$ for $G = 0$ dB and $k = 0.5$ (k measures the asymmetry of the hierarchical 16-QAM constellation; $k = 0.5$ corresponds to the uniform constellation). This case is more spectrally efficient than the reference QPSK modulation without multiresolution, presented in Figure 6.3, because it uses a higher SF. An iterative receiver based on the one described in Chapter 4 is employed for decoding both blocks of bits. For the reference value of BLER $= 10^{-2}$ the difference of total transmitted power between the strong and the weak blocks is about 5.5 dB for either Vehicular A or Pedestrian B. This means that there is a substantial difference in coverage between the two data streams to assure the reference BLER.

Figure 6.9 presents average throughput and corresponds to Figure 6.5. It is obvious that multiresolution using hierarchical modulation provides higher throughput than with single resolution. Taking the same reference $E_c/I_{or} = 80\%$, the achieved throughput for the Pedestrian B channel is 240 kbps. There is an obvious increase in throughput obtained with hierarchical modulation compared to multiresolution by multicode. The throughput gain is 28.3%, but there is the disadvantage of requiring more complex receivers.

6.4.1.3 MIMO Results

Figure 6.10 presents an alternative way of offering multiresolution where the bit rate is 256 kbps using SM, two transmitting and two receiving antennas, 2×2 MIMO, and a single spreading code with $SF = 16$ for $G = -3$ dB. The complex correlation

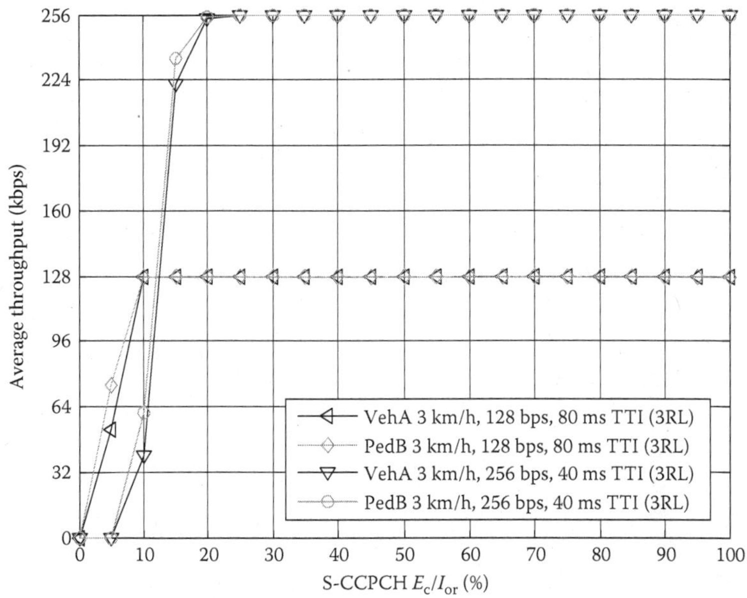

Figure 6.7 S-CCPCH average throughput versus Tx. power (soft combining 3RL).

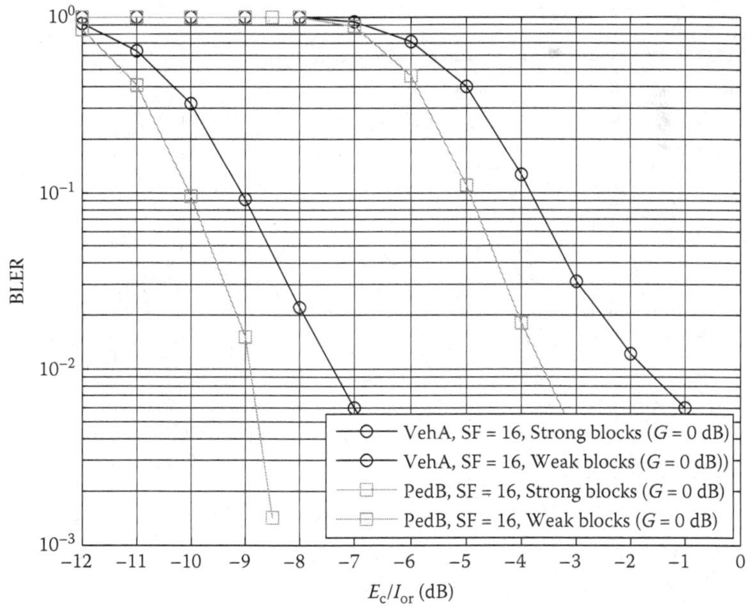

Figure 6.8 BLER versus Tx. power for 16-QAM strong and weak blocks of bits ($SF = 16$), $k = 0.5$.

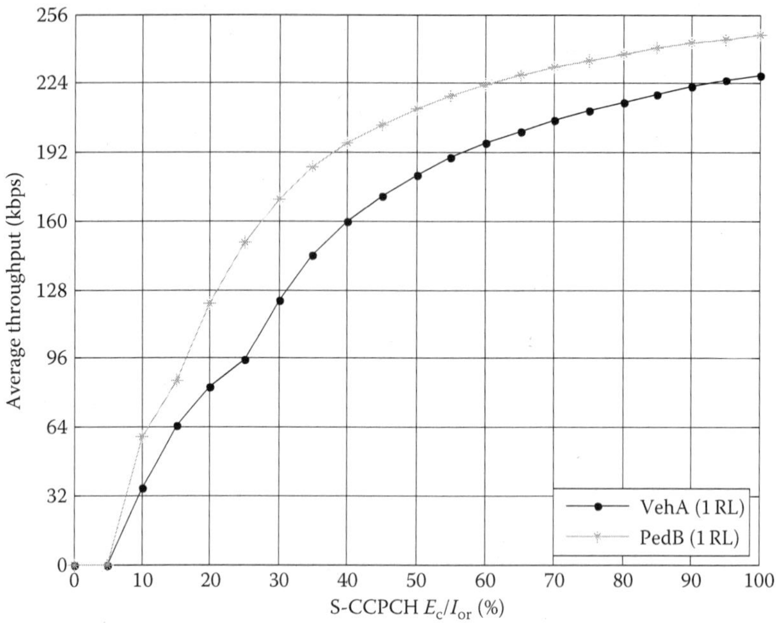

Figure 6.9 16-QAM average throughput versus Tx. power (1 RL).

coefficients between antennas were taken from Ref. [3GPP 2008a]. This case has the same spectral efficiency than the previous one with multiresolution (using hierarchical constellations), presented in Figures 6.8 and 6.9, because it uses the same SF. However, it provides higher coverage because the geometry is lower. The base and the enhancement layers are transmitted by different antennas. However, the results here presented only consider the case where the transmitted power per antenna is the same and equal to half of the single transmitting antenna case. If we had considered different transmitted powers per antenna, then we would have a better BLER performance for the base layer compared to the enhancement layer stream.

The receiver for both single-resolution SISO and multiresolution MIMO is the same described in Chapter 5. For the reference value of BLER $= 10^{-2}$, the difference of total transmitted power between the single-resolution SISO (1×1) and the multiresolution MIMO (2×2) schemes is less than 1 dB (Vehicular A) or equal to 1 dB (Pedestrian B). Considering that we are transmitting at 128 kbps with single resolution and 256 kbps with multiresolution, we would expect the double of the transmitted power for the latter. However, we can conclude that the multiresolution scheme is much more power efficient than the single-resolution one for BLER $< 10^{-2}$.

Figure 6.11 presents average throughput and corresponds to Figures 6.5 and 6.9. It is obvious that multiresolution using MIMO provides higher throughput than the previous multiresolution techniques. However, the observation of the curves indicates that there is a decrease in the throughput of the enhanced layer compared to the base

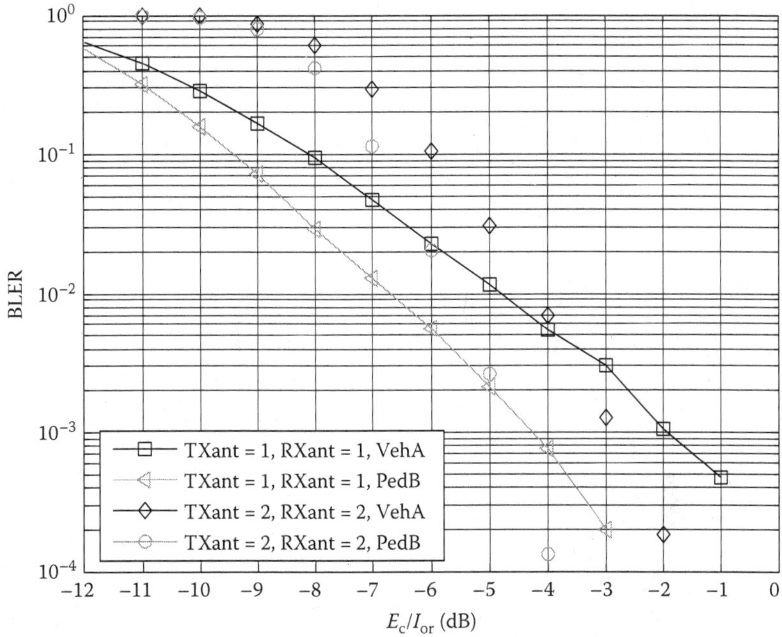

Figure 6.10 BLER versus Tx. power for SISO (1 x 1) and MIMO (2 x 2), the geometry is $G = -3$ dB.

layer. For instance, let us consider the Pedestrian B channel and $E_c/I_{or} = 40\%$. The base layer throughput is 128 kbps but the total throughput (base + enhancement) is around 240 kbps. Considering again the previous reference $E_c/I_{or} = 80\%$, the achieved throughput for the Pedestrian B channel is 256 kbps. The throughput gain is 32.82% relative to QPSK single resolution ($SF = 8$).

6.4.1.4 Conclusions

In this section we have analyzed several multiresolution broadcast systems for WCDMA cellular mobile networks, namely multicode, hierarchical 16-QAM constellations, and MIMO systems. In addition to the obvious advantage of multiresolution compared to single resolution in terms of graceful degradation of quality, each technique presented here offered coverage and throughput performance gains over conventional single-resolution broadcast systems. We have applied each multiresolution broadcast system to the PMP mode of MBMS in SC-PMP scenarios.

A comparison between the three specific broadcast multiresolution systems indicates that multicode is the one with less performance gain and has no spectral efficiency gain compared to single resolution (single code). Hierarchical 16-QAM modulation has the double spectral efficiency and higher performance gains than multicode. MIMO with two transmitting and two receiving antennas has also the double spectral efficiency

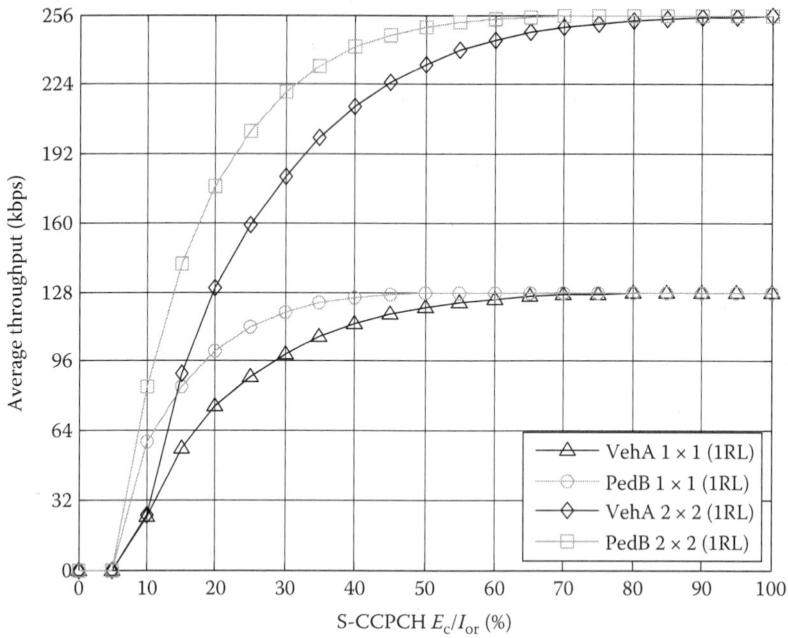

Figure 6.11 Average throughput versus Tx. power (1RL).

and offers the highest gains. However, the expected capacity gains that MIMO and hierarchical 16-QAM schemes provide require more complex receivers than multicode or single resolution.

6.4.2 OFDMA Broadcast Multiresolution Results

For 16QAM, two classes of bits with different error protection are used (for 64QAM, three classes of bits are used). Each information stream was encoded with a different block size depending on the coding rate. We have considered two different coding rates, namely 1/2 and 3/4. This leads to block sizes of each information stream with 2400 and 3600 bits, respectively, during each slot with a duration of 0.5 ms. The total corresponding transmitted information bit rates per cell sector are 4800 kbps and 7200 kbps, respectively.

6.4.2.1 Frequency Reuse 1/3 Results

As a reference for measuring the capacity, we will consider the number of transmitted TV channels, each with a bit rate of 256 kbps. For the chosen fractional frequency reuse of 1/3, we get 18.75 and 28.125 TV channels, respectively. We want to compare directly the OFDM/OFDMA results with those obtained previously with the WCDMA technology [Correia et al. 2007]. All the parameters used for OFDM during these simulations were based on 3GPP documents [3GPP 2006b; 3GPP 2007b; 3GPP 2008a].

About one-third of the total PRBs is transmitted in each sector. This corresponds to an instantly occupied BW of 3.0 MHz, where we have considered an average of 16.67 PRBs per sector, each with 180 kHz of adjacent BW (corresponding to 12 subcarriers with a frequency spacing of 15 kHz). The number of adjacent subcarriers in each PRB was chosen according to 3GPP specifications. With the dynamic allocation of the resources per sector, sectors 2 and 3 have 17 PRBs while sector 1 has 16 PRBs in the first subframe duration. Sectors 1 and 3 have 17 PRBs while sector 2 has 16 PRBs in the second subframe duration. Finally, sectors 1 and 2 have 17 PRBs while sector 3 has 16 PRBs in the third subframe duration. On average there are 16.67 PRBs per sector. We can conclude that the transmission of each TV channel with the LTE technology requires less than one PRB for any analyzed coding rates.

We have evaluated in the link-level simulations the hierarchical 16QAM (H16-QAM) and 64QAM (H64-QAM) with two different coding rates and $k = 0.5$. In Figures 6.12 and 6.13, we consider the Vehicular A propagation channel to be used in the SC-PMP scenario and we present the BLER versus E_s/N_0 for the hierarchical 16QAM and 64QAM, respectively. In the legend, for 16QAM, H1 means strong bit blocks, H2 means weak bit blocks. For 64QAM, H1 means strong bit blocks, H2 means medium bit blocks, and H3 means weak bit blocks. As expected, the coding rate 1/2 provides smaller E_s/N_0 compared to rate 3/4 resulting in higher coverage than with rate 3/4. However, the coding rate 3/4 provides bit rates that are 1.5 higher than rate 1/2. There is a trade-off between bit rate (or throughput) and coverage. We will consider later on macrodiversity combining to increase the coverage (and throughput) at the cell borders.

When we compare Figures 6.12 and 6.13 we observe that due to the higher bit rates offered by 64QAM, exactly 1.5 times the bit rates of 16QAM, the corresponding E_s/N_0

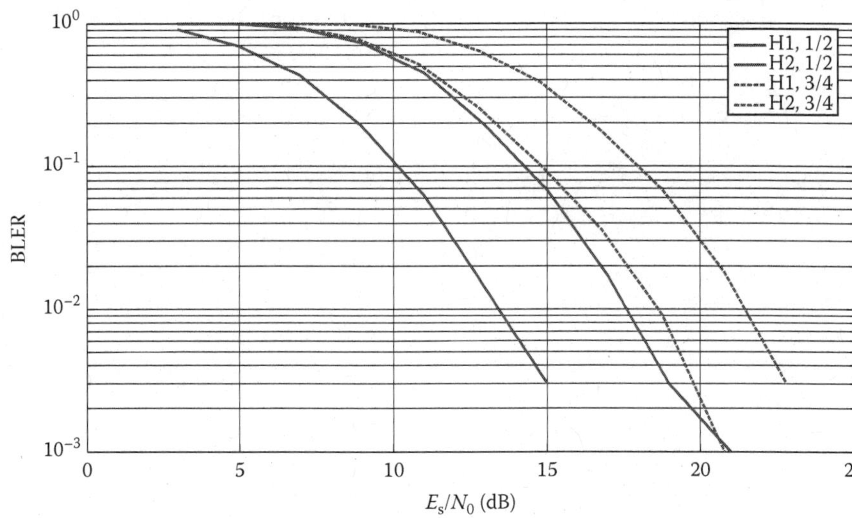

Figure 6.12 BLER versus E_s/N_0 for H16-QAM, Vehicular A 30 km/h, reuse 1/3.

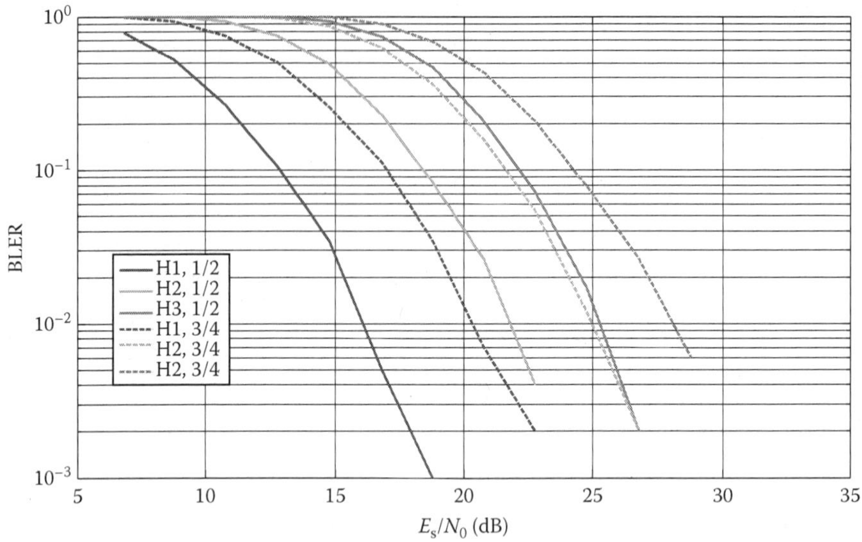

Figure 6.13 BLER versus E_s/N_0 for H64-QAM, Vehicular A 30 km/h, reuse 1/3.

are higher than 16QAM resulting in less coverage for 64QAM. It seems that 64QAM with coding rate 1/2 has around 3 dB higher E_s/N_0 for BLER = 0.01, compared to 16QAM rate 3/4 (both provide the same maximum bit rates). In addition, the sensitivity of 64QAM to channel estimation errors is a feature that should not be forgotten, especially for hierarchical 64QAM. The introduction of macrodiversity combining will increase the coverage of 16QAM H2 blocks and 64QAM H2, H3 blocks (medium and weak bits).

In Figures 6.14 and 6.15, we present the BLER versus E_s/N_0 for the hierarchical 16QAM and 64QAM in the MBSFN scenario with the MBSFN propagation channel. Comparison between Figures 6.12 and 6.13 with Figures 6.14 and 6.15 indicates that the MBSFN channel, due to the longer multipath PDP, provides higher multipath diversity. This can be confirmed by the clear increase in the BLER performance of both coding rates compared to the Vehicular A, especially with 16QAM. There is also higher inherent ISI in the MBSFN channel, which is more evident for 16QAM rate 3/4 (H2) and 64QAM (H2 and H3 blocks). In spite of this, there is no significant loss in the BLER performance for 16QAM rate 3/4 due to the redundancy of the channel coding. However, 64QAM has a substantial increase in E_s/N_0 for H3 blocks of bits. Recall that, as the MBSFN scenario has lower intercell interference, it is expected to compensate the lower achievable bit rates resulting from the use of a longer guard time to avoid the effects of ISI.

In the system-level simulations, mobile users receive strong and weak bit blocks transmitted from BSs. Each block undergoes small- and large-scale fading and multicell interference. In terms of coverage or throughput, the SNR of each block is computed taking into account all the above impairments and based on the comparison between the reference SNR at a BLER of 1% and the evaluated SNR, it is decided whether or not the

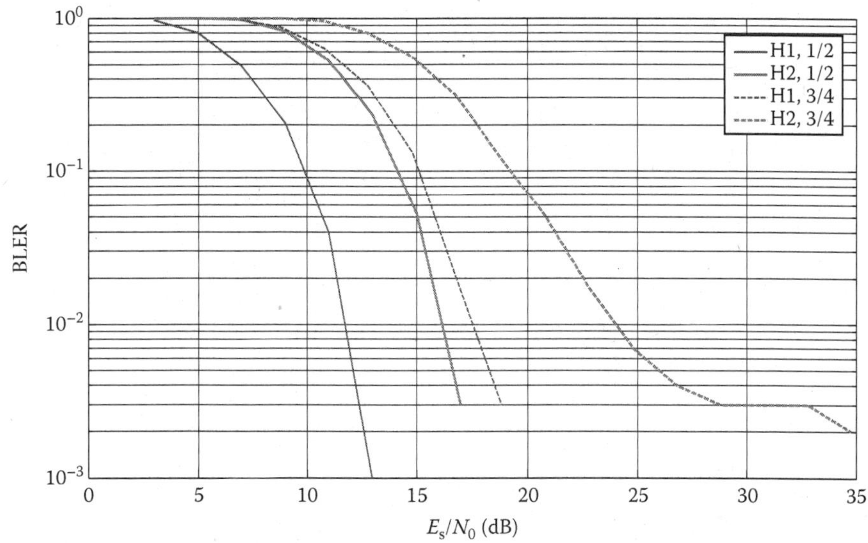

Figure 6.14 BLER versus E_s/N_0 for H16-QAM, MBSFN 30 km/h, reuse 1/3.

block is correctly received. This is done for all the transmitted blocks and for all users in all sectors of the 19 cells, during typically 10 min.

Figure 6.16 presents the coverage versus the fraction of the total transmitted power (denoted as E_c/I_{or}), for SC-PMP scenario where there is interference only from

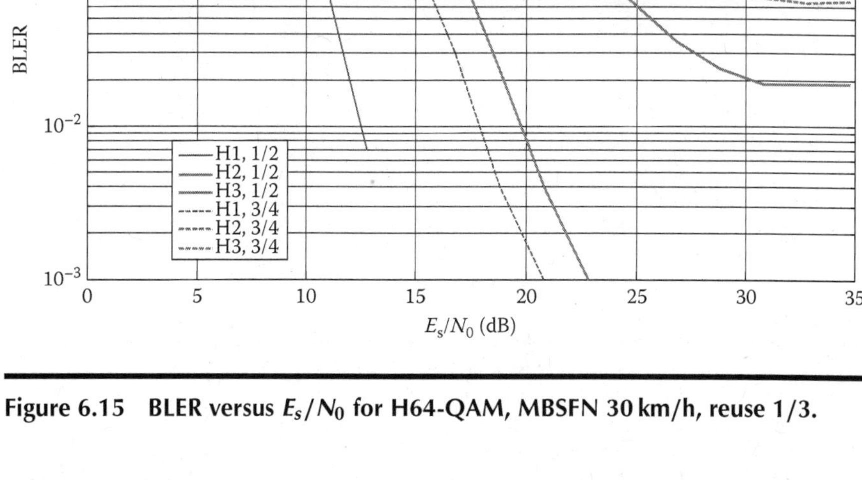

Figure 6.15 BLER versus E_s/N_0 for H64-QAM, MBSFN 30 km/h, reuse 1/3.

one-third of the sectors due to the frequency reuse of 1/3 (see Figure 6.2). All interfering sites transmit with the maximum power of 90% according to the parameters indicated in Table 6.2.

The cell radius is 750 or 1500 m, and we have separated strong blocks (H1) from weak blocks (H2) without including macrodiversity combining, denoted as 1RL, and also with macrodiversity combining the two best RLSs (2RL). Recall that the basic scenario SC-PMP does not include macrodiversity. The multicell interference is 90% of the maximum transmitted power in each site. Both coding rates considered were 1/2 and rate 3/4. Recall that it is necessary to assure the coverage of 95% for the strong bit blocks (H1). The only case that never reaches such coverage is when the cell radius has 1500 m and the coding rate is 3/4, even with a macrodiversity of 2RL. For coding rate 1/2, it is equivalent to have 1RL and cell radius of 750 m or 2RL and cell radius of 1500 m. For $E_c/I_{or} = 50\%$ and rate 1/2, the coverage of H1 is 98% and H2 coverage is around 89%. For the same E_c/I_{or} but with rate 3/4, the coverage values of H1 and H2 are 77% and 57%, respectively. According to the coverage results of this figure we can conclude that with coding rate 1/2 we can increase the cell radii from 750 to 1500 m as long as we include macrodiversity combining of the two best RLs. However for coding rate 3/4 we cannot double the cell radius and assure the designed coverage, even if we add macrodiversity of 2RL. We must choose between increasing the number of TV channels (or the TV channel bit rates) or increasing the cell radius.

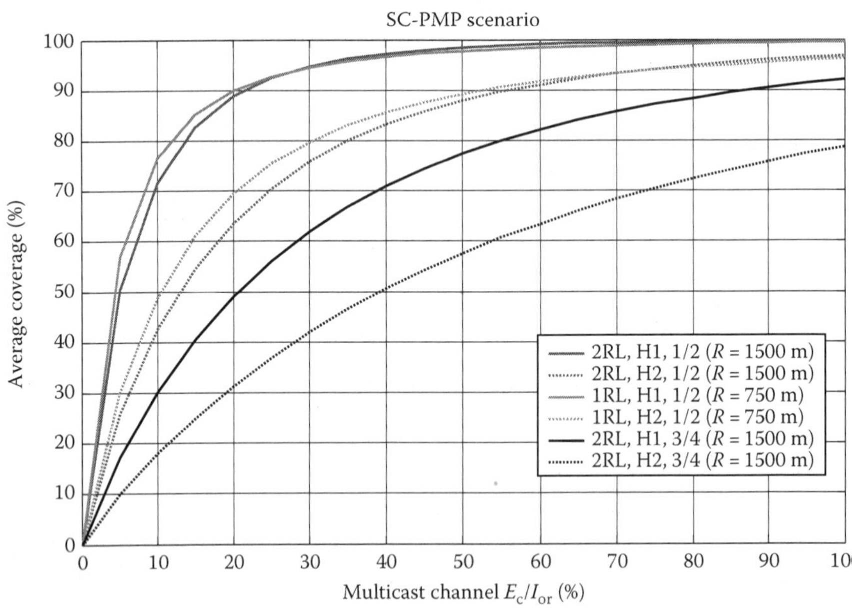

Figure 6.16 Average coverage (%) versus E_c/I_{or} for SC-PMP.

Figure 6.17 presents the average throughput distribution as a function of the E_c/I_{or} for the SC-PMP scenario network with and without macrodiversity for both cell radii 750 and 1500 m. We observe a considerable gain in throughput when macrodiversity (2RL) is considered compared to the single RL case. This is particularly true for the high coding rate 3/4. For E_c/I_{or} above 50% and, coding rate 3/4, the average throughput is above 256 kbps, which is the maximum throughput for rate 1/2. However, not all UEs are able to achieve such high throughput since users located at the cell borders never reach such high values of throughput.

Figure 6.18 considers the throughput distribution as a function of the distance between UEs and BS for the $E_c/I_{or} = 90\%$, with and without macrodiversity for the same cell radius of 1500 m and different coding rates. For the chosen E_c/I_{or}, both macrodiversity (2RL) and 1RL assure the maximum throughput for users located near the BS. As the distance between UEs and BS increases, then the throughput of 1RL decreases significantly. However, the decrease in throughput for rate 3/4 and mobile users at the cell borders is more obvious. It is observed that with macrodiversity (2RL), only for the rate 1/2, the throughput is almost the maximum, independent of the distance. For the high coding rate 3/4, a single RL only offers high throughput for users close to the BS. This figure complements the previous one where the average throughput for the cell area was presented. When we consider that approximately 3/4 of the users are located in the area between the cell border and half radius of the cell, then we realize the importance (for the average throughput) of the achieved throughput in this area. For users located above 1200 m, when there is macrodiversity 2RL, the coding rate 1/2 provides higher throughput than the rate 3/4. Again, we must opt between increasing the ISD (coverage) or increasing the number of TV channels (capacity).

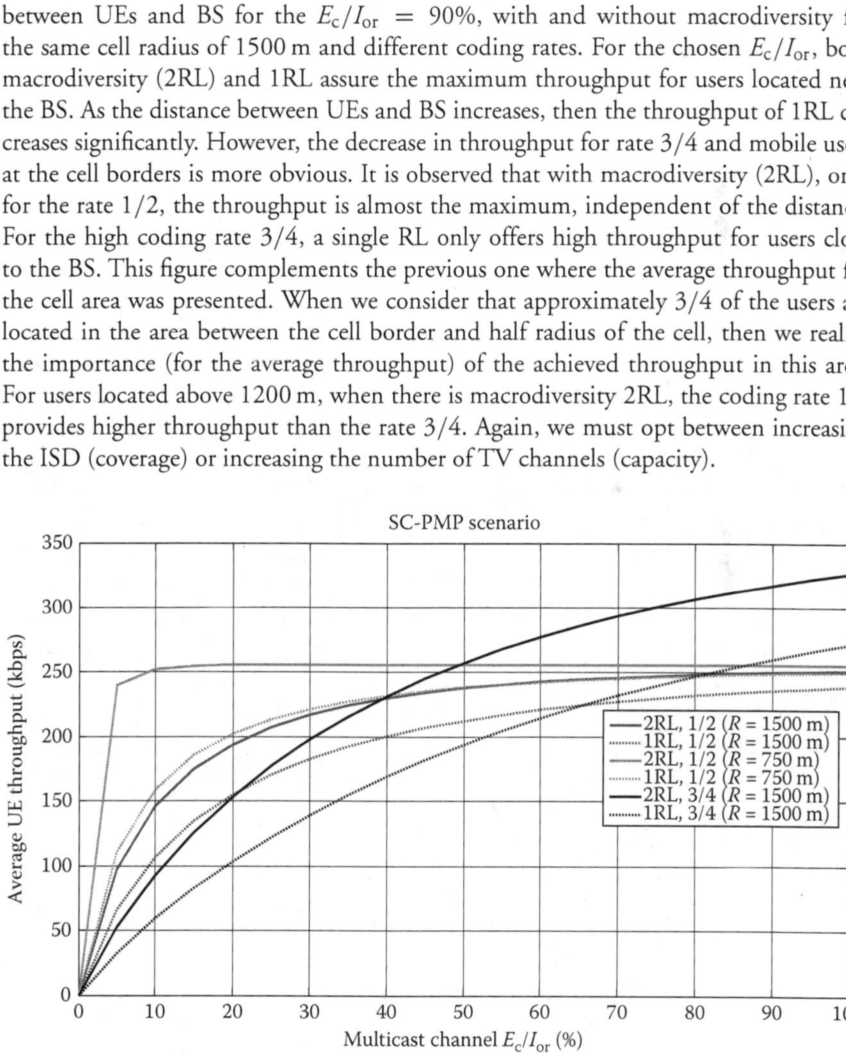

Figure 6.17 Average UEs throughput versus E_c/I_{or} for SC-PMP.

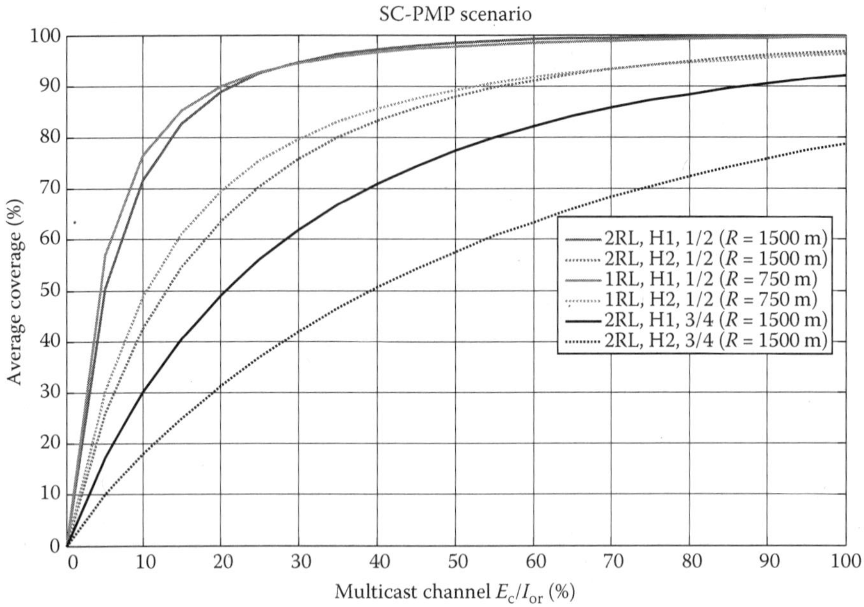

Figure 6.18 Throughput versus distance for SC-PMP.

In Figure 6.19, the coverage performance curves for MBSFN scenario, versus E_c/I_{or}, are presented for both cell radii, 750 and 1500 m. These results should be compared against the corresponding results of Figure 6.16 for the SC-PMP scenario. As expected there is a difference in the coverage between the two scenarios where MBSFN takes advantage of its lower intercell interference. The coverage values are above 95%, even for small values of E_c/I_{or} such as 25%. The only exception is the weak bit blocks (H2) of coding rate 3/4 and cell radius $R = 1500$ m. There is a coverage similarity between rate 1/2 with $R = 1500$ m and rate 3/4 with $R = 750$ m. This means that we can choose between increasing the coding rate (the average throughput) or increasing the coverage. When we increase both coverage and throughput at the same time, then we observe a decrease in coverage, in particular, H2 blocks.

Figure 6.20 presents the average throughput distribution as a function of E_c/I_{or} for the MBSFN scenario for cell radii of 750 and 1500 m. As expected we observe that, for the high coding rate 3/4, the throughput performance is higher for $R = 750$ m compared to $R = 1500$ m due to the higher coverage of H2 blocks. However, for E_c/I_{or} values equal to 90% (dedicated carrier to MBSFN) there is almost no difference between the two performance curves. The average throughput of 256 kbps, which is the maximum throughput for rate 1/2, is achievable for $E_c/I_{or} = 45\%$. This means that we can have two MBSFN carriers, each transmitting at least 18 TV channels. The operator must always choose between increasing the capacity, $2 \times 18 = 36$ TV channels, keeping $R = 750$ m, or increase the coverage to $R = 1500$ m with only 28 TV channels.

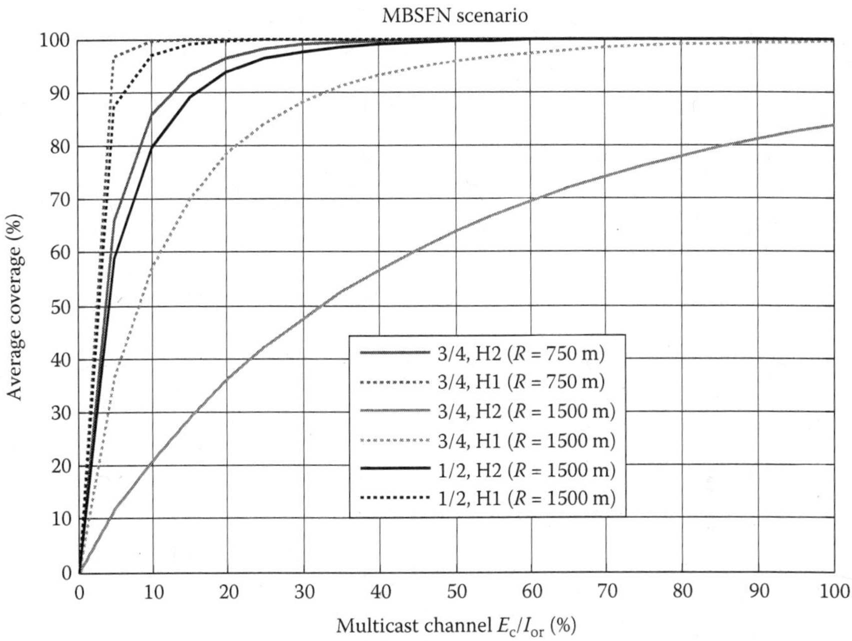

Figure 6.19 Average coverage (%) versus E_c/I_{or} for MBSFN.

In Figure 6.21, the throughput distribution as a function of the distance between UEs and BS is presented for $E_c/I_{or} = 90\%$ and considering both scenarios. For the chosen E_c/I_{or}, both scenarios assure the maximum throughput for users located near the BS. As the distance between UEs and BS increases, the throughput of SC (single cell with macrodiversity of 2RL) decreases significantly. However, the decrease in throughput for coding rate 3/4 than with rate 1/2 is more obvious. The SFN ensures not only higher coverage than SC due to lower intercell interference but also higher throughput. For coding rate 3/4, the throughput gain of MBSFN compared against SC-PMP indicates a value close to 1.5 considering the use of 16QAM multiresolution scheme. This corresponds to the situation where we measure the gain as the aggregate throughput in all the cell area (weighted by the distribution of users) under the two throughput performance curves. Note that, for broadcasting mobile TV channels, it is also important to increase the ISD to 3000 m in order to reduce the number of sites. This is only easily assured with MBSFN. Recall that, with ISD = 2R, when we double R, we also double the ISD, allowing for 50% reduction in the number of sites.

To obtain the 16QAM multiresolution gain over the single resolution with QPSK used by MBMS with the WCDMA technology specified in Release 6, we should compute the aggregate throughput in all the cell area with multiresolution and divide by the single-resolution aggregate throughput in the cell area. As the coverage of QPSK blocks becomes the same as the strong bit blocks of hierarchical 16QAM due to macrodiversity

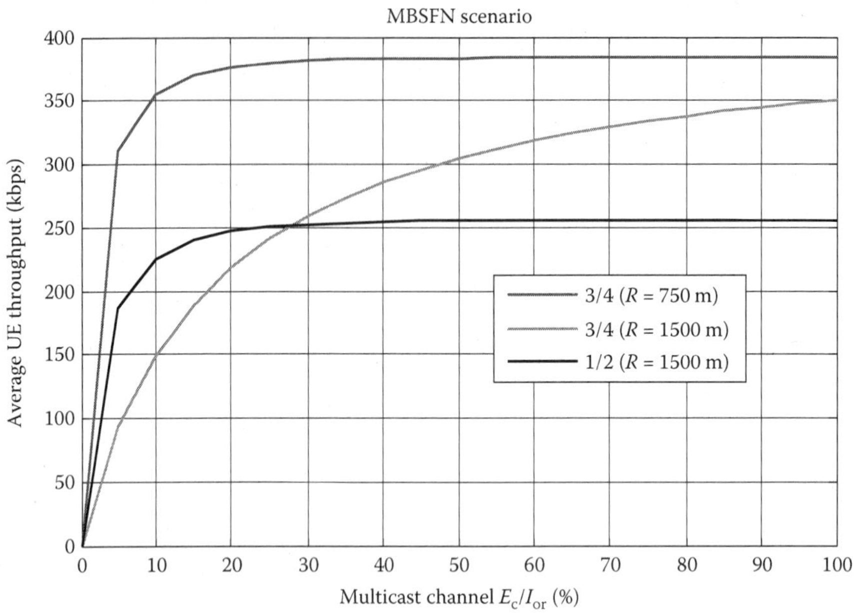

Figure 6.20 Average UEs throughput versus E_c/I_{or} for MBSFN.

combining, the comparison of the aggregate throughput is based on the different coverage of the weak bit blocks.

It is clear that the smallest throughput gain is achieved for coding rate $=1/2(256\text{ kbps})$. For this case, taking into account that the single-resolution throughput of QPSK (which is 128 kbps), the throughput gain is 2. The highest throughput gain is achieved for coding rate 3/4 (384 kbps) and the MBSFN scenario. For this case, the throughput gain is around 2.7. However, for the SC-PMP scenario with macrodiversity and the same coding rate 3/4, the throughput gain stays close to $2.7/1.2 = 2.25$ (see Figure 6.21).

6.4.2.2 Frequency Reuse 1 Results

We will now consider the frequency reuse 1, where all the available band is used everywhere. The number of TV channels per sector reaches 18.75×3 and 28.125×3 TV channels, with coding rate $1/2$ and $3/4$, respectively. We want to check whether, due to the higher intercell interference in the cellular OFDM/OFDMA system, the QoS degrades.

In Figures 6.22 and 6.23, we present the BLER versus E_s/N_0 for the hierarchical 16QAM and 64QAM in the MBSFN scenario with the MBSFN propagation channel. Comparing Figures 6.22 and 6.23 with Figures 6.14 and 6.15 indicates that, for the MBSFN, reuse 1 provides slightly better BLER performance results than the frequency reuse of $1/3$. This is due to higher blocks of bits allowing a better turbo iterative decoding.

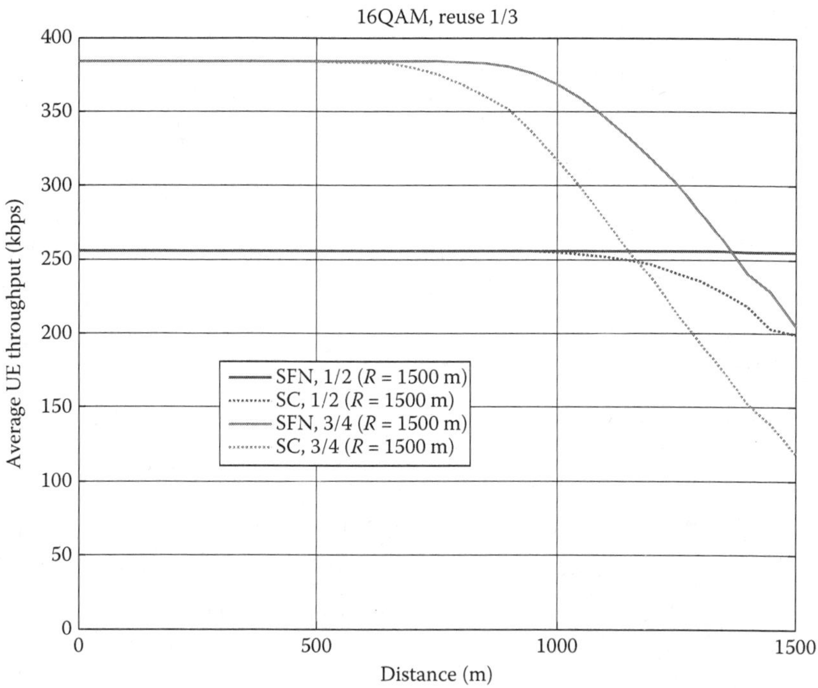

Figure 6.21 Throughput versus distance for both scenarios.

However, in terms of SNR required to achieve the BLER $= 0.01$, reuse 1 demands for higher SNR.

Figures 6.24 and 6.25 present the coverage versus the fraction of the total transmitted power (E_c/I_{or}), for cell radius $R = 750$ m, hierarchical 64QAM, rate 1/2, and both scenarios and frequency reuse of 1/3 and 1, respectively. 64QAM with rate 1/2 has maximum bit rates of 384 kbps per TV channel such as 16QAM with rate 3/4. In Figure 6.24 there is interference only from one-third of the sectors due to the frequency reuse of 1/3. However, the coverage of weak bit blocks is very low, independent of the scenario but especially for the SC scenario. As a consequence of this very low coverage, the maximum throughput achievable with 64QAM rate 1/2 will be lower than with 16QAM rate 3/4 (see corresponding results of Figure 6.19). Recall that all interfering sites transmit with the maximum power of 90%, independent of the scenario. As expected, the MBSFN scenario provides always better performance results than the SC scenario.

The comparison between Figures 6.25 and 6.24 indicates a substantial decrease in the coverage of all blocks of bits of 64QAM with frequency reuse 1, due to the higher intercell interference levels. Weak bit blocks have null coverage independent of the scenario, but especially for SC. As a consequence, the maximum throughput achievable with 64QAM rate 1/2 with reuse 1 will be much lower than with reuse 1/3, for the same modulation and coding rate.

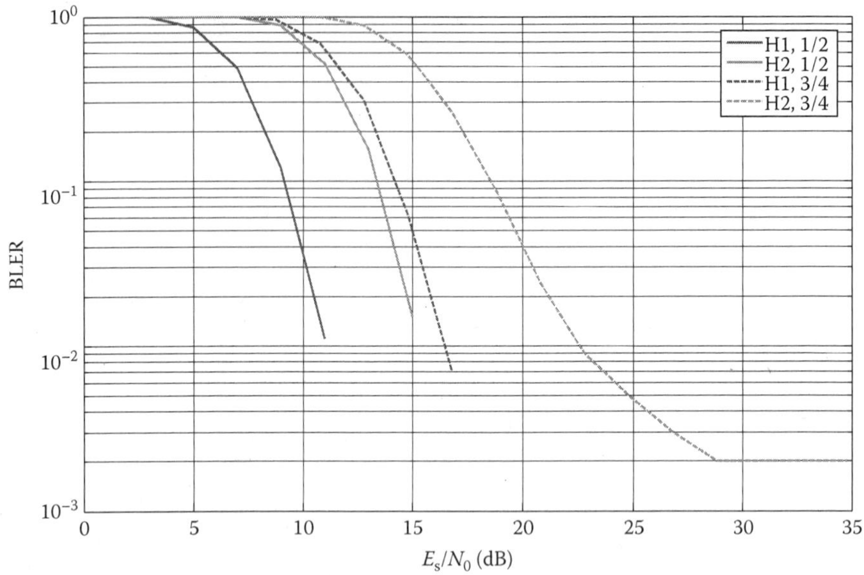

Figure 6.22 BLER versus E_s/N_0, H16QAM and reuse 1.

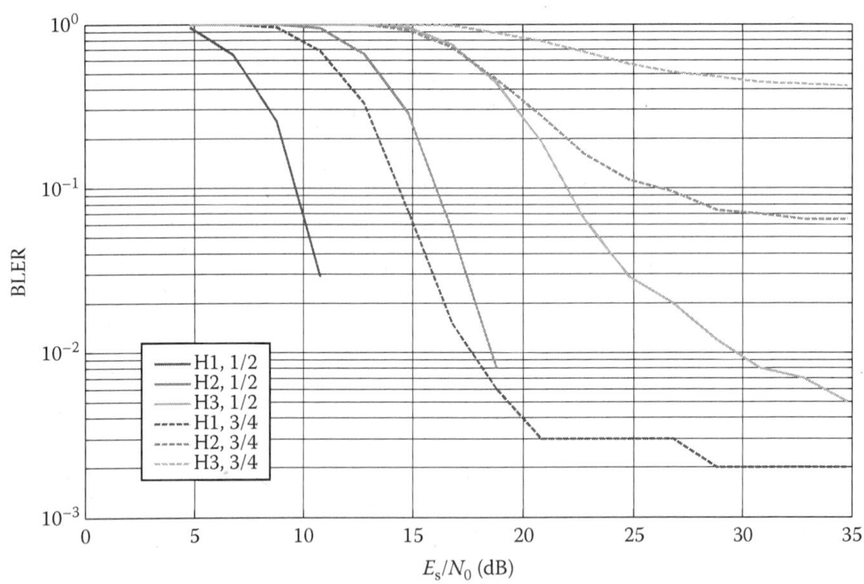

Figure 6.23 BLER versus E_s/N_0, H64QAM and reuse 1.

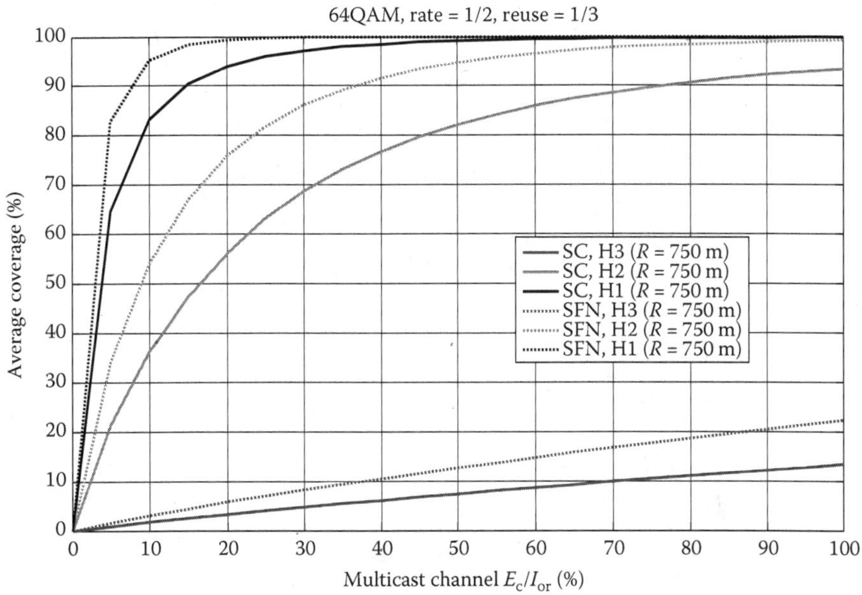

Figure 6.24 Coverage versus E_c/I_{or}, 64QAM, rate 1/2, and reuse 1/3.

In Figure 6.26, the throughput distribution as a function of the distance between UEs and BS is presented for $E_c/I_{or} = 90\%$ and considering both modulations and frequency reuse. The chosen ISD is 3000 m ($R = 1500$ m). As expected, due to the very low 64QAM coverage results, for $R = 750$ m, the throughput achieved with 16QAM is higher than that with 64QAM, even for mobile users close to the BS. However, for users close to the BS, the throughput reduction of 16QAM is higher, when going from reuse 1/3 to reuse 1, than with 64QAM. This is explained by the presence of H3 blocks in 64QAM that, in spite of having reduced coverage, these blocks are less susceptible to the additional intercell interference of reuse 1. Anyway, 16QAM provides always better throughput than 64QAM for reuse 1/3 (the only exception is the cell border). Nevertheless, with reuse factor 1, 64QAM offers higher throughput for users located between the middle and the cell border.

6.5 Summary and Conclusions

We have evaluated the use of multiresolution with hierarchical modulations for the evolved-MBMS (E-MBMS), to be standardized in the next Release for the LTE-Advanced. In this chapter, link- and system-level simulations of LTE multicellular networks considering broadcast/multicast transmissions using OFDM/OFDMA-based LTE technology were presented. We evaluate the capacity, in terms of number of TV channels, with given bit rates or total spectral efficiency and coverage.

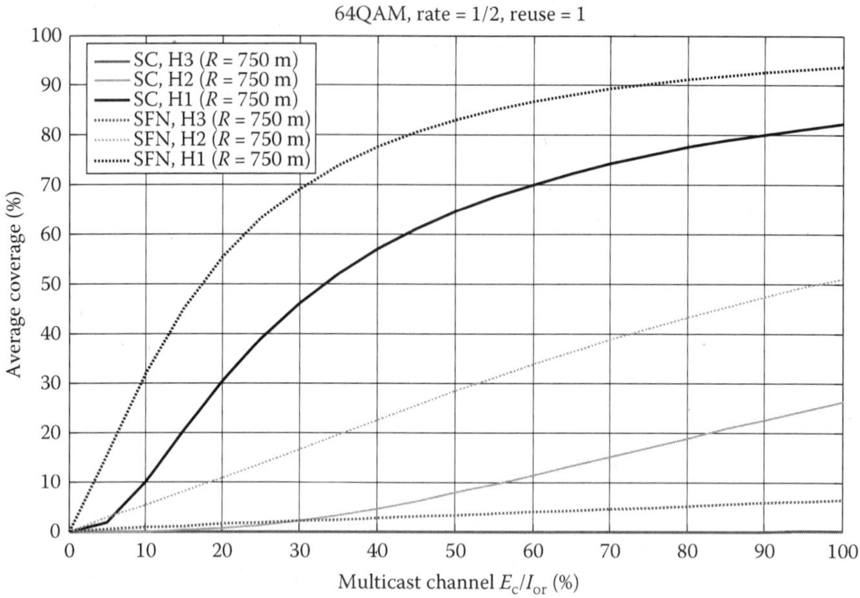

Figure 6.25 Coverage versus E_c/I_{or}, 64QAM, rate 1/2, and reuse 1.

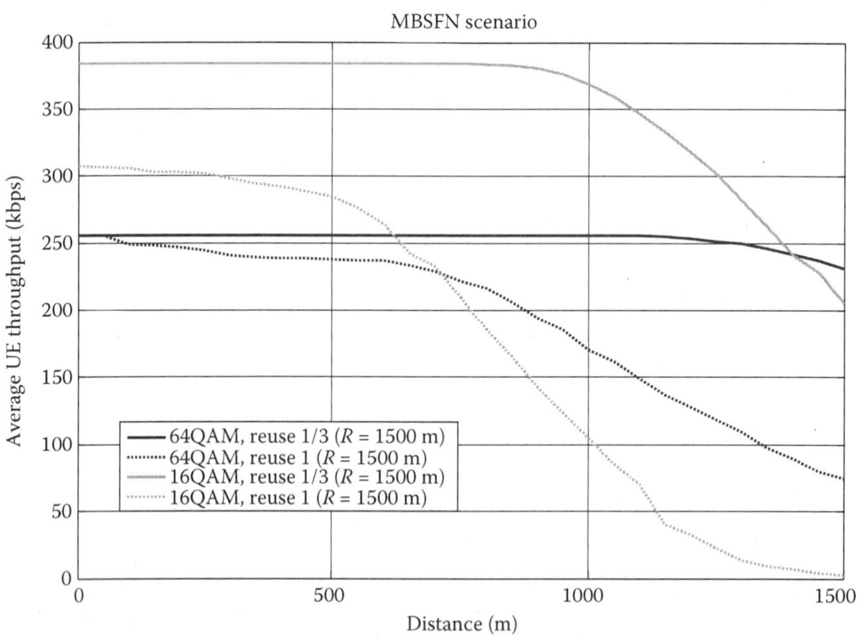

Figure 6.26 Throughput distribution for different modulations and reuses.

Taking the 95% coverage as reference, the evaluation of the achievable capacity gain (number of transmitted mobile TV channels for WCDMA and LTE) is next presented and discussed. The ISD gain is also presented allowing for a substantial reduction in the number of cell sites when LTE will replace WCDMA.

The scenarios based on the use of SFN with the MBSFN channel and the SC-PMP network considering the Vehicular A propagation channel were both evaluated for 16QAM/64QAM hierarchical modulations, and compared against the present MBMS network, that is, WCDMA based. In general, we can state that multiresolution is suitable for any of the analyzed MBSFN and SC-PMP scenarios. Indeed, it works well in any single-cell scenario without macrodiversity combining or in multicells with macrodiversity.

Table 6.3 shows that the SC-PMP scenario without macrodiversity (1RL), due to multiresolution, increases the channel bit rate of each TV channel to 256 kbps (compared to 128 kbps provided by QPSK with single resolution) for users close to the BS and ISD equal to 1500 m. When we add macrodiversity (2RL) of the best two RLs, the multiresolution schemes become less sensitive to the used channel bit rates and it is possible to increase the channel coding rate, keeping the same ISD, or to increase the ISD to 3000 mt, keeping the channel bit rate. The operator must choose between the trade-off of increasing capacity or coverage (see Table 6.3). It is not possible to increase both at the same time.

In the MBSFN scenario, due to smaller intercell interference and substantial macrodiversity combining, in order to achieve higher multiresolution gains, it is suggested to increase both the channel bit rates from 256 to 384 kbps (i.e., to change the channel coding rate from 1/2 to 3/4). In addition, the ISD can also be extended from 1500 to 3000 m. Considering the transmission of 18.75 TV channels using all the 10 MHz BW, for the high channel bit rate of 384 kbps, the spectral efficiency achieved per cell sector is 0.72 bps/Hz/cell. This value of spectral efficiency is valid for users at the cell edge. The ISD associated to this spectral efficiency is 3000 m. Alternatively, 28.125 TV channels with 256 kbps could be transmitted at the same time, as indicated in Table 6.3.

Table 6.4 shows the capacity of MBMS single resolution, taking into account results for the standard MBMS normalized in Release 6 presented in Refs. [Correia et al. 2007;

Table 6.3 Capacity Values for MBSFN and SC-PMP Scenarios using 16QAM Hierarchical Multiresolution OFDMA (BW = 10 MHz)

QoS (kbps)	*#TV Channels*	*Spectral Efficiency (bps/Hz/cell)*	*ISD (m)*	*Scenario*
256	18.75	0.48	1500	SC-PMP 1RL
256	18.75	0.48	3000	SC-PMP 2RL
256	28.125	0.72	3000	MBSFN
384	18.75	0.72	3000	MBSFN

Table 6.4 Capacity Values for QPSK Single Resolution, WCDMA Scheme for BW = 10 MHz

QoS (kbps)	*#TV Channels*	*Spectral Efficiency (bps/Hz/cell)*	*ISD (m)*	*Scenario*
256	14	0.358	1000	SC-PMP 2RL (2 × 5 MHz)

Hartung et al. 2007], for the same SC-PMP scenario with macrodiversity of two RLs, with two carriers, each one having BW = 5 MHz.

The comparison between Tables 6.3 and 6.4 is not straightforward due to the different ISDs employed. However, it is possible to draw a capacity gain of at least 2 between hierarchical 16QAM and QPSK (note that higher ISD is an advantage for broadcasting).

The ISD gain is related to the decrease in the number of sites due to longer ISD. We have checked in this chapter that, for the MBSFN scenario, we can increase the cell radius (the ISD is the double of the cell radius) from 750 to 1500 m, without any decrease in the average throughput, as long as the fraction of the total transmitted power dedicated to E-MBMS reaches 90%. This corresponds to a reduction in the number of sites of 50%. An alternative is to use the same carrier for E-MBMS traffic and other unicast type of traffic, without increasing the ISD.

Appendix A

WCDMA Channel Model

A.1 Channel Characterization

In mobile communications, the channel is one of the most limiting factors for achieving a reliable transmission. Owing to the mobility of the transmitter, of the receiver, or both, the resulting channel affects the transmitted signal that will basically undergo two effects: slow fading (shadowing) and fast fading. The slow fading is mainly caused by the terrain contour between the transmitter and the receiver, being directly related to the presence of obstacles in the path of the signal. This effect can be compensated with power control schemes that take into account the mean signal strength over a small period of time and will not be studied in this thesis.

Fast fading is caused by the reflection of the signal in various objects (buildings, trees, vehicles, etc.) that originate multiple replicas of the signal reaching the receiver through different paths. These replicas arrive at the receiver with different delays and attenuations, superimposed in such a way that they will interfere with each other, either constructively or destructively. Owing to the mobility of the transmitter or receiver and of the surrounding objects themselves, the replicas are subject to variations on their paths, and hence in their delays and attenuations, leading to great oscillations on the envelope of the received signal.

Since the multiple replicas of the signal arrive with different delays there will be temporal dispersion of the received signal. This means that if an impulse is transmitted, the received signal will have a noninfinitesimal duration and thus will not be an impulse. This temporal dispersion can be represented using a PDP, $P(\tau)$, which represents the average received power as a function of the delay τ. Figure A.1 shows an example of a PDP, where the parameter P_{max} denotes the maximum value. The other parameters shown in the figure will be described later.

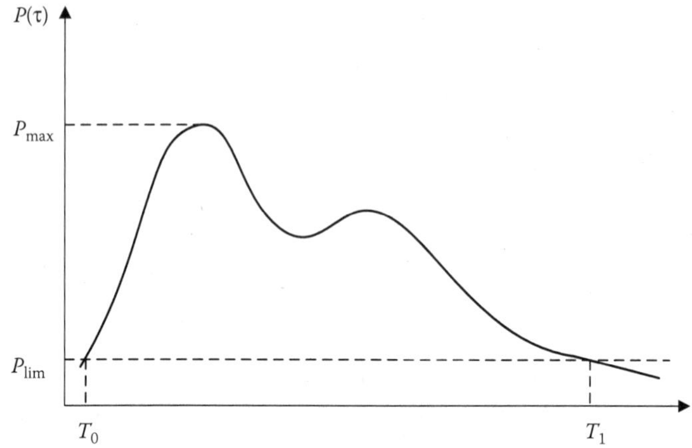

Figure A.1 Continuous power delay profile.

In order to better describe the behavior of the channel, a few important parameters are defined, such as the total received power, the mean delay, and the delay spread. The total received power is formulated as

$$P_{tot} = \int_0^{\infty} P(\tau)\,d\tau \cong \int_{T_0}^{T_1} P(\tau)\,d\tau, \tag{A.1}$$

where T_0 is the time instant when the average power rises above the threshold values P_{lim} while T_1 is the time instant when the average power drops below the threshold. The mean delay is given by

$$\bar{\tau} = \frac{1}{P_{tot}} \int_0^{\infty} \tau P(\tau)\,d\tau, \tag{A.2}$$

and finally, the delay spread is computed with

$$\sigma_{\tau} = \sqrt{\frac{1}{P_{tot}} \int_0^{\infty} \tau^2 P(\tau)\,d\tau - \bar{\tau}^2}. \tag{A.3}$$

The temporal dispersion of the signal leads to the existence of ISI that can be opposed by lowering the transmission rate and using equalizers in the receiver.

Two other quantities can be defined for describing the channel, one is the time domain and the other is the frequency domain. The first is the coherence time, T_{coe}, and is related to the mobility of the terminal. This mobility refers to the fact that two signals in different time instants are statistically independent if their separation is greater than

T_{coe}. The other parameter is the coherence BW, B_{coe}, which is related to multipath. The multipath implies that two signals with different frequencies are statistically independent if their separation is greater than B_{coe}.

The Doppler frequency shift and T_{coe} form a dual pair that allows studying the rapid variations of the channel. According to Ref. [Steele 1994], if T_{coe} is defined as the time over which the time correlation function is above 0.5, then it can be expressed as

$$T_{coe} \cong \frac{9}{16\pi f_D}, \tag{A.4}$$

where f_D is the maximum Doppler frequency caused by the existence of movement. In Table A.1, the maximum Doppler shifts and T_{coe} values for typical velocities in UMTS for the 2000 MHz band are shown. Since in UMTS the chip time is approximately 260 ns [3GPP 2004c], which is smaller than any T_{coe} presented in the table, we can conclude that this system behaves as a slow-varying channel at any realistic velocity.

As for the coherence BW, it is inversely proportional to the delay spread and both can be used to study the time dispersion caused by the channel. According to Ref. [Lee 1989], the coherence BW depends on the correlation coefficient between two sinusoids separated by B_{coe}. Although there is no exact expression, an approximate value of B_{coe} for a correlation coefficient of 0.9 is presented in Ref. [Lee 1989], which is expressed as

$$B_{coe} \cong \frac{1}{50\sigma_\tau}. \tag{A.5}$$

It is necessary to take into account that since the channel is time varying, σ_τ changes with time, and thus Equation A.5 is an estimate for the mean value of B_{coe}.

The coherence BW can be used to classify a channel in wideband or narrowband. All channels whose BW is greater than B_{coe} are defined as wideband (or frequency selective) and a narrowband (or frequency flat) channel is defined if otherwise. Using the temporal dispersion values for each typical environment described in Ref. [ETSI 1998a], and applying Equation A.5 for computing B_{coe}, we obtain for the UMTS system the values presented in Table A.2. In this table, two different types of channels are

Table A.1 Values for f_D and T_{coe}, for Different Mobility Scenarios

Velocity (km/h)	*Doppler Frequency f_D (Hz)*	*Coherence Time T_{coe} (μs)*
3	6	32,230
10	19	9669
50	93	1934
120	222	806
250	463	387
500	926	193

Table A.2 B_{coe} Values for Scenarios Defined in Ref. [ETSI 1998a]

	Channel A		Channel B	
Environment	Delay Spread, σ_τ (ns)	Coherence BW B_{coe} (kHz)	Delay Spread, σ_τ (ns)	Coherence BW B_{coe} (kHz)
Indoor	35	571	100	200
Pedestrian	45	444	750	27
Vehicular	370	54	4000	5

defined for each environment, representing two typical delay spread cases that can occur frequently for each environment. Channel A corresponds to the low delay spread case while Channel B is a medium delay spread case. Since the UMTS BW is approximately 5 MHz [3GPP 2004c], we can conclude that according to the obtained values it is a wideband system.

In a frequency-selective channel it is possible to resolve the multipath components and provide the receiver with various signal replicas undergoing independent fadings. In this way diversity is obtained. In fact, taking into account that the temporal resolution is the inverse of the signal's BW ($1/B_{wd}$), and the maximum temporal delay is T_m (corresponding to the width of the PDP), it is possible to distinguish $T_m \cdot B_{wd}$ components of the multipath signal. In order to take advantage of the frequency diversity, the signal processing can be done using a RAKE receiver [Price and Green 1958] as will be shown later on.

A.2 Channel Model Construction

To analyze or simulate the performance of a system in a specific channel, it is necessary to have some kind of model for that channel. There are two types of channel models: directional and nondirectional. In the former case, temporal and angular dispersion are considered, which is essential for adaptive antennas, while in the latter case, only temporal dispersion is taken into account. In this thesis, only nondirectional channels are used.

The physical structure of a certain environment can be defined as a set of reflectors/re-radiators arranged throughout the terrain at a certain distance from the mobile, which originates a set of waves that reach the mobile terminal with random power and delays. When the physical structure varies, so does the channel's impulse response. For the characterization of the channel, the Wide-Sense Stationary Uncorrelated Scaterers (WSSUS) model can be used [Bello 1963]. In this model, it is assumed that the correlation properties of the channel are stationary, that is, the fading properties remain unaltered for a determined period of time and signal replicas arriving with different delays can be considered as being subject to uncorrelated fading coefficients.

Although the PDP of a channel is continuous, the radio channel can also be modeled by a linear filter with a discrete time-varying impulse response. According to this model,

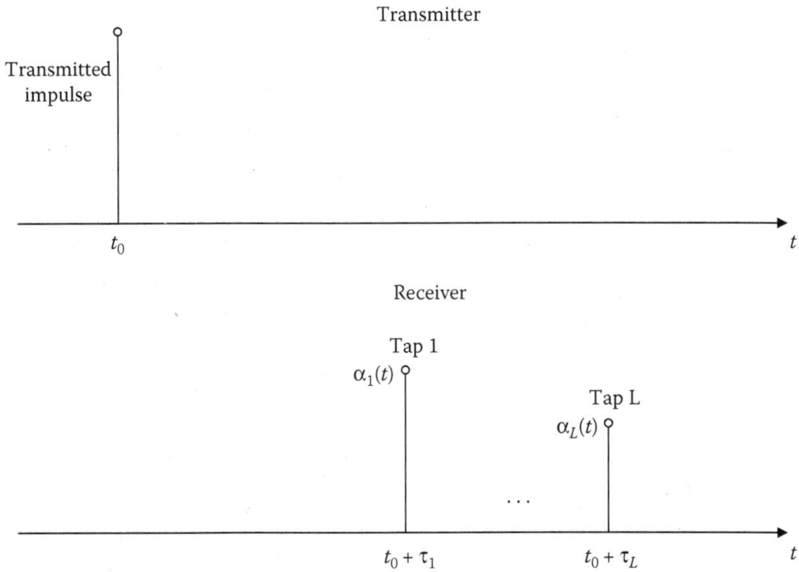

Figure A.2 Discrete time channel model.

shown in Figure A.2, there are several different taps, corresponding to replicas of the transmitted impulse, arriving with different delays and suffering different fadings.

Hence, if an impulse, $\delta(\tau)$, is applied at time t_0, then the received signal can be written as

$$h_c(t, t_0) = \sum_{l=1}^{L(t)} \alpha_l(t)\delta(t - \tau_l(t) - t_0), \tag{A.6}$$

where $\alpha_l(t) = |\alpha_l(t)|e^{j\phi_l(t)}$ represents the amplitude and phase shift of the received impulse in path l, $\tau_l(t)$ the respective delay, and $L(t)$ is the number of total received discrete impulses (which in the most general case can be variable).

Considering that for each delay, τ_l, a large number of scattered waves arrive from random directions then, from the Central Limit Theorem, $\alpha_l(t)$ can be modeled as a complex Gaussian process with zero mean. This means that the phase $\phi_l(t)$ will follow a uniform distribution in the interval $[0\ 2\pi]$ and the fading amplitude, $|\alpha_l(t)|$, will follow a Rayleigh distribution whose PDF is expressed as

$$p_R(|\alpha|) = \frac{2|\alpha|}{E\left[|\alpha|^2\right]} \exp\left(-\frac{|\alpha|^2}{E\left[|\alpha|^2\right]}\right). \tag{A.7}$$

If there is also a significant nonfading LOS component with amplitude A arriving at the receiver, then $\alpha_l(t)$ will be a complex Gaussian process with nonzero mean and thus

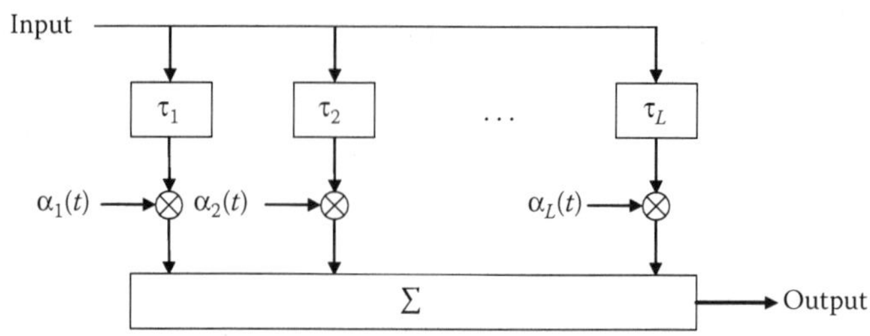

Figure A.3 Wideband channel modeled using a TDL.

the fading amplitude, $|\alpha_l(t)|$, will have a Rice PDF, which is expressed as

$$p_R(|\alpha|) = \frac{2|\alpha|}{E\left[|\alpha|^2\right]} I_0\left(\frac{2|\alpha|A}{E\left[|\alpha|^2\right]}\right) \exp\left(-\frac{|\alpha|^2 + A^2}{E\left[|\alpha|^2\right]}\right). \tag{A.8}$$

If the impulse response of the channel is considered time invariant, its baseband representation can be written as

$$h_c(t) = \sum_{l=1}^{L} \alpha_l(t)\delta(t - \tau_l). \tag{A.9}$$

This type of model can be represented using a TDL, as shown in Figure A.3.

The time-invariant model can be used for studying WCDMA systems similar to UMTS since, as already shown, for realistic velocities the channel has a very slow variation. This model is employed in this thesis. Using the TDL of Figure A.3, the input will be a WCDMA signal modulated at the base station or mobile terminal. The number of taps, respective average fading powers, and delays are generated according to the scenario being studied, which can be one of the reference environments defined in Ref. [ETSI 1998a]. In these reference environments, the delays may not coincide with simulation sampling time instants and thus some conversion of the models has to be performed, as was studied in Ref. [Silva et al. 2003].

According to the WSSUS model, the fading coefficients are uncorrelated between different taps. The fading in each tap can be modeled using a Rayleigh distribution and their time variation can be described using a Doppler power spectrum based on Jakes' model [Jakes 1974] (shown in Figure A.4), which can be described using

$$A_0(f) = \begin{cases} \frac{1}{\pi f_D \sqrt{1-(f/f_D)^2}}, & |f| \leq f_D \\ 0, & |f| > f_D \end{cases} \tag{A.10}$$

where f_D is the maximum Doppler frequency caused by the existence of movement. This frequency is given by

$$f_D = \frac{v}{\lambda_c} \tag{A.11}$$

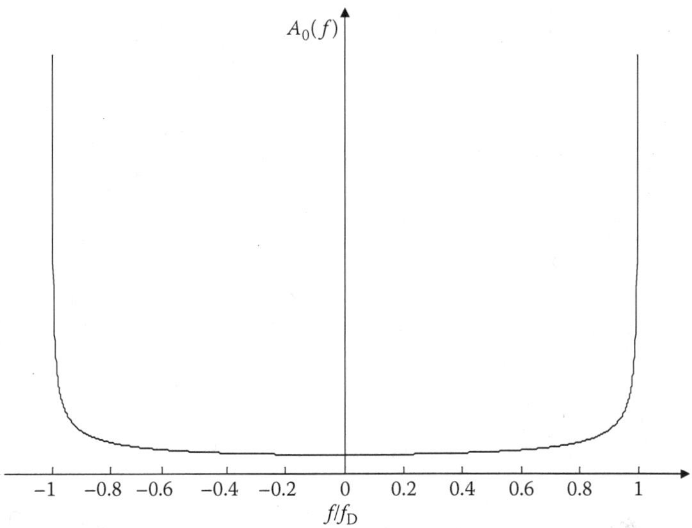

Figure A.4 Jakes' spectrum.

with λ_c being the wavelength of the carrier (whose frequency is f_c) and v is the vehicle speed.

It is important to note that it is possible to employ other types of channel models such as a finite state channel model, as discussed in Ref. [Silva et al. 2004b]. In this type of model, the channel is described by a set of probabilities that relate the input signal to the output signal. Although in terms of simulation this model is more efficient, the computation of the probabilities required for its description is computationally time consuming and is dependent on the specific environment.

Table A.3 Delay and Power Profile of Vehicular A Environment [ETSI 1998a]

Tap, i	*Delay, τ_i (ns)*	*Power, Ω_i^2 (dB)*
1	0	0
2	310	−1
3	710	−9
4	1090	−10
5	1730	−15
6	2510	−20

Table A.4 Delay and Power Profile of Indoor A environment [ETSI 1998a]

Tap, i	*Delay,* τ_i *(ns)*	*Power,* Ω_i^2 *(dB)*
1	0	0
2	50	−3
3	110	−10
4	170	−18
5	290	−26
6	310	−32

Several performance results presented in this book were obtained for some of the 3GGP reference environments published in Ref. [ETSI 1998a], namely Vehicular A and Indoor, A whose tap delays and respective powers are shown in Tables A.3 and A.4.

References

3GPP, 25.101-v6.6.0,User equipment (UE) radio transmission and reception (FDD), December 2004c.

3GPP, 25.211-v6.3.0, Physical channels and mapping of transport channels onto physical channels (FDD), December 2004a.

3GPP, 25.212-v5.2.0, Multiplexing and channel coding (FDD), 2003b.

3GPP, Deployment aspects, 3GPP 25.943 v5.1.0, Sophia Antipolis, France, 2002.

3GPP, 25.213-v6.1.0, Spreading and modulation (FDD), December, 2004b.

3GPP, 25.214-v5.5.0, Physical layer procedures (FDD), 2003c.

3GPP, 25.803, V6.0.0, S-CCPCH performance for MBMS, September 2005b.

3GPP, 25.814 v7.1.0, Release 7, Technical specification group radio access network; physical layers aspects for evolved (UTRA), September 2006d, http://www.3gpp.org

3GPP, 25.892 v6.0.0, Release 6, Feasibility study for orthogonal frequency division multiplexing (OFDM) for UTRAN enhancement, June 2004, http://www.3gpp.org

3GPP, 25.905 version 7.2.0, Release 7, Feasibility study on improvement of the multimedia broadcast multicast service (MBMS), January 2008, http://www.3gpp.org

3GPP, 25.913 version 7.3.0, Release 7, Requirements for evolved UTRA (E-UTRA) and evolved UTRAN (E-UTRAN), March 2006, http://www.3gpp.org

3GPP, 36.913, V8.0.0, Requirements for further advancement for E-UTRA (LTEAdvanced), June 2008a.

3GPP, 36.942 Release 8, Evolved universal terrestrial radio access (E-UTRA); radio frequency (RF) system scenarios, 2008b, http://www.3gpp.org

Abe, T. and Matsumoto, T., Space–time turbo equalization in frequency-selective MIMO channels, *IEEE Transactions on Vehicle Technology*, 52, 3, 2003.

Ahmadi, S., An overview of next-generation mobile WiMAX technology, *IEEE Communications Magazine*, 47, 6, 84–98, 2009.

Al-Dhahir, N. and Sayed, A. H., The finite-length multi-input multi-output MMSE-DFE, *IEEE Transactions on Signal Processing*, 48, 2921–2936, 2000.

Alamouti, S. M., A simple transmitter diversity scheme for wireless communications, *IEEE JSAC*, 1451–1458, 1998.

Andrews, J. G., Gosh, A., and Muhamed, R., *Fundamentals of WiMAX: Understanding BroadbandWireless Networking*, Prentice-Hall, NJ, 2007.

Ansari, A., Dutta, S., and Tseytlin, M., S-WiMAX: Adaptation of IEEE 802.16e for mobile satellite services, *IEEE Communications Magazine*, 47, 6, 150–155, 2009.

Araújo, T. and Dinis, R., Efficient detection of zero-padded OFDM signals with large blocks *IASTED SIP'06*, Honolulu, HI, 2006.

Armstrong, J., New OFDM peak-to-average power reduction scheme, *IEEE VTC'2001(Spring)*, 1, 756–760, 2001.

Astély, D., et al., A future radio-access framework, *IEEE JSAC*, 24, 3, 2006.

Astély, D., et al., LTE: The evolution of mobile broadband, *IEEE Communications Magazine*, 44–51, 2009.

Bahl, L. R., Cocke, J., Jeinek, F., and Raviv, J., Optimal decoding of linear codes for minimizing symbol error rate, *IEEE Transactions on Information Theory*, IT-20, 248–287, 1974.

Bello, P. A., Characterization of randomly time-variant linear channels, *IEEE Transactions on Communications System*, *CS.11*, 5, 360–393, 1963.

Benvenuto, N. and Tomasin, S., On the comparison between OFDM and single carrier with DFE using a frequency domain feed-forward filter, *IEEE Transactions on Communications*, 50, 947–955, 2002a.

Benvenuto, N. and Tomasin, S., Block iterative DFE for single carrier modulation, *IEEE Electronic Letters*, 39, 19, 1144–1145, 2002b.

Benvenuto, N. and Tomasin, S., Iterative design and detection of a DFE in the frequency domain, *IEEE Transactions on Communications*, 53, 11, 1867–1875, 2005.

Berrou, C., Glavieux, A., and Thitimajshima, P., Near Shannon limit error correcting coding and decoding: Turbo-codes, *Proceedings of the IEEE International Conference on Communications*, pp. 1064–1070, Geneva, Switzerland, May 23–26, 1993.

Bingham, J. A. C., Multicarrier modulation for data transmission: An idea whose time has come, *IEEE Communication Magazine*, 28, 5, 5–14, 1990.

Bogineni, K., et al., LTE Part II: Radio access, *IEEE Communications Magazine*, 40–42, 2009.

Bölcskei, H., Gesbert, D., and Paulraj, A. J., On the capacity of OFDM-based spatial multiplexing systems, *IEEE Transactions on Communications*, 50, 225–234, 2002.

Boudreau, G., et al., Interference coordination and cancellation for 4G networks, *IEEE Communications Magazine*, 47, 4, 74–81, 2009.

Cavers, J. K., An analysis of pilot symbol assisted modulation for Rayleigh fading channels, *IEEE Transactions on Vehicular Technology*, 40, 4, 686–693, 1991.

Chiurtu, N., Rimoldi, B., and Telatar, E., Dense multiple antenna systems, *Proceedings of the IEEE Information TheoryWorkshop*, 2001, pp. 108–109.

Chu, D., Poly phase codes with good periodic correlation properties, *IEEE Transactions and Information Theory*, 18, 4, 531–532, 1972.

Cimini, L., Analysis and simulation of a digital mobile channel using orthogonal frequency division multiplexing, *IEEE Transactions on Communication*, 33, 1985.

Cimini, L., Jr. and Sollenberger, N., Peak-to-average power reduction of an OFDM signal using partial transmit sequences, *IEEE Communication Letters*, 4, 3, 86–88, 2000.

Coldrey, M. (Tapio) and Bohlin, P., Training-based MIMO systems: PartI—performance comparison, *IEEE Transactions on Signal Processing*, 12, 3, 226–229, 2005.

Cooley, J. W. and Tukey, J. W., An algorithm for the machine calculation of complex Fourier series, *Mathematics of Computation*, 19, 90, 297–301, 1965.

Correia, A., Silva, J., Souto, N., Silva, L., Boal, A., and Soares, A., Multi-resolution broadcast/multicast systems for MBMS, *IEEE Transactions on Broadcasting*, 53, 1, 224–234, 2007.

Correia, A., Souto, N., Silva, J., and Soares, A., Air interface enhancements for MBMS, Chapter 17, in *Handbook on Mobile Broadcasting*, B. Furht and S. Ahson (Eds), CRC Press, Taylor & Francis Group, New York, 2008.

Cover, T., Broadcast channels, *IEEE Transactions on Information Theory*, IT-18, 2–14, 1972.

da Silva, M. M., Correia, A. M. C., and Dinis, R., Interference suppression consisting of pre-distortion filtering and selective transmit diversity,*Wireless Personal Communications*, 47, 2, 2008, Springer, the Netherlands, DOI: 10.1007/s11277-008-9456-7.

Dahlman, E., et al., *3G Evolution: HSPA and LTE for Mobile Broadband*, 2nd edition, Academic Press, New York, 2008.

Dahlman, E., Frenger, P., Guey, J., Klang, G., Ludwig, R., Meyer, M., Wiberg, N., and Zangi, K., A framework for future radio access, *IEEE VTC'05* (*Spring*), May–June 2005.

Damosso, E., European Commission COST 231, Digital mobile radio towards future generation systems, 1999.

Da-Shan, S., Foschini, G. J., Gans, M. J., and Kahn, J. M., Fading correlation and its effect on the capacity of multi-element antenna systems, *IEEE Transactions on Communications*, 48, 3, 502–513, 2000.

Deneire, L., Gyselinckx, B., and Engels, M., Training sequence versus cyclic prefix—A new look on single carrier communications, *IEEE Communication Letters*, 5, 7, 292–294, 2001.

Dinis, R. and Gusmão, A., A class of nonlinear signal processing schemes for bandwidth efficient OFDM transmission with low envelope fluctuation, *IEEE Transactions on Communications*, November 2004.

Dinis, R., Gusmão, A., and Esteves, N., On broadband block transmission over strongly frequency-selective fading channels, *15th International Conference on Wireless Communications (Wireless 2003)*, pp. 261–269, Calgary, Canada, 2003.

Dinis, R., Kalbasi, R., Falconer, D., and Banihashemi, A., Iterative layered space-time receivers for single-carrier transmission over severe time-dispersive channels, *IEEE Comm. Letters*, 8, 9, 579–581, 2004.

Dinis, R., Silva, P., and Gusmão, A., An iterative frequency-domain decision-feedback receiver for MC-CDMA schemes, *IEEE VTC'05 (Spring)*, pp. 271–275, Stockholm, Sweden, 2005.

Divsalar, D., Simon, M., and Raphaeli, D., Improved parallel interference cancellation for CDMA, *IEEE Transactions on Communications*, 46, 2, 258–268, 1998.

Dogan, S., et al., Video content adaptation using transcoding for enabling UMA over UMTS, *Proceedings of theWiamis 2004*, Lisbon, Portugal, April 2004.

Douillard, C., Jezequel, M., Berrou, C., Picart, A., Didier, P., and Glavieux, A., Iterative correction of intersymbol interference: Turbo-equalization, *European Transactions on Telecommunications*, 6, 507–511, 1995.

Driessen, P. F. and Foschini, G. J., On the capacity formula for multiple-input multiple output wireless channels: A geometric interpretation, *IEEE Transactions on Communications*, 47, 173–176, 1999.

Durgin, G. D. and Rappaport, T. S., Effects of multipath angular spread on the spatial cross-correlation of received voltage envelopes, *49th IEEE Vehicular Technology Conference (VTC) 1999*, 2, 996–1000.

Eklund, C., et al., IEEE Standard 802.16: A technical overview of the Wireless MAN air interface for broadband wireless access, *IEEE Communications Magazine*, 40, 6, 98–107, 2002.

Engels, V. and Rohling, H., Multi-resolution 64-DAPSK modulation in a hierarchical COFDM transmission system, *IEEE Transactions on Broadcasting*, 44, 1, 139–149, 1998.

ETSI, Channel models for HiperLAN/2 in different indoor scenarios, ETSI EP BRAN 3ERI085B; pp. 1–8, 1998a.

ETSI, ETS 300 744: Digital video broadcasting (DVB) framing structure, channel coding and modulation for digital terrestrial television (DVB-T) V1.5.1, European Telecommunication Standard, November, 2004.

ETSI, TR 101.112 v3.2.0, Selection procedures for the choice of radio transmission technologies of UMTS, Sophia Antipolis, France, 1998b.

Hartung, F., et al., Delivery of broadcast services in 3G networks, *IEEE Transactions on Broadcasting*, 53, 1, 188–199, 2007.

Falconer, D. and Ariyavisitakul, S., Broadband wireless using single carrier and frequency domain equalization, *IEEEWCNC*, pp. 27–36, Honolulu, USA, 2002.

Falconer, D., Ariyavisitakul, S., Benyamin-Seeyar, A., and Eidson, B., Frequency domain equalization for single-carrier broadband wireless systems, *IEEE Communications Magazine*, 4, 4, 58–66, 2002.

Farhang-Boroujeny, B., Pilot-based channel identification: A proposal for semi-blind identification of communication channels, *Electronics Letters*, 31, 13, 1044–1046, 1995.

Ferro, E. and Potorti, F., Bluetooth and Wi-Fi wireless protocols: A survey and a comparison, *IEEE Wireless Communications*, 12, 1, 12–26, 2005.

Fonollosa, J. R., et al., The IST METRA project, *IEEE Communications Magazine*, 78–86, 2002.

Foschini, G. J. and Gans, M. J., On limits of wireless communications in fading environments when using multiple antennas, *Wireless Personal Communications*, 6, 315–335, 1998.

Foschini, G. J., Layered space–time architecture for wireless communication in a fading environment when using multiple antennas, *Bell Laboratories Technical Journal*, 1, 2, 41–59, 1996.

Geier, J., Wireless LANs, SAMs Publishing, Indianapolis, IN, 2002.

George, A. and Joseph L. L., *Computer Solution of Large Sparse Positive Definite Systems*, Prentice-Hall, Englewood Cliffs, New Jersey, 1981.

Gesbert, D., Christophersen, N., and Ekman, T., Capacity limits of dense palm-sized MIMO arrays, in *Proceedings of the Globecom Conference*, 2002.

Ghogho, M., McLernon, D., Hernandez, E. A., and Swami, A., Channel estimation and symbol detection for block transmission using data-dependent superimposed training, *IEEE Signal Processing Letters*, 12, 3, 226–229, 2005.

Glisic, S. and Vucetic, B., *Spread Spectrum CDMA Systems for Wireless Communications*, Artech House Publishers, Norwood, Massachusetts, 1997.

Goldsmith, A. and Chua, S. G., Variable-rate variable power M-QAM for fading channels, *IEEE Transactions on Communications*, 45, 10, 1218–1230, 1997.

Gorokhov, A. Transmit diversity versus SDMA: Analytic and numerical comparisons, *Proceedings of the IEEE ICC 2000*, New Orleans, USA, Spring 2000.

Gusmão, A., Dinis, R., Conceição, J., and Esteves, N., Comparison of two modulation choices for broadband wireless communications, *IEEEVTC'00* (Spring), pp. 1300–1305, Tokyo, Japan, 2000.

Gusmão, A., Torres, P., Dinis, R., and Esteves, N., A turbo FDE technique for reduced-CP SC-based block transmission systems, *IEEE Transactions on Communications*, 55, 1, 16–20, 2007.

Gusmão, A., Dinis, R., and Esteves, N., On frequency-domain equalization and diversity combining for broadband wireless communications, *IEEE Transactions on Communications*, 51, 7, 1029–1033, 2003.

Gusmão, A., Dinis, R., and Torres, P., Low-PMEPROFDM transmission with an iterative receiver technique for cancelation of nonlinear distortion, *IEEEVTC'05* (Fall), pp. 2367–2371, Dallas, Texas, 2005.

Gusmão, A., Torres, P., Dinis, R., and Esteves, N., A class of iterative FDE techniques for reduced-CP SC-based block transmission, *International Symposium on Turbo Codes*, Munich, April 2006.

Hall, E. K. and Wilson, S. G., Turbo codes for noncoherent channels. *Proceedings IEEE GLOBECOM'97*, pp. 66–70, Phoenix, Arizona, November 1997.

Higuchi, K., Fujiwara, A., and Sawahashi, M., Multipath interference canceller for highspeed packet transmission with adaptive modulation and coding scheme in WCDMA forward link, *IEEE Journal on Selected Areas in Communications*, 20, 2, 419–432, 2002.

Ho, C., Farhang-Boroujeny, B., and Chin, F., Added pilot semi-blind channel estimation scheme for OFDM in fading channels, *IEEE GLOBECOM'01*, November 2001.

Hochwald, B., Marzetta, T., and Papadias, C., A transmitter diversity scheme for wideband CDMA systems based on space–time spreading, *IEEE Journal on Selected Area in Communications*, 19, 1, 48–60, 2001.

Hoeher, P. and Tufvesson, F., Channel estimation with superimposed pilot sequence, *Proceedings of the IEEE GLOBECOM 1999*, 4, 2162–2166.

Hoher, P., Kaiser, S., and Robertson, P., Pilot-symbol-aided channel estimation in time and frequency, *IEEE Communication Theory Mini-Conference (CTMC), IEEE GLOBECOM'97*, 90–96, 1997.

Holma, H. and Toskala, A., *WCDMA for UMTS*, JohnWiley & Sons, New York, 2000.

Holma, H. and Toskala, A., *WCDMA for UMTS: HSPA Evolution and LTE*, 4th edition, John Wiley & Sons, New York, 2007.

Hottinen, A., Tirkkonen, O., and Wichman, R., *Multi-antenna Transceiver Techniques for 3G and Beyond*, John Wiley & Sons, Chichester, UK, 2003.

IEEE 802.16 Relay Task Group, 2008, http://www.ieee802.org/16/relay

IEEE Standard 802.16–2004, Air interface for fixed broadband wireless access systems, 2004.

IEEE Standard 802.16e-2005, Amendment to air interface for fixed and mobile broadband wireless access systems—physical and medium access control layers for combined fixed and mobile operations in licensed bands, 2005.

IEEE Standard 802.16m-07/002r8, IEEE 802.16m System Requirements, January 2009, http://ieee802.org/16/tgm/index.html

ITU-R M.2134—Requirements related to technical system performance for IMTAdvanced Radio Interface(s) (IMT-TECH), draft new report, November 2008.

ITU-R Recommendation M. 1645—Framework and overall objectives for the future development of IMT-2000 and systems beyond IMT-2000, 2003.

Jakes, W. C., *Microwave Mobile Communications*, John Wiley & Sons, New York, 1974.

Jiang, H. and Wilford, P. A., A hierarchical modulation for upgrading digital, *IEEE Transactions on Broadcasting*, 51, 2, 223–229, 2005.

Johansson, A. L. and Svensson, A., On multi-rate DS/CDMA schemes with interference cancellation, *Journal on Wireless Personal Communications, Kluwer*, 9, 1, 1–29, 1999.

Jones, A. and Wilkinson, T., Combined coding for error control and increased robustness to system nonlinearities in OFDM, *IEEE VTC'96*, Vol. 2, pp. 904–908, Atlanta, May 1996.

Josiam, K. and Rajan, D., Bandwidth efficient channel estimation using super-imposed pilots in OFDM systems, *IEEE Transactions on Wireless Communication*, 6, 6, 2234–2245, 2007.

Kalbasi, R., Dinis, R., Falconer, D., and Banihashemi, A., Layered space-time receivers for single-carrier transmission with iterative frequency-domain equalization, *IEEE VTC'04 (Spring)*, pp. 575–579, May, Milan, Italy, 2004.

Kay, S. M., *Fundamentals of Statistical Signal Processing: Estimation Theory*, Prentice-Hall, Englewood Cliffs, New Jersey, 1993.

Kay, Y., Bengtsson, M., Ottersten, B., McNamara, D., Karlsson, P., and Beach, M., Second order statistics of NLOS indoor MIMO channels based on 5.2 GHz measurements, *IEEE GLOBECOM 2001*, 1, 156–160.

Kim, Y.-S., Kim, C.-J., Jeong, G.-Y., Bang, Y.-J., Park, H.-K., and Choi, S. S., New Rayleigh fading channel estimator based on PSAM channel sounding technique, *Proceedings of the IEEE International Conference on Communications*, pp. 1518–1520, Montreal, Canada, June 1997.

Klein, A., Data detection algorithms specially designed for the downlink of CDMA mobile radio systems, *Proceedings of the Vehicular Technology Conference VTC'97*, pp. 203–207, Phoenix, AZ, May 1997.

Koffman, I. and Roman, V., Broadband wireless access solutions based on OFDM access in IEEE 802.16, *IEEE Communication Magazine*, 40, 4, 96–103, 2002.

Kudoh, E. and Adachi, F., Transmit power efficiency of a multi-hop virtual cellular system, *Proceedings of IEEE Vehicular Technology Conference, 2003 (VTC2003-Fall)*, 5, 2910–2914, October 6–9, 2003.

Lam, C., Falconer, D., and Danilo-Lemoine, F., Iterative frequency domain channel estimation for DFT-precoded OFDM systems using in-band pilots, *IEEE Journal on Selected Areas in Communications*, 26, 2, 348–358, 2008.

Lam, C., Falconer, D., Danilo-Lemoine, F., and Dinis, R., Channel estimation for SCFDE systems using frequency domain multiplexed pilots, *IEEE VTC'06* (Fall), Montreal, Canada, September 2006.

Larmo, A., et al., The LTE link layer design, *IEEE Communications Magazine*, April 2009.

Latva-aho, M. and Juntti, M., LMMSE detection for DS-CDMA systems in fading channels, *IEEE Transactions on Communications*, 48, 2, 2000.

Lee, W. C. Y., *Mobile Cellular Telecommunications Systems*, McGraw-Hill, NewYork, 1989.

Li, W., Overview of fine granularity scalability in MPEG-4 video standard, *IEEE Transactions on CSVT*, 11, 3, 301–317, 2001.

Liang, X. and Xia, X., On the nonexistence of rate-one generalized complex orthogonal designs, *IEEE Transactions on Information Theory*, 49, 11, 2003.

Liu, H. and Li, G., *OFDM-Based Broadband Wireless Networks*, John Wiley & Sons, New Jersey, 2005.

Liu, J., Li, B., and Zhang, Y.-Q., Adaptive video multicast over the Internet, *IEEE Multimedia*, 10, 1, 22–33, 2003.

Lodge, J. H. and Moher, M. L., Time diversity for mobile satellite channels using Trellis coded modulations, *IEEE Global Telecommunications Conference*, Tokyo, Japan, 1987.

Lozano, A. and Papadias, C., Layered space–time receiver for frequency-selective wireless channels, *IEEE Transactions on Communications*, 50, 65–73, 2002.

Lozano, F. R. and Farrokhi Valenzuela, R. A., Lifting the limits on high-speed wireless data access using antenna arrays, *IEEE Communication Magazine*, 39, 156–162, 2001.

Marques da Silva, M., PhD Thesis Emergent Radio Systems, Instituto Superior Tecnico, Lisbon, Portugal, September 2005.

Marques da Silva, M. and Correia, A., Combined transmit diversity and beamforming for WCDMA, *IEEE EPMCC'03*, Glasgow, Scotland, April 2003b.

Marques da Silva, M. and Correia, A., Interference cancellation using joint pre-distortion filtering and selective transmit diversity, *4th Conference on Telecommunications*, Aveiro, Portugal, June 18–20, 2003d.

Marques da Silva, M. and Correia, A., Interference cancellation with combined predistortion filtering and transmit diversity, *14th IEEE International Symposium on Personal Indoor and Mobile Radio Communications*, Beijing, China, September 7–10, 2003e.

Marques da Silva, M. and Correia, A., Joint beamforming and transmit diversity for WCDMA, *Joint RTO IST and SET Symposium on Smart Antennas*, RTO-MP-119, Chester, UK, April 7–9, 2003c.

Marques da Silva, M. and Correia, A., Joint multi-user detection and intersymbol interference cancellation for WCDMA satellite UMTS, *International Journal of Satellite Communications and Networking—Special Issue on Interference Cancellation—Wiley*, 21, 1 January, pp. 93–117, 2003a (Invited paper).

Marques da Silva, M. and Correia, A., MAI cancellation with commutation signaling, *IEEE—Vehicular Technology Conference, VTC2000*, Spring, Tokyo, Japan, May 15–18, 2000a.

Marques da Silva, M. and Correia, A., Parallel interference cancellation with commutation signaling, *IEEE—International Conference on Communications—ICC2000*, Spring, New Orleans, USA, June 18–22, 2000b.

Marques da Silva, M. and Correia, A., Space time block coding for 4 antennas with coding rate 1, *IEEE—International Symposium on Spread Spectrum Techniques and Application—ISSSTA*, Prague, Czech Republic, September 2–5, 2002a.

Marques da Silva, M. and Correia, A., Space time diversity for the downlink of WCDMA, *IEEE—Wireless Personal and Mobile Communications—WPMC'01*, Aalborg, Denmark, September 9–12, 2001.

Marques da Silva, M. and Correia, A., Space–time coding schemes for 4 or more antennas, *IEEE—International Symposium on Personal Indoor and Mobile Radio Communications—PIMRC'02*, Lisbon, Portugal, September 16–18, 2002b.

Marques da Silva, M. and Correia, A., A combined precoding scheme with selective transmit diversity, *International Conference on Telecommunications 2004 (ICT'04)*, Fortaleza, Brazil, August 2004b.

Marques da Silva, M. and Correia, A., A pre-distortion filtering for MAI reduction with selective transmit diversity—PIMRC 2004, Barcelona, Spain, September 2004c.

Marques da Silva, M. and Correia, A., Joint MIMO and MAI suppression for the HSDPA, *International Conference on Telecommunications 2004 (ICT'04)*, Fortaleza, Brazil, August 2004a.

Marques da Silva, M., Correia, A., and Dinis, R., A decorrelating MUD approach for the downlink of UMTS considering a RAKE in the receiver, *Proceedings of the 16th IEEE Personal Indoor and Mobile Radio Communications 2005 (PIMRC'05)*, Berlin, Germany, September 11–14, 2005a.

Marques da Silva, M., Correia, A., and Dinis, R., On pre-processing for MIMOWCDMA, *11th International Symposium on Wireless Personal Multimedia Communications (WPMC 2008)*, Lapland, Finland, September 2008.

Marques da Silva, M., Correia, A., and Dinis, R., On transmission techniques for multiantenna W-CDMA systems, *European Transactions on Telecommunications*, 20, 1, 107–121, 2009a.

Marques da Silva, M., Correia, A., Silva, J. C., and Souto, N., Joint MIMO and parallel interference cancellation for the HSDPA, *Proceedings of IEEE International Symposium on Spread Spectrum Techniques and Applications 2004 (ISSSTA'04)*, Sydney, Australia, September 2004a.

Marques da Silva, M., Correia, A., Silva, J. C., and Souto, N., Interference reduction consisting of pre-distortion filtering with selective transmit diversity, *ISSSTA 2004*, Sydney, Australia, August 31–September 02, 2004c.

Marques da Silva, M., Correia, A., Souto, N., Silva, J. C., Francisco, C., and António, R., Interference suppression consisting of pre-distortion filtering and beamforming with joint transmit diversity—SEACORNWorkshop on Simulation of Enhanced UMTS Access and Core Networks, Cambridge, UK, March 23, 2004b.

Marques da Silva, M., Dinis, R., and Correia, A., A V-BLAST detector approach for W-CDMA signals with frequency-selective fading, *Proceedings of the 16th IEEE Personal Indoor and Mobile Radio Communications 2005 (PIMRC'05)*, Berlin, Germany, September 11–14, 2005b.

Marques da Silva, M., Dinis, R., and Correia, A., Joint detection and channel estimation for SC-FDE with STBC, *Proceedings of 2009 15th Asia-Pacific Conference on Communications (APCC)*, Shanghai, China, October 08–10, 2009b.

Marques da Silva, M., Dinis, R., and Correia, A., Iterative frequency-domain receivers for STBC schemes, *Proceedings ofVTC2009 Fall*, Anchorage, Alaska (USA), September 20–23, 2009c.

Mazzenga, F., Channel estimation and equalization for M-QAM transmission with a hidden pilot sequence, *IEEE Transactions on Broadcasting*, 46, 2, 170–176, 2000.

Meng, X., Tugnait, J., and He, S., Iterative joint channel estimation and data detection using superimposed training: Algorithms and performance analysis, *IEEE Transactions on Vehicular Technology*, 56, 4, 1873–1880, 2007.

Montezuma, P. and Gusmão, A., A pragmatic coded modulation choice for future broadband wireless communications, *Proceedings of the IEEE VTC'01 (Spring)*, Rhodes, Greece, May 2001.

Müller, S., Bäuml, R., Fischer, R., and Huber, J., OFDM with reduced peak-to-average power ratio by multiple signal representation, *Annals of Telecommunications*, 52, 58–67, 1997.

Müller, S. and Huber, J., OFDM with reduced peak-to-average power ratio by optimum combination of partial transmit sequences, *Electronic Letters*, 33, 5, 368–369, 1997.

Muquet, B., Courville, M., Dunamel, P., and Giannakis, G., OFDM with trailing zeros versus OFDM with cyclic prefix: Links, comparisons and application to the HiperLAN/2 system, *IEEE ICC'00*, 1049–1053, Santa-Barbara, California, June 2000a.

Muquet, B., Courville, M., Giannakis, G., Wang, Z., and Dunamel, P., Reduced complexity equalizers for zero-padded OFDM transmission, *IEEE ICASSP'00*, pp. 2973–2976, Istanbul, Turkey, June 2000b.

Nam, S. H. and Lee, K. B., Transmit power allocation for an extended V-BLAST system, PIMRC'2002, Lisbon, Portugal, September 2002.

Nefedov, N. and Pukkila, M., Iterative channel estimation for GPRS, *Proceedings of the IEEE PIMRC'2000*, London, UK, September, 2000.

NGMN, Next generation mobile networks beyond HSPA and EVDO, v.1.3, December 2006, www.ngmn.org

NGMN, NGMN radio access performance evaluation methodology, v.1.2, June 2007; www.ngmn.org

Ohno, S. and Giannakis, G., Optimal training and redundant precoding for block transmissions with application to wireless OFDM, *IEEE Transactions on Communications*, 50, 12, 2113–2123, 2002.

Ohrtman, F., *WiMAX Handbook*, McGraw-Hill Communications, New York, 2008.
Ohrtman, F. and Roeder, K., *Wi-Fi Handbook: Building 802.11b Networks*, McGraw-Hill, New York, 2003.
Ojanperä, T. and Prasad, R., *Wideband CDMA for Third Generation Mobile Communications*, Artech House Publishers, Norwood, Massachusetts, 1998.
Orozco-Lugo, A., Lara, M., and McLernon, D., Channel estimation using implicit training, *IEEE Transactions on Signal Processing*, 52, 1, 2004.
Osseiran, A., et al., The road to IMT-advanced communication systems: state-of-the-art and innovation areas addressed by the WINNER+project, *IEEE Communications Magazine*, 47, 6, 38–47, 2009.
Oyman, O., Laneman, J. N., and Sandhu, S., Multi hop relaying for broadband wireless mesh networks: From theory to practice, *IEEE Communications Magazine*, 45, 11, 116–122, 2007.
Patel, P. and Holtzman, J., Analysis of simple successive interference cancellation scheme in DS/CDMA, *IEEE Journal on Selected Areas in Communications*, 12, 5, 796–807, 1994.
Peters, S. W. and Heath, R. W. The future of WiMAX: Multihop relaying with IEEE 802.16j, *IEEE Communications Magazine*, 104–111, 2009.
Price, R. and Green, P. E. A communication technique for multipath channels, *Proceedings of the IRE*, 46, 3, 555–570, 1958.
Proakis, J., Digital *Communications*, 4th edition, McGraw-Hill, New York, NY, 2001.
Pursley, M. B. and Shea, J. M., Non-uniform phase-shift-key modulation for multimedia multicast transmission in mobile wireless networks, *IEEE Journal on Selected Areas in Communications*, 17, 5, 774–783, 1999.
Raleigh, G. and Cioffi, J., Spatio-temporal coding for wireless communication, *IEEE Transactions on Communications*, 46, 3, 357–366, 1998.
Raleigh, G. and Jones, V., Multivariate modulation and coding for wireless communication, *IEEE Journal on Selected Areas in Communications*, 17, 5, 851–866, 1999.
Ramchandran, K., Ortega, A., Uz, K. M., and Vetterli, M., Multiresolution broadcast for digital HDTV using joint source/channel coding, *IEEE Journal on Selected Areas in Communications*, 11, 1, 6–23, 1993.
Rapajic, P. B. and Popescu, D., Information capacity of a random signature multiple-input multiple-output channel, *IEEE Transactions on Communications*, 48, 1245–1248, 2000.
Raphaeli, D. and Zarai, Y., Combined turbo equalization and turbo decoding, *IEEE Communications Letters*, 2, 4, 107–109, 1998.
Reynolds, D. and Wang, X., Low complexity turbo-equalization for diversity channels, *Signal Processing*, 81, 5, 989–995, 2001.
Robertson, P., Villebrun, E., and Hoeher, P., A comparison of optimal and sub-optimal MAP decoding algorithms operating in the log domain, *Proceedings of the IEEE International Conference on Communications (ICC'95)*, 2, 1009–1013, 1995.
Rooyen, P. V., Lötter, M., and Wyk, D., *Space-Time Processing for CDMA Mobile Communications*, Kluwer Academic Publishers, Boston, 2000.

Roy, S. and Falconer, D., Optimum infinite-length MMSE multi-user decision-feedback space–time processing in broadband cellular radio, *Wireless Personal Communications*, 27, 1–32, 2003.

Sampei, S. and Sunaga, T., Rayleigh fading compensation for QAMin land mobile radio communications, *IEEE Transactions on Vehicular Technology*, 42, 2, 137–147, 1993.

Sandell, M., Iterative channel estimation using soft decision feedback, *Proceedings of IEEE GLOBECOM'98*, 6, pp. 3728–3733, Sydney, Australia, November 1998.

Sanzi, F. and Speidel, J., An adaptive two-dimensional channel estimator for wireless OFDM with application to mobile DVB-T, *IEEE Transactions on Broadcasting*, 46, 128–133, 2000.

Sari, H., et al., An analysis of orthogonal frequency-division multiplexing for mobile radio applications, *IEEE VTC'94*, Stockholm, Sweden, 1994.

Sengupta, M. and Mitra, P. P., Capacity of multivariate channels with multiplicative noise: I. Random matrix techniques and large-n expansions for full transfer matrices, *Physics Archives*, no. 0 010 081, 2000.

Sfar, S., Murch, R., and Letaief, K., Layered space–time multiuser detection over wireless uplink systems, *IEEE Transactions on Communications*, 2, 4, 653–668, 2003.

Shannon, C. E., A mathematical theory of communication, *Bell System Technical Journal*, 27, 623–656, 1948.

Shiu, D., Wireless communication using dual antenna arrays, *International Series in Engineering and Computer Science*, Kluwer, Norwell, MA, 1999.

Shoumin, L. and Zhi, T., Near-optimum soft decision equalization for frequency selective MIMO channels, *IEEE Transactions on Signal Processing*, 52, 3, 721–733, 2004.

Silva, J. C., et al., Enhanced MMSE detection for MIMO systems, *Proceedings of Conf-Tele'2005*, Tomar, Portugal, April 2005a.

Silva, J. C., et al., Equalization based receivers for wideband MIMO/BAST systems, *Proceedings of WPMC'2005*, Aalgorg, Denmark, September 2005b.

Silva, J. C., Souto, N., Correia, A., Cercas, F., and Rodrigues, A., Multipath interference canceller for high speed downlink packet access in enhanced UMTS networks, *Proceedings of the IEEE International Symposium on Spread Spectrum Techniques and Applications—ISSSTA 2004*, pp. 609–612, Sydney, Australia, August 30–September 2, 2004a.

Silva, J. C., Souto, N., Rodrigues, A., Correia, A., Cercas, F., and Jesus, S., Finite state channel modeling for UMTS HSDPA link level performance evaluation, *Proceedings of the IEEE International Symposium on Wireless Personal Multimedia Communications WPMC 2004*, 1, pp. 416–420, Padua, Italy, September 12–15, 2004b.

Silva, J. C., Souto, N., Rodrigues, A., Cercas, F., and Correia, A., Conversion of reference tapped delay line channel models to discrete time channel models, *Proceedings of the IEEE Vehicular Technology Conference—VTC2003 Fall*, Orlando, USA, October 2003.

Silva, J. C., Souto, N., Rodrigues, A., Correia, A., Cercas, F., and Dinis, R., A L-MMSE DS-CDMA detector for MIMO/BLAST systems with frequency selective fading, *Proceedings of the IEEE Ist Mobile & Wireless Communications Summit*, Dresden, Germany, June 19–23, 2005.

Silverstein, J. W., Strong convergence of the empirical distribution of eigenvalues of large dimensional random matrices, *Journal of Multivariate Analysis*, 55, 2, 331–339, 1995.

Soares, A., Silva, J., Leitão, F., Correia, A., and Souto, N., MIMO based radio resource management for UMTS multicast broadcast multimedia services, *Wireless Personal Communications Journal*, 42, 2, 225–246, 2007b.

Soares, A., Souto, N., Silva, J., Eusébio, P., and Correia, A., Effective radio resource management for MBMS in UMTS networks, *Wireless Personal Communications Journal*, 42, 2, 185–211, 2007a.

Souto, N., et al., Iterative turbo multipath interference cancellation for WCDMA systems with non-uniform modulations, *Proceedings of the IEEE Vehicular Technology Conference—VTC2005-Spring*, Stockholm, Sweden, May–June, 2005a.

Souto, N., et al., Non-uniform constellations for broadcasting and multicasting services in WCDMA systems, *Proceedings of the IEEE 1st Mobile & Wireless Communications Summit*, Dresden, Germany, June 19–23, 2005b.

Souto, N., Cercas, F., Dinis, R., and Silva, J. C., On the BER performance of hierarchical M-QAM constellations with diversity and imperfect channel estimation, *IEEE Transactions on Communications*, 55, 10, 1852–1856, 2007a.

Souto, N., Correia, A., Dinis, R., Silva, J., and Abreu, L., Multi resolution MBMS transmissions for MIMO UTRA LTE systems, *Proceedings of the IEEE International Symposium on Broadband Multimedia Systems and Broadcasting—BTS 2008*, Las Vegas, Nevada, March 31–April 2, 2008.

Souto, N., Dinis, R., and Silva, J. C., Iterative decoding and channel estimation of MIMO-OFDM transmissions with hierarchical constellations and implicit pilots, *International Conference on Signal Processing and Communications (ICSPC'07)*, Dubai (United Arab Emirates), November 24–27, 2007b.

Souto, N., Dinis, R., Cercas, F., Silva, J., and Correia, A., An iterative receiver for WCDMA systems with MIMO transmissions and hierarchical constellations, *Proceedings of the IEEE International Symposium on Spread Spectrum Techniques and Applications—ISSSTA 2006*, August 2006, Manaus, Brazil.

Souto, N., Silva, J. C., Correia, A., Cercas, F., da Silva, M. M., and Rodrigues, A., Multiuser detector schemes for the UMTS uplink transmission, *Proceedings of the IEEE International Symposium on Spread Spectrum Techniques and Applications—ISSSTA 2004*, pp. 285–289, Sydney, Australia, August 30–September 2, 2004.

Steele, R., *Mobile Radio Communications*, IEEE Press, New York, 1994.

Stuber, G. L. *Principles of Mobile Communication*, Kluwer, Norwell, Massachusetts, 2001.

Sydir, J. and Taori, R., An evolved cellular system architecture incorporating relay stations, *IEEE Communications Magazine*, 47, 6, 150–155, 2009.

Tarokh, V., et al., Space–time block codes from orthogonal designs, *IEEE Transactions on Information Theory*, 1456–1467, 1999.

Telatar, I. E., Capacity of multiantenna Gaussian channels, *AT&T Bell Laboratories, Technical Memorandum*, 1995.

Telatar, I. E., Capacity of multiantenna Gaussian channels, *European Transactions on Communications*, 10, 6, 585–595, 1999.

Tidestav, C., Sternad, M., and Ahlèn, A., Reuse within a cell—interference rejection or multiuser detection?, *IEEE Transactions on Communications*, 47, 10, 1999.

Torrance, J. M. and Hanzo, L., Comparative study of pilot symbol assisted modem schemes, *Proceedings of the Radio Receivers and Associated Systems Conference—RRAS'95*, pp. 36–41, Bath, UK, September 26–28, 1995.

Tubbax, J., Vanderperre, L., Donnay, S., and Engels, M., Single-carrier communications using decision-feedback equalization for multiple antennas, *Proceedings of IEEE ICC*, Alaska, USA, May 2003.

Tüchler, M. and Hagenauer, J., Linear time and frequency domain turbo equalization, *IEEE VTC'01 (Fall)*, pp. 2773–2777, Atlantic City, USA, 2001.

Tüchler, M. and Hagenauer, J., Turbo equalization using frequency domain equalizers, *38th Annual Allerton Conference on Communication*, pp. 1234–1243, Monticello, Illinois, 2000.

Tüchler, M., Koetter, R., and Singer, A., Turbo equalization: Principles and new results, *IEEE Transactions on Communication*, 50, 5, 754–767, 2002.

Tugnait, J. and He, S., Doubly-selective channel estimation using data-dependent superimposed training and exponential bases models, *Proceedings of the 40th Annual Conference on Information Sciences and Systems*, 2006.

Tugnait, J. and Meng, X., Superimposed training for channel estimation: Performance, analysis, training power allocation and frame synchronization, *IEEE Transactions on Signal Processing*, 54, 2, 752–765, 2006.

Valenti, M. C., Iterative channel estimation and decoding of pilot symbol assisted turbo codes over flat-fading channels, *IEEE Journal on Selected Areas in Communications*, 19, 9, 1697–1705, 2001.

Varanasi, M. K. and Aazhang, B., Multistage detection in asynchronous CDMA communications, *IEEE Transactions on Communications*, 38, 4, 509–519, 1990.

Vetro, A., Christopoulos, C., and Sun, H., Video transcoding architectures and techniques: An overview, *IEEE Signal Proccessing Magazine*, 20, 2, 18–29, 2003.

Viterbi, J., Very low rate convolutional codes for maximum theoretical performance of spread-spectrum multiple-access channels, *IEEE Journal on Selected Areas in Communications*, 8, 4, 641–649, 1990.

Vitthaladevuni, P. K. and Alouini, M.-S., A recursive algorithm for the exact BER computation of generalized hierarchical QAM constellations, *IEEE Transactions on Information Theory*, 49, 297–307, 2003.

Vitthaladevuni, P. K. and Alouini, M.-S., A closed-form expression for the exact BER of generalized PAM and QAM constellations, *IEEE Transactions on Communications*, 52, 5, 698–700, 2004.

Vitthaladevuni, P. K. and Alouini, M.-S., BER computation of 4/M-QAM hierarchical constellations, *IEEE Transactions on Broadcasting*, 47, 3, 228–239, 2001.

Vucetic, B. and Yuan, J., *Turbo Codes: Principles and Applications*, Kluwer Academic Publishers, Norwell, Massachusetts, 2002.

Wang, Z. and Giannakis, G. B. Wireless multi carrier communications—where Fourier meets Sannon, *IEEE Signal Processing Magazine*, 17, 3, 29–48, 2000.

Wang, S., Kwon, S., and Yi, B. K., On enhancing hierarchical modulation, *Proceedings of the IEEE International Symposium on Broadband Multimedia Systems and Broadcasting—BTS 2008*, Las Vegas, Nevada, March 31–April 2, 2008.

Webb, W. T. and Hanzo, L., *Modern Quadrature Amplitude Modulation: Principles and Applications for Fixed and Wireless Channels*, IEEE Press, New York, 1994.

Webb, W. T. and Steele, R., Variable rate QAM for mobile radio, *IEEE Transactions on Communications*, 43, 7, 2223–2230, 1995.

Wei, L.-F., Coded modulation with unequal error protection, *IEEETransactions on Communications.*, 41, 10, 1439–1449, 1993.

Wei, S., Goeckel, D., and Janaswami, R., On the capacity of fixed length linear arrays under band limited correlated fading, in *Proceedings of the CISS*, Princeton, New Jersey, 2002.

Winters, J., Salz, J., and Gitlin, R., The capacity of wireless communication systems can be substantially increased by the use of antenna diversity, *Conference on Information Science and Systems*, Princeton, New Jersey, 1992.

Wireless LANMedium Access Control (MAC) and Physical Layer (PHY) specifications, ANSI/IEEE Std 802.11; 1999 (E) Part 11, ISO/IEC 8802-11, 1999.

Wolniasky, P. W., Foschini, G. J., Golden, G. D., and Valenzuela, R. A., V-BLAST: An architecture for realizing very high rates over rich-scattering wireless channel, *Proceedings of the ISSSE'98*, Valenzuela, 1998.

Yarali, A. and Rahman, S., WiMAX broadband wireless access technology: Services, architecture and deployment models, *IEEE CCECE2008,* Niagara Falls, Ontario, pp. 77–82, May 4–7, 2008.

Zhu, H., Farhang-Boroujeny, B. and Schlegel, C., Pilot embedding for joint channel estimation and data detection in MIMO communication systems, *IEEE Communication Letters*, 7, 1, 30–32, 2003.

Index

D

G

H

I

P

Q

R

S

Z